INORGANIC SYNTHESES

Volume 31

Editor-in-Chief
ALAN H. COWLEY
The University of Texas at Austin

• •

INORGANIC SYNTHESES

Volume 31

A Wiley-Interscience Publication
JOHN WILEY & SONS, INC.

New York   Chichester   Brisbane   Toronto   Singapore   Weinheim

Library of Congress Catalog Number: 39-23015

ISBN 0-471-15288-9

Printed in the United States of America

10 9 8 7 6 5 4 3 2 1

PREFACE

The field of inorganic chemistry continues to thrive, as evidenced by the substantial volume of published literature as well as the large number of meetings and symposia that are devoted to the diverse aspects of this subject. *Inorganic Syntheses*, now in its thirty-first volume, plays an extremely important role in the burgeoning literature of inorganic chemistry. Indeed, in a time span exceeding half a century, *Inorganic Syntheses* has become recognized worldwide as *the* primary source for the preparation of a host of useful inorganic compounds. In large measure, this enviable reputation has come about because each preparation is checked completely independently in a laboratory other than that of the submitter(s).

The present volume feautres over 150 different preparations and the subject matter is divided into four chapters. By a slender margin, the largest number of preparations relate to main-group compounds, and these procedures have been grouped together in Chapter 1. Until relatively recently, main-group chemistry had been a somewhat neglected area of endeavor—particularly in the United States. Happily, this situation is now changing rapidly. One of the driving forces in the renaissance of main-group chemistry is the demand posed by the field of materials science. This trend is represented in Chapter 1 of this volume. The majority of the compounds for which preparations are described represent candidates for molecules-to-materials conversions. *A distinctive feature of this volume, therefore, is the presence of synthetic procedures for a number of important precursors.*

Chapter 2 is devoted both to the synthesis of important ligands and also to the preparation of useful reagents. The syntheses described range from a porphine and catalytically relevant phosphine ligands to relatively complex heteropolytungstate synthons. Organometallic chemistry remains a quintessentially strong area of chemistry and this interesting field is represented in Chapter 3. Finally, and in keeping with previous volumes of *Inorganic Syntheses*, the preparation of several complexes of the d- and f-block elements have been included in Volume 31 (Chapter 4).

Mindful of the many distinguished inorganic chemists who have served as Editor-in-Chief for *Inorganic Syntheses*, it has been an honor to function in this capacity. My sincere thanks are due to many individuals who have rendered invaluable assistance in the preparation of this volume. The list includes the authors and checkers of each article, the Editorial Board for their many helpful comments, and Tom Sloan of *Chemical Abstracts* for fastidious work on nomenclature. Last, but certainly not least, I am indebted to Ms. Melissa Cabal and Ms. Jennifer Fernandez for their outstanding organizational and secretarial assistance.

ALAN H. COWLEY

Austin, Texas

NOTICE TO CONTRIBUTORS
AND CHECKERS

The *Inorganic Syntheses* series is published to provide all users of inorganic substances with detailed and foolproof procedures for the preparation of important and timely compounds. Thus the series is the concern of the entire scientific community. The Editorial Board hopes that all chemists will share in the responsibility of producing *Inorganic Syntheses* by offering their advice and assistance in both the formulation of and the laboratory evaluation of outstanding syntheses. Help of this kind will be invaluable in achieving excellence and pertinence to current scientific interests.

There is no rigid definition of what constitutes a suitable synthesis. The major criterion by which syntheses are judged is the potential value to the scientific community. An ideal synthesis is one that presents a new or revised experimental procedure applicable to a variety of related compounds, at least one of which is critically important in current research. However, syntheses of individual compounds that are of interest or importance are also acceptable. Syntheses of compounds that are readily available commercially at reasonable prices are not acceptable. Corrections and improvements of syntheses already appearing in *Inorganic Syntheses* are suitable for inclusion.

The Editorial Board lists the following criteria of content for submitted manuscripts. Style should conform with that of previous volumes of *Inorganic Syntheses*. The introductory section should include a concise and critical summary of the available procedures for synthesis of the product in question. It should also include an estimate of the time required for the synthesis, an indication of the importance and utility of the product, and an admonition if any potential hazards are associated with the procedure. The Procedure should present detailed and unambiguous laboratory directions and be written so that it anticipates possible mistakes and misunderstandings on the part of the person who attempts to duplicate the procedure. Any unusual equipment or procedure should be clearly described. Line drawings should be included when they can be helpful. All safety measures should be stated clearly. Sources of unusual starting materials must be given, and, if possible, minimal standards of purity of reagents and solvents should be stated. The scale should be reasonable for normal laboratory operation, and any problems involved in scaling the procedure either up or down should be discussed. The criteria for judging the purity of the final product should be delineated clearly. The Properties section should supply and discuss those physical and chemical characteristics that are relevant to judging the purity of the product and to permitting its handling and use in an intelligent manner. Under References, all pertinent literature citations should be listed in order. A style sheet is available from the Secretary of the Editorial Board.

The Editorial Board determines whether submitted syntheses meet the general specifications outlined above. Every procedure will be checked in an independent laboratory, and publication is contingent upon satisfactory duplication of the syntheses.

Each manuscript should be submitted in duplicate to the Secretary of the Editorial Board, Professor Jay H. Worrell, Department of Chemistry, University of South Florida, Tampa, FL 33620. The manuscript should be typewritten in English. Nomenclature should be consistent and should follow the recommendations presented in *Nomenclature of Inorganic Chemistry*, 2nd ed., Butterworths & Co, London, 1970 and in *Pure and Applied Chemistry*, Volume 28, No. 1 (1971). Abbreviations should conform to those used in publications of the American Chemical Society, particularly *Inorganic Chemistry*.

Chemists willing to check syntheses should contact the editor of a future volume or make this information known to Professor Worrell.

TOXIC SUBSTANCES AND
LABORATORY HAZARDS

Chemicals and chemistry are by their very nature hazardous. Chemical reactivity implies that reagents have the ability to combine. This process can be sufficiently vigorous as to cause flame, an explosion, or, often less immediatley obvious, a toxic reaction.

The obvious hazards in the syntheses reported in this volume are delineated, where appropriate, in the experimental procedure. It is impossible, however, to foresee every eventuality, such as a new biological effect of a common laboratory reagent. As a consequence, *all* chemicals used and *all* reactions described in this volume should be viewed as potentially hazardous. Care should be taken to avoid inhalation or other physical contact with all reagents and solvents used in this volume. In addition, particular attention should be paid to avoiding sparks, open flames, or other potential sources which could set fire to combustible vapors or gases.

A list of 400 toxic substances may be found in the *Federal Register*, Volume 40, No. 23072, May 28, 1975. An abbreviated list may be obtained from *Inorganic Syntheses*, Vol. 18, p. xv, 1978. A current assessment of the hazards associated with a particular chemical is available in the most recent edition of *Threshold Limit Values for Chemical Substances and Physical Agents in the Workroom Environment* published by the American Conference of Governmental Industrial Hygienists.

The drying of impure ethers can produce a violent explosion. Further information about this hazard may be found in *Inorganic Syntheses*, Volume 12, p. 317.

CONTENTS

Chapter Two   LIGANDS AND REAGENTS

Chapter Three ORGANOMETALLIC COMPOUNDS

Chapter One

MAIN GROUP COMPOUNDS

1. VOLATILE β-DIKETONATE COMPLEXES OF CALCIUM(II), STRONTIUM(II), AND BARIUM(II)

SUBMITTED BY DOUGLAS L. SCHULZ,* DEBORAH A. NEUMAYER,*
and TOBIN J. MARKS*
CHECKED BY HIEP LY,[†] GERTROD E. KRÄUTER,[†]
and WILLIAM S. REES, Jr.[†]

Efficient, reproducible metal–organic chemical vapor deposition (MOCVD) processes[1] depend crucially upon the availability of high-purity metal–organic precursors with high and stable vapor pressure as well as predictable gas phase reactivity. For MOCVD routes to thin films of high-T_c superconducting (HTS) cuprate materials, the development of effective metal–organic alkaline earth (Ca^{2+}, Sr^{2+}, Ba^{2+}) precursors has presented a major challenge. The small charge-to-radius ratios and kinetic labilities of these ions has rendered encapsulation in sterically encumbered, nonpolar ligation environments, a prerequisite for favorable volatility, particularly difficult.[2] Initial MOCVD studies employed β-diketonates (acetylacetonates, dipivaloyl-methanates, heptafluorodimethyloctanedionates) which, in most cases, have oligomeric structures and less than satisfactory, unstable vapor pressure characteristics.[2, 3]

Complexes of the type $M(hfa)_2 \cdot L$ [hfa = 1,1,1,5,5,5-hexafluoro-2,4-pentanedionato; M = Ca, Sr, and Ba; L = 2,5,8,11,14-pentaoxapentadecane-(tetraglyme), 2,5,8,11-tetraoxadodecane(triglyme), and 2,5,8,11,14,17,20-heptaoxaheneicosane(hexaglyme)] were reported by Meinema et al.[4]

* Department of Chemistry, Science and Technology Center for Superconductivity, Northwestern University, Evanston, IL 60208-3113.
† Department of Chemistry and Materials Research and Technology Center, The Florida State University, Tallahassee, FL 323060.

These complexes exhibit high, stable vapor pressures compared to the aforementioned alkaline earth β-diketonates.[5] The superior vapor pressure characteristics and thermal stability of the $M(hfa)_2 \cdot$ tetraglyme complexes have been exploited in the MOCVD growth of high-quality thin films of the Y–Ba–Cu–O,[6] Bi–Sr–Ca–Cu–O,[7] and Tl–Ba–Ca–Cu–O[8] HTS phases as well as of the ferroelectric perovskite $BaTiO_3$.[9] The reported syntheses of these complexes require the use of Schlenk techniques, dried hydrocarbon solvents, and expensive commercial alkaline earth reagents [$M(hfa)_2 \cdot H_2O$ or MH_2] of limited purity.[4, 5] High metal purity is of paramount importance in the formation of defect-free materials in the solid state.[10] The MOCVD-derived films of $BaTiO_3$, grown using a low-metal-purity (99%) $Ba(hfa)_2 \cdot$ tetraglyme precursor, displayed semiinsulating electrical behavior ($\rho = 200$–$1000 \ \Omega \, cm$). In contrast, a film grown with the high-metal-purity (99.999%) $Ba(hfa)_2 \cdot$ tetraglyme precursor displayed insulating electrical behavior ($\rho = 10^6$–$10^7 \ \Omega \, cm$) comparable to reported resistivity values of 10^7–$10^8 \ \Omega \, cm$.[11] Meinema et al.[4] reported 75–88% yields for the synthesis of $M(hfa)_2 \cdot L$ complexes (M = Sr and Ba, L = tetraglyme; M = Ba, L = hexaglyme; M = Ca, L = triglyme) by reaction of commercial $M(hfa)_2 \cdot H_2O$ complexes with the appropriate L in toluene solution, followed by recrystallization of the product from pentane. Kirlin et al.[5] reported a 40% yield of $Ba(hfa)_2 \cdot$ tetraglyme by treating uncomplexed Hhfa with a suspension of commercial BaH_2 in tetraglyme and rinsing the product with pentane for purification. We present here an efficient (total synthesis time < 48 h), aqueous synthesis of $M(hfa)_2 \cdot$ tetraglyme complexes (M = Ca, Sr, and Ba) from Hhfa and the metal chlorides or nitrates. This economical procedure affords high yields and offers the possibility of incorporating high-purity (99.999% or greater) metal salts as starting reagents.

A. BIS(1,1,1,5,5,5-HEXAFLUORO-2,4-PENTANEDIONATO-O,O'-)-CALCIUM(II) (2,5,8,11,14-PENTAOXAPENTADECANE-O,O',O'',O''',O'''')

$$CaCl_2 \cdot 2H_2O + \text{tetraglyme} + 2Hhfa + 2H_2N\text{-}n\text{-}C_3H_7 \xrightarrow[H_2O]{DMF}$$

$$Ca(hfa)_2 \cdot \text{tetraglyme} + 2H_3N\text{-}n\text{-}C_3H_7^+ Cl^-$$

Procedure

■ **Caution.** *1,1,1,5,5,5-Hexafluoro-2,4-pentanedione (Hhfa) and $Ba(NO_3)_2$ are considered hazardous. Avoid skin contact and inhalation. Hhfa is a volatile, noxious liquid and must be handled in a fume hood. $Ba(NO_3)_2$ is toxic; $LD_{50} = 355 \ mg/kg$, OSHA (PEL) $= 0.5 \ mg \ Ba/m^3$; ACVGIH (TLV) $= 0.5 \ mg \ Ba/m^3$.*

To a 1-L, round-bottomed flask containing 150 mL dimethylformamide (DMF)* and a magnetic stirring bar is added 6.15 g (0.042 mol) of 99.997% $CaCl_2 \cdot xH_2O$†, 9.30 g (0.042 mol) of 99% 2,5,8,11,14-pentaoxapentadecane (tetraglyme),* and 100 mL deionized, distilled water. This mixture is stirred until the calcium salt has dissolved. In a separate 500-mL Erlenmeyer flask containing 100 mL DMF and a magnetic stirring bar is added 17.40 g (0.084 mol) freshly distilled 1,1,1,5,5,5-hexafluoro-2,4-pentanedione (Hhfa) $[bp_{760} = 60\text{-}70°C]$‡ and 4.94 g (0.084 mol) 98% n-propylamine* with stirring. The Hhfa plus amine solution is added to the stirring Ca^{2+} solution over a period of 30 s. The solvent is then removed using a rotary evaporator until a yellow slurry remains (~ 30 mL volume). Approximately 350 mL deionized, distilled water is then added to the slurry, and the resulting mixture is stirred vigorously for 5 min, at which time the precipitate is collected via suction filtration. This rinse/stir/filtration step is repeated twice. The collected solid is set aside and allowed to dry in air overnight.

Purification

A standard Dailey Pyrex sublimer§ equipped with a water-cooled cold finger is charged with 19.57 g (0.029 mol) crude pale-yellow solid. The sublimer is evacuated to 4×10^{-3} torr and then heated to 90°C with an oil bath. After 8 h, the white solid on the cold finger is scraped off and subjected to a second sublimation cycle. After the second sublimation cycle, the white solid on the cold finger is collected (scraped off) under ambient conditions, yielding 18.17 g (0.027 mol), 64% Ca(hfa)$_2$·tetraglyme, based on 6.15 g (0.042 mol) of $CaCl_2 \cdot 2H_2O$.

Anal. Calcd. for $C_{20}H_{24}O_9F_{12}Ca$: C, 35.51; H, 3.58. Found: C, 35.49; H, 3.48.

Properties

The colorless crystals (mp 94–95°C) of Ca(hfa)$_2$·tetraglyme are air-stable and nonhygroscopic. The ^1H NMR spectrum (CDCl$_3$ solution) exhibits a singlet at δ 3.25 ppm corresponding to the terminal methyl groups of tetraglyme, multiplets at δ 3.41, 3.60, and 3.76 ppm corresponding to the internal methylene groups of tetraglyme, and a singlet at δ 5.92 ppm corresponding to the methine proton of hexafluoroacetylacetonate. A TGA scan (50-mg sample, N$_2$ atmosphere; temperature ramp, 10°C/min) gives $T_{50} = 248°C$ with 3%

* Aldrich, 1001 West Saint Paul Avenue, Milwaukee, WI 53233.
† AESAR, Johnson Matthey, P.O. Box 8247, Ward Hill, MA 01835-0747.
‡ Genzyme, Hollands Road, Haverhill Suffolk, CB9 8PU, England.
§ Catalog number CG-2100, Chemglass, 3861 North Mill Road, Vineland, NJ 08360.

residue to 350°C.[†] (Note: T_{50} is defined as the temperature at which 50% of the initial weight of the sample is lost.)

B. BIS(1,1,1,5,5,5-HEXAFLUORO-2,4-PENTANEDIONATO-*O,O'*-)-STRONTIUM(II) (2,5,8,11,14-PENTAOXAPENTADECANE-*O,O',O'',O''',O''''*)

$$Sr(NO_3)_2 + \text{tetraglyme} + 2Hhfa + 2H_2N\text{-}n\text{-}C_3H_7 \xrightarrow[\text{H}_2\text{O}]{\text{DMF}}$$

$$Sr(hfa)_2 \cdot \text{tetraglyme} + 2H_3N\text{-}n\text{-}C_3H_7^+ \, NO_3^-$$

Procedure

To a 1-L round-bottomed flask containing 200 mL DMF* and a magnetic stirring bar is added 9.56 g (0.045 mol) of 99.9965% $Sr(NO_3)_2$,[†] 10.04 g (0.045 mol) of 99% 2,5,8,11,14-pentaoxapentadecane (tetraglyme),* and 100 mL distilled, deionized water. This mixture is stirred until the strontium salt has dissolved. In a separate 500-mL Erlenmeyer flask containing 100 mL DMF and a magnetic stirring bar is added 18.79 g (0.090 mol) distilled 1,1,1,5,5,5-hexafluoro-2,4-pentanedione (Hhfa) [$bp_{760} = 69$–$70°C$][‡] and 5.34 g (0.090 mol) 98% *n*-propylamine* with stirring. The Hhfa plus amine solution is added to the stirring Sr^{2+} solution over a period of 30 s. The solvent is then removed using a rotary evaporator until a yellow slurry remains (~ 30 mL volume). Approximately 350 mL deionized, distilled water is then added to the slurry, and the resulting mixture is stirred vigorously for 5 min, at which time the precipitate is collected via suction filtration. This rinse/stir/filtration step is repeated twice. The collected solid is set aside and allowed to dry in air overnight.

Purification

A standard Dailey Pyrex sublimer[§] equipped with a water-cooled cold finger is charged with 24.88 g (0.034 mol) crude pale-yellow solid. The sublimer is evacuated to 4×10^{-3} torr and then heated to 110°C with an oil bath. After 8 h, the white solid on the cold finger is scraped off and subjected to a second sublimation cycle. After the second sublimation cycle, the white solid on

---

[†] For a 5-mg sample, the checkers found that $T_{50} = 181°C$ with 1% residue to 350°C.
* Aldrich, 1001 West Saint Paul Avenue, Milwaukee, WI 53233.
[†] AESAR, Johnson Matthey, P.O. Box 8247, Ward Hill, MA 01835-0747.
[‡] Genzyme, Holland Road, Haverhill Suffolk, CB9 8PU, England.
[§] Catalog number CG-2100, Chemglass, 3861 North Mill Road, Vineland, NJ 08360.

the cold finger is collected (scraped off) under ambient conditions, yielding 23.62 g (0.033 mol), 72% Sr(hfa)$_2$·tetraglyme, based on 9.56 g (0.045 mol) Sr(NO$_3$)$_2$.

Anal. Calc. for C$_{20}$H$_{24}$O$_9$F$_{12}$Sr: C, 33.18; H, 3.34. Found: C, 33.21; H, 3.24.

Properties

The colorless crystals (mp 134–135°C) of Sr(hfa)$_2$·tetraglyme are air-stable and nonhygroscopic. The ^{1}H NMR spectrum (CDCl$_3$ solution) exhibits a singlet at δ 3.32 ppm corresponding to the terminal methyl groups of tetraglyme, multiplets at δ 3.42, 3.56, and 3.75 ppm corresponding to the internal methylene groups of tetraglyme, and a singlet at δ 5.86 ppm corresponding to the methine proton of hexafluroacetylacetonate. A TGA scan (50 mg, N$_2$ atmosphere; temperature ramp, 10°C/min) gives $T_{50} = 260$°C with 4% residue to 350°C.*

C. BIS(1,1,1,5,5,5-HEXAFLUORO-2,4-PENTANEDIONATO-*O,O′*)-BARIUM(II) (2,5,8,11,14-PENTAOXAPENTADECANE-*O,O′,O″,O‴,O⁗*)

$$\text{Ba(NO}_3)_2 + \text{tetraglyme} + 2\text{Hhfa} + 2\text{H}_2\text{N-}n\text{-C}_3\text{H}_7 \xrightarrow[\text{H}_2\text{O}]{\text{DMF}}$$

$$\text{Ba(hfa)}_2 \cdot \text{tetraglyme} + 2\text{H}_3\text{N-}n\text{-C}_3\text{H}_7^+ \text{NO}_3^-$$

Procedure

To a 1-L, round bottomed flask containing 200 mL DMF[†] and a magnetic stirring bar is added 14.92 g (0.057 mol) of 99.999% Ba(NO$_3$)$_2$,[‡] 12.69 g (0.057 mol) of 99% 2,5,8,11,14-pentaoxapentadecane (tetraglyme),[†] and 100 mL distilled, deionized water. This mixture is stirred until the barium salt has dissolved. In a separate 500-mL Erlenmeyer flask containing 100 mL DMF and a magnetic stirring bar is added 23.75 g (0.114 mol) distilled 1,1,1,5,5,5-hexafluoro-2,4-pentanedione (Hhfa) [bp$_{760}$ = 69–70°C][§] and 6.75 g (0.114 mol) 98% *n*-propylamine[†] with stirring. The Hhfa plus amine solution is added to the stirring Ba^{2+} solution over a period of 30 s. The solvent is then removed using a rotary evaporator until a yellow slurry

---

* For a 5-mg sample, the checkers found that $T_{50} = 194$°C with 1% residue to 350°C.
† Aldrich, 1001 West Saint Paul Avenue, Milwaukee, WI 53233.
‡ AESAR, Johnson Matthey, P.O. Box 8247, Ward Hill, MA 01835-0747.
§ Genzyme, Hollands Road, Haverhill Suffolk, CB9 8PU, England.

remains ($\sim$ 30 mL volume). Approximately 350 mL deionized, distilled water is then added to the slurry, and the resulting mixture is stirred vigorously for 5 min, at which time the precipitate is collected via suction filtration. This rinse/stir/filtration step is repeated twice. The collected solid is set aside and allowed to dry in air overnight.

Purification

A standard Pyrex sublimer* equipped with a water-cooled cold finger is charged with 29.01 g (0.037 mol) crude pale-yellow solid. The sublimer is evacuated to 4×10^{-3} torr and then heated to 130°C with an oil bath. After 8 h, the white solid on the cold finger is scraped off and subjected to a second sublimation cycle. After the second sublimation cycle, the white solid on the cold finger is collected (scraped off) under ambient conditions, yielding 27.92 g (0.036 mol), 63% Ba(hfa)$_2$ tetraglyme, based on 14.92 g (0.057 mol) Ba(NO$_3$)$_2$.

Anal. Calcd. for $C_{20}H_{24}O_9F_{12}Ba$: C, 31.05; H, 3.13. Found: C, 31.19; H, 2.99.

Properties

The colorless crystals (mp 151–152°C) of Ba(hfa)$_2$·tetraglyme are air-stable and nonhygroscopic. The ^1H NMR spectrum (CDCl$_3$ solution) exhibits a singlet at δ 3.34 ppm corresponding to the terminal methyl groups of tetraglyme, multiplets at δ 3.45, 3.56, 3.73, and 3.80 ppm corresponding to the internal methylene groups of tetraglyme, and a singlet at δ 5.82 ppm corresponding to the methine proton of the hexafluoroacetylacetonate. A TGA scan (50 mg, N$_2$ atmosphere; temperature ramp, 10°C/min) gives $T_{50} = 260°C$ with 16% residue to 350°C.[†]

Acknowledgment

This research was supported by the National Science Foundation (DMR 91-20000) through the Science and Technology Center for Superconductivity.

References

1. K. F. Jensen and W. Kern, in *Thin Film Processes II*, J. L. Vossen and W. Kern (eds), Academic, New York, 1991, pp. 283–368.

---

* Catalog number CG-2100, Chemglass, 3861 North Mill Road, Vineland, NJ 08360.
† For a 5-mg sample, the checkers found that $T_{50} = 209°C$ with 2% residue to 350°C.

2. (a) L. M. Tonge, D. S. Richeson, T. J. Marks, J. Zhao, J. Zhang, B. W. Wessels, H. O. Marcy, and C. R. Kannewurf, *Adv. Chem. Ser.*, **226**, 351 (1990). (b) T. Hirai and H. Yamane, *J. Cryst. Growth*, **107**, 683 (1991) and references therein. (c) K. Zhang and A. Erbil, *Mater. Sci. Forum.* **130-2**, 255 (1993). (d) T. J. Marks *Pure Appl. Chem.* **67**, 313 (1995).

3. (a) J. Zhao, K.-H. Dahmen, H. O. Marcy, L. M. Tonge, T. J. Marks, B. W. Wessels, and C. R. Kannewurf, *Appl. Phys. Lett.*, **53**, 1750 (1988). (b) J. Zhang, J. Zhao, H. O. Marcy, L. M. Tonge, B. W. Wessels, T. J. Marks, and C. R. Kannewurf, *Appl. Phys. Lett.*, **54**, 1166 (1989). (c) D. S. Richeson, L. M. Tonge, J. Zhao, J. Zhang, H. O. Marcy, T. J. Marks, B. W. Wessels, and C. R. Kannewurf, *Appl. Phys. Lett.*, **54**, 2154 (1989).

4. K. Timmer, C. I. M. A. Spee, A. Mackor, H. A. Meinema, A. L. Spek, and P. van der Sluis, *Inorg. Chim. Acta*, **190**, 109 (1991) and references therein.

5. R. Gardiner, D. W. Brown, P. S. Kirlin, and A. L. Rheingold, *Chem. Mater.*, **3**, 1053 (1991).

6. (a) S. J. Duray, D. B. Buchholz, S. N. Song, D. S. Richeson, J. B. Ketterson, T. J. Marks, and R. P. H. Chang, *Appl. Phys. Lett.*, **59**, 1503 (1991). (b) C. I. M. A. Spee, E. A. Vander Zouwen-Assink, K. Timmer, A. Mackor, and H. A. Meinema, *J. Phys. IV*, **1**, C2/295 (1991). (c) S. J. Duray, D. B. Buchholz, H. Zhang, V. P. Dravid, D. L. Schulz, T. J. Marks, J. B. Ketterson, and R. P. H. Chang, *J. Vac. Sci. Technol. A*, **11**, 1346 (1993). (d) K. A. Dean, D. B. Buchholz, L. D. Marks, R. P. H. Chang, B. V. Vuchic, K. L. Merkle, D. B. Studebaker, and T. J. Marks *J. Mater. Res.* **10**, 2700 (1995).

7. (a) J. M. Zhang, B. W. Wessels, D. S. Richeson, T. J. Marks, D. C. DeGroot, and C. R. Kannewurf, *J. Appl. Phys.*, **69**, 2743 (1991). (b) B. W. Wessels, J. M. Zhang, F. DiMeo, Jr., D. S. Richeson, T. J. Marks, D. C. DeGroot, and C. R. Kannewurf, *SPIE Proc.*, **1394**, 232 (1990). (c) H. A. Lu, J. Chen, B. W. Wessels, D. L. Schulz, and T. J. Marks *J. Appl. Phys.* **73**, 3886 (1993).

8. (a) G. Malandrino, D. S. Richeson, T. J. Marks, D. C. DeGroot, and C. R. Kannewurf, *Appl. Phys. Lett.*, **58**, 182 (1991). (b) D. L. Schulz, D. S. Richeson, G. Malandrino, D. A. Neumayer, T. J. Marks, D. C. DeGroot, J. L. Schindler, T. Hogan, and C. R. Kannewurf, *Thin Solid Films*, **216**, 45 (1992). (c) D. A. Neumayer, D. L. Schulz, D. S. Richeson, T. J. Marks, D. C. DeGroot, J. L. Schindler, and C. R. Kannewurf, *Thin Solid Films*, **216**, 41 (1992). (d) B. J. Hinds, D. L. Schulz, D. A. Nevmayer, B. Han, T. J. Marks, Y. Y. Wang, V. P. Dravid, J. L. Schindler, T. P. Hogan, and C. R. Kannewurf *Appl. Phys. Lett* **63**, 231 (1994).

9. (a) L. A. Wills, B. W. Wessels, D. S. Richeson, and T. J. Marks, *Appl. Phys. Lett.*, **60**, 41 (1992). (b) H. A. Lu, L. A. Wills, B. W. Wessels, W. P. Lin, P. G. Zhang, G. K. Wong, D. A. Neumayer, T. J. Marks, *Appl. Phys. Lett.*, **62**, 12 (1993).

10. (a) F. Batllo, B. Jannot, J.-C. Jules, J.-C. Niepce, and A. Beauger, *Ferroelectrics*, **94**, 195 (1989). (b) B. A. Wechsler and M. B. Klein, *J. Opt. Soc. Am. B*, **5**, 1711 (1988) and references therein.

11. B. W. Wessels, L. A. Wills, S. R. Gilbert, D. A. Neumayer, and T. J. Marks, *Proc. Electrochem. Soc.*, **93-2**, 291 (1993).

2. BIS(1,1,1,3,3,3-HEXAMETHYLDISILAZANATO)-BIS(TETRAHYDROFURAN)BARIUM

SUBMITTED BY ROGER L. KUHLMAN,* BRIAN A. VAARTSTRA,*
and KENNETH G. CAULTON*
CHECKED BY PAMELA S. TANNER† and TIMOTHY P. HANUSA†

$$Ba(s) + 2HN(SiMe_3)_2 \xrightarrow[NH_{3(g)}]{THF} Ba\{N(SiMe_3)_2\}_2(THF)_2 + H_2 (g)$$

Metal alkoxides are of interest as precursors to oxide materials with ceramic or electronic applications.[1] Barium incorporation in such a precursor is particularly interesting because of the known high-T_c superconductor, $YBa_2Cu_3O_7$. The title compound has vast possibilities for reactions of the type

$$M_n(OR)_m(ROH)_2 + Ba(NR'_2)_2 \rightarrow BaM_n(OR)_{m+2} + 2HNR'_2$$

and has already been used to form the mixed-metal alkoxides $\{BaZr_2(O^iPr)_{10}\}_2$ and $BaZr_4(O^iPr)_{18}$ in a reaction with $Zr_2(O^iPr)_8(^iPrOH)_2$.[2] This source of Ba^{2+} is ideal for these reactions, and generally in organometallic chemistry where an anhydrous, hydrocarbon-soluble reagent is required. The bulky disilazanide group was chosen to help stabilize barium in its atypically low coordination number. This ligand has been used successfully to stabilize other metals with low coordination number.[3] The $NH_3(g)$ catalyst apparently oxidizes the barium to Ba^{2+}, and the resulting barium amide is more finely divided than Ba(s), and is perhaps soluble in tetrahydrofuran (THF). A subsequent acid–base reaction with the silylamine regenerates $NH_3(g)$, as the product forms, completing the catalytic cycle. The original report of this synthesis contains a more detailed description of this catalysis.[4] Alternative syntheses involve transmetallation[5] (a redox reaction between Ba^0 and $Sn[N(SiMe_3)_2]_2$) and preactivation of the elemental alkaline earth by its formation as a finely divided powder from BaI_2 and K.[6] Subsequently, it has been reported that even the halide metathesis reaction [$BaCl_2$ and $NaN(SiMe_3)_2$] succeeds in good yield, although this product was reported to be only slightly soluble in hydrocarbon solvents.[7] The synthesis reported here is generally the most cost-effective, because it uses hexamethyldisila*zane* as starting material, which is much cheaper than commercially available sila*zides.*

* Chemistry Department, Indiana University, Bloomington, IN 47401.
† Department of Chemistry, Vanderbilt University, Bloomington, IN 47401.

Procedure

All manipulations are done under a nitrogen atmosphere using standard Schlenk techniques unless otherwise noted.[8] Tetrahydrofuran is dried and distilled before use. Barium granules (Alfa, up to ~0.8 cm on an edge), and 1,1,1,3,3,3-hexamethyldisilazane (Aldrich) are used as received without further purification. In a dry box, 4.00 g (29.1 mmol) of barium granules are added to a 300-mL, round-bottomed, 3-neck flask equipped with stir bar, vacuum adapter, and rubber septum. The flask is then removed from the dry box and attached to a Schlenk line in a well-ventilated hood. The barium is then covered with 24 mL (113.8 mmol) of $HN(SiMe_3)_2$ via syringe, followed by 24 mL of THF. Access to nitrogen is closed off, and $NH_3(g)$ ■ (**Caution!** *Use an efficient hood!*) is bubbled into the liquid via a syringe needle through the septum for 5 min with constant stirring. The flow of $NH_3(g)$ is monitored by allowing it to escape into the hood through a bubbler filled with Nujol. The evolution of bubbles at the surface of the barium indicates that the reaction is initiated, and the $NH_3(g)$ flow is no longer needed. In the event that no H_2 bubbles evolve, the treatment with $NH_3(g)$ can be repeated. The $NH_3(g)$ flow is then reduced to a slow trickle, and the mixture is stirred for 18 h. The mixture typically turns a dark purple color within 30 min of beginning the $NH_3(g)$ flow. The mixture is again treated with a more rapid flow of $NH_3(g)$, at 5-min intervals, until no more bubbles evolve (i.e., until the reaction is complete). Typically, four or five of these final $NH_3(g)$ treatments are employed, at 20-min intervals. The volume is then reduced to about 10 mL in vacuo, and 100 mL of pentane (or hexanes) is added. The mixture is warmed and stirred to aid dissolution. At this point the reaction mixture is a light yellow solution, with insoluble purple Ba and $Ba(NH_2)_2$ impurities. These impurities are removed by filtration into a flame-dried Schlenk flask that is fitted with a filter frit. The mixture is transferred via cannula through a septum fitted to the top of a sintered glass filter stick. The residue is then disposed of appropriately ■ (**Caution!** *Barium is toxic! Wear gloves!*). The solvent of the filtrate is removed in vacuo, leaving 14.1 g (80.2%) of a white powder which is the title compound. If necessary, the product can be recrystallized from cold pentane with only minimal loss in yield.

Properties

The product is a white powder, soluble in benzene, toluene, pentane, and hexane. It is most conveniently recognized by its ^1H NMR spectrum (toluene-d_8): δ 3.52 (m, 8 H), 1.40 (m, 8 H), 0.28 (s, 36 H).[9] The product also shows three peaks in its $^{13}C\{^1H\}$ NMR (toluene-d_8): δ 68.62(s), 25.46(s),

5.60(s). When redissolved in toluene (the solvent must have a higher boiling point than THF) and stripped to dryness, the compound loses one coordinated THF molecule to yield the dimer $\{Ba[N(SiMe_3)_2]_2(THF)\}_2$. ^1H NMR (toluene-d_8): δ 3.49 (m, 4 H), 1.38 (m, 4 H), 0.26 (s, 36 H). $^{13}C\{^1H\}$ NMR (toluene-d_8): δ 68.61 (s), 25.35(s), 5.52(s). Alternately, the product can be isolated with no coordinated THF by sublimation onto a water-cooled cold finger, using heat from an IR lamp[8] and a vacuum of about 50 mtorr. This product is also a dimer, $\{Ba[N(SiMe_3)_2]_2\}_2$. ^1H NMR (toluene-d_8): δ 0.19(s). ^{13}C NMR (toluene-d_8): δ 5.39(s). The structures of all three compounds,[4] variable temperature and multinuclear NMR, and vibrational and mass spectra[5] have been reported. The title compound is perhaps the most convenient in terms of "shelf life," since progressive removal of solvent molecules makes subsequent products increasingly susceptible to decomposition by air and moisture. This variable solvent loss and extreme hydrolytic sensitivity have frustrated the use of elemental analysis as a proof of purity.

References

1. (a) L. G. Hubert-Pfalzgraf, *New J. Chem.*, **11**, 663 (1987). (b) K. G. Caulton and L. G. Hubert-Pfalzgraf, *Chem. Rev.*, **90**, 969 (1990).
2. (a) B. A. Vaartstra, J. C. Huffman, W. E. Streib and K. G. Caulton, *J. Chem. Soc., Chem. Commun.*, **1990**, 1750. (b) B. A. Vaartstra, J. C. Huffman, W. E. Streib and K. G. Caulton, *Inorg. Chem.*, **30**, 3068 (1991).
3. (a) M. F. Lappert, P. P. Power, A. R. Sanger and R. C. Srivastava, *Metal and Metalloid Amides*, Wiley, New York, 1980. (b) K. F. Tesh, T. P. Hanusa, and J. C. Huffman, *Inorg. Chem.*, **29**, 1584 (1990).
4. B. A. Vaartstra, J. C. Huffman, W. E. Streib, and K. G. Caulton, *Inorg. Chem.*, **30**, 121 (1991).
5. M. Westerhausen, *Inorg. Chem.*, **30**, 96 (1991). For the calcium analog, see M. Westerhausen and W. Schwarz, *Z. Anorg. Allg. Chem.*, **604**, 127 (1991) and P. B. B. Hitchcock, M. F. Lappert, G. A. Lawless, and B. Royo, *J. Chem. Soc., Chem. Commun.*, **1990**, 1141.
6. (a) M. J. McCormick, K. B. Moon, S. R. Jones and T. P. Hanusa, *J. Chem. Soc., Chem. Commun.*, **1990**, 778. (b) 198th National Meeting of the American Chemical Society, September 1989, Abstract I14.
7. J. M. Boncella, C. J. Coston, and J. K. Cammack, *Polyhedron*, **10**, 769 (1991).
8. A. L. Wayda and M. Y. Darensbourg, *Experimental Organometallic Chemistry*, American Chemical Society, Washington, DC, 1987, p. 30.
9. Integration of the ^1H NMR spectrum by the checkers showed 1.92 THF molecules per barium.

3. AMMONIUM AND BARIUM SALTS OF THE TRIS[1,2-BENZENEDIOLATE(2−)-*O,O′*]TITANIUM(IV) DIANION

SUBMITTED BY JULIAN A. DAVIES* and SYLVAIN G. DUTREMEZ[†]

CHECKED BY CORTLANDT G. PIERPONT[‡]

Alkaline earth metal salts of the tris(catecholate)titanium(IV)[§] dianion have been used as precursors to titanate ceramics with controlled stoichiometries.[1, 2] The acid, $H_2[Ti(cat)_3]$ (cat = dianion of catechol)[§], has been prepared from titanium tetrachloride in a nonaqueous medium and has been shown[1] to react with barium carbonate to generate $Ba[Ti(cat)_3]$, which is a precursor to $BaTiO_3$. Other salts of the tris(catecholate)titanium(IV) dianion (e.g., Sr, Ca, and Pb) have also been used as precursors to the corresponding titanate ceramics. The patent literature[2] contains a description of the synthesis of $[NH_4]_2[Ti(cat)_3]$ from tetraisopropoxytitanium(IV) and the subsequent formation of $BaTiO_3$ via generation and pyrolysis of the barium salt. Syntheses of the complex $[NH_4]_2[Ti(cat)_3]$ (and closely related hydrates) have been described where both "hydrous titanium dioxide" [formally $Ti(OH)_4$] and $TiCl_4$ have been employed as starting materials.[3, 4] The source of titanium for the syntheses described here is titanium dioxide, which is depolymerized by acid digestion. Acid-digestion techniques are routinely used by ceramists and present less of an obstacle than the use of a volatile, reactive material such as $TiCl_4$. The following procedures for the preparation of $[NH_4]_2[Ti(cat)_3] \cdot 2H_2O$ and its conversion to $Ba[Ti(cat)_3] \cdot 3H_2O$ have been developed[5, 6] to provide the ceramist with a route to these ceramic precursors that employs no organic solvents and that avoids the use of ill-defined reagents such as hydrous titanium oxide or volatile reagents such as $TiCl_4$ and $Ti(O-i-Pr)_4$. The precursors described here are air-stable solids under normal conditions and require no special handling techniques. Similar methods to those described may be employed[7] to generate other salts of the $[Ti(cat)_3]^{2-}$ ion and related strategies lead to catecholate precursors of ceramics based on Zr, Bi, Sn, and so on.[6]

* Department of Chemistry, University of Toledo, Toledo, OH 43606.

[†] Department of Chemistry, Indiana University, Bloomington, IN 47405.

[‡] Department of Chemistry and Biochemistry, University of Colorado at Boulder, Boulder, CO, 80309.

[§] Throughout this manuscript catechol = 1,2-benzenediol and catecholate = 1,2-benzenediolate (2−)-*O,O′*.

A. AMMONIUM TRIS[1, 2-BENZENEDIOLATE(2 −)-*O,O'*] TITANIUM(IV) DIHYDRATE

$$TiO_2 \xrightarrow{H_2SO_4/(NH_4)_2SO_4} Ti(IV)*$$

$$Ti(IV) + 3\ 1,2\text{-}C_6H_4(OH)_2 \xrightarrow{NH_4OH} [NH_4]_2 [Ti(1,2\text{-}C_6H_4O_2)_3] \cdot 2H_2O$$

Procedure

Titanium dioxide (1.04 g, 13 mmol), ammonium sulfate (16.8 g, 127 mmol), and concentrated sulfuric acid (43 mL; 95.6% w/w; 35.87 N) are loaded into a 100-mL round-bottomed flask equipped with an air condenser and heated using a sand bath and Bunsen burner until a clear, yellow solution is obtained (note that appropriate safety measures must be employed when handling sulfuric acid at high temperatures). The addition of ammonium sulfate to elevate the boiling point of the sulfuric acid solvent is critical for effective dissolution of the titanium dioxide reagent. With a sand bath temperature of about 500°C, dissolution takes 4–8 h depending upon the nature of the TiO_2 employed. Anatase, rutile, and mixtures of anatase and rutile may be depolymerized effectively by this method. The solution is then allowed to cool to room temperature, at which point it is clear and colorless. The concentrated acid solution is then added carefully to distilled water (130 mL) and the diluted solution is transferred to a 250-mL bypass-dropping funnel.

Separately, concentrated ammonium hydroxide solution (400 mL) is taken in a 1-L, three-necked, round-bottomed flask equipped with the bypass-dropping funnel, a nitrogen inlet, and a glass stopper, and containing a magnetic stirrer bar. The ammonium hydroxide solution is deaerated by bubbling with nitrogen for 1 h. Catechol (4.31 g, 39 mmol) is added and the contents are stirred under nitrogen until a clear solution results. The acid solution is then added from the bypass-dropping funnel over a period of 5 min while the contents of the flask are stirred under nitrogen. During the addition, a red suspension is produced and this is stirred for an additional 4 h at room temperature. Stirring is then discontinued and the precipitate is allowed to settle. The solid product is isolated by filtration at a water pump and is allowed to air dry. The yield is 5.22 g (90%). If trace amounts of sulfate are present, they can be removed by washing the solid with cold isopropyl alcohol.

Anal. Calcd. for $C_{18}H_{24}O_8N_2Ti$: C, 48.66; H, 5.54; N, 6.31; Ti, 10.78; S, 0.0. Found: C, 47.80, 47.89; H, 5.35, 5.39; N, 6.60, 6.62; Ti, 11.01; S, 0.0.%.

---

* Ti(IV) is presumably present as a sulfate complex such as $[Ti(SO_4)_3]^{2-}$.

Properties

Ammonium tris(catecholate)titanium(IV) dihydrate is a rust-red solid that is soluble in water and in dimethyl sulfoxide. It is air-stable as a solid but requires storage under an inert atmosphere when in solution. It is insoluble in hexane, diethyl ether, and other nonpolar organic solvents. Identification of the product as a dihydrate is based upon thermogravimetric analysis where the complex is converted to titanium dioxide (identified by X-ray powder diffraction) by heating in air at 800°C for 12 h. The ^{1}H NMR spectrum of a DMSO-d_6 solution exhibits resonances due to the aromatic protons of the catecholate ligands (6.0–6.4 ppm; multiplet), the ammonium ion protons (7.2 ppm, singlet, broad), and the protons of the water of crystallization (3.3 ppm, singlet, broad). The ^{13}C{^{1}H} NMR spectrum in the same solvent exhibits three resonances at 109.8, 115.8, and 160.0 ppm due to the carbon atoms of the coordinated catecholate ligands. The IR spectrum[4] exhibits broad features due to the ammonium ion and the water of crystallization at about 3700–2800 cm^{-1} and sharp bands due to the catecholate ligands in the region below 1600 cm^{-1}. A cyclic voltammogram of the complex measured in aqueous 1M KCl exhibits a single reversible wave at -1.38 V (vs. Ag/AgCl), indicating the absence of detectable dinuclear species in solution.[4]

B. BARIUM TRIS[1, 2-BENZENEDIOLATE(2−)-O,O']TITANIUM(IV) TRIHYDRATE

$$[NH_4]_2[Ti(cat)_3] \cdot 2H_2O + Ba(OH)_2 \cdot 8H_2O \longrightarrow$$

$$Ba[Ti(cat)_3] \cdot 3H_2O + 9H_2O + 2NH_3$$

Procedure

Ammonium tris(catecholate)titanium(IV) dihydrate (0.41 g, 0.92 mmol) is dissolved in distilled water (50 mL) in a 100-mL, round-bottomed flask containing a magnetic stirrer bar. Ba(OH)$_2 \cdot$ 8H$_2$O (0.29 g, 0.92 mmol) is added to the flask, which is then fitted with a water-cooled condenser and heated by an oil bath at 60–70°C for 2 h with stirring. The resulting brown suspension is allowed to cool at room temperature and is filtered at a water pump to produce a clear solution. The solution is then evaporated to dryness, at room temperature, using a vacuum-line, producing a brown solid. The yield is 0.37 g (72%).

Anal. Calcd. for C$_{18}$H$_{18}$O$_9$TiBa: C, 38.36; H, 3.22; N, 0.00; Ti, 8.50. Found: C, 38.34; H, 3.01; N, 0.50; Ti, 8.39%.

Properties

Barium tris(catecholate)titanium(IV) trihydrate is a brown solid that is soluble in water and in dimethyl sulfoxide. It is air-stable as a solid but requires storage under an inert atmosphere when in solution. It is insoluble in hexane, diethyl ether, and other nonpolar organic solvents. Identification of the product as a trihydrate is based upon thermogravimetric analysis where the complex is converted to barium titanate (identified by X-ray powder diffraction) by heating in air at 800°C for 12 h. Despite variable, low levels of nitrogen in the combustion analysis of the barium salt, the X-ray powder diffraction pattern of the calcined material conforms exactly to that of tetragonal $BaTiO_3$. No other phases such as BaO, TiO_2, or other compounds in the BaO/TiO_2 system were detected. The presence of low levels of nitrogen in the combustion analyses of the barium salt thus does not interfere with its use as a precursor to $BaTiO_3$. The ^1H NMR spectrum of a D_2O solution exhibits resonances due to the aromatic protons of the catecholate ligands (6.2–6.4 ppm; multiplet). No resonance due to the ammonium ion protons is observed with either D_2O or DMSO-d_6 as the solvent. The $^{13}C\{^1H\}$ NMR spectrum in D_2O exhibits three resonances at 111.8, 117.4, and 159.8 ppm due to the carbon atoms of the coordinated catecholate ligands. The IR spectrum shows that the broad band at about 3000–2600 cm^{-1} due to the ammonium ions is no longer present. There is a broad band at about 3500 cm^{-1} that is attributed to water of crystallization. Sharp bands due to the catecholate ligands appear in the region below 1600 cm^{-1} (the checkers found that the product sometimes showed a weak band in the 650-cm^{-1} region that may be associated with a $Ba_2[TiO(cat)_2]_2$ contaminant present at low level). A cyclic voltammogram of the complex measured in aqueous 1M KCl exhibits a single reversible wave at −1.40 V (vs. Ag/AgCl), indicating the absence of detectable dinuclear or titanyl species in solution.[4]

References

1. N. J. Ali and S. J. Milne, *Br. Ceram. Trans. J.*, **86**, 113 (1987).
2. R. H. Heistand and L. G. Duquette, *U.S. Patent* 4946810 (1990).
3. A. Rosenheim and O. Sorge, *Chem. Ber.*, **53**, 932 (1920).
4. B. A. Borgias, S. R. Cooper, Y. B. Koh, and K. N. Raymond, *Inorg. Chem.*, **23**, 1009 (1984).
5. J. A. Davies and S. G. Dutremez, *J. Am. Ceram. Soc.*, **73**, 1429 (1990).
6. J. A. Davies and S. G. Dutremez, *U.S. Patent* 5082812 (1992).
7. J. A. Davies and S. G. Dutremez, *J. Am. Ceram. Soc.*, **73**, 2570 (1990).

4. N-DONOR ADDUCTS OF DIMETHYLZINC

SUBMITTED BY ANTHONY C. JONES,* PAUL O'BRIEN,[†]
and JOHN R. WALSH[†]
CHECKED BY BRIDGETT KILLION,[‡] JOHN CANBELL,[‡]
STEPHEN SKOOG,[‡] and WAYNE GLADFELTER[‡]

Although adducts between dimethylzinc (Me_2Zn) and nitrogen donor ligands have been known for many years,[1, 2] they have only recently found a technological application as precursors in the growth of zinc chalcogenides by MOCVD,[3-5] and as the dopant sources in the growth of III–V materials[6] by MOCVD. For instance, adducts between Me_2Zn and triethylamine (NEt_3) or hexahydro-1,3,5-trimethyl-1,3,5-triazine (triazine) have been used successfully in combination with hydrogen selenide (H_2Se) or hydrogen sulfide (H_2S) for the growth of epitaxial ZnSe or ZnS layers respectively.[4]

The use of such adducts eliminates homogeneous reaction (prereaction) between the group II precursor and the group VI hydride. Layers of II–VI materials grown using such adducts have greatly improved structural electrical and optical properties as compared[4] with equivalent layers grown using base-free Me_2Zn.

The mechanisms underlying the improved quality of the layers has not precisely been determined, although the presence of nitrogen donor ligands has been shown to be essential.[7] The use of such adducts is likely to be of increasing importance, especially with the potential application of II–VI alloys in devices such as blue LEDs.[8] Adducts also have considerable potential as *p*-dopant sources for GaAs, InP, and related III–V semiconductors grown by MOCVD. The *N,N,N',N'*-tetramethyl-1,2-ethanediamine adduct of dimethylzinc ($ZnMe_2$ TMED) allows[6] the reproducible *p*-doping of GaAs over the range $p = 10^{17}$–10^{19} cm^{-3}. Such adducts are stable single sources and avoid the need for gas mixtures such as Me_2Zn/H_2.

Traditionally, the adducts have been prepared by adding stoichiometric amounts of the ligand to a solution of dimethylzinc (in diethyl ether or petroleum spirit) followed by removal of the solvent *in vacuo*. The compounds presented here were synthesized by direct reaction of the ligand and the electronic grade metal alkyl in the absence of any solvent.

* Epichem Ltd., Power Road, Bromborough, Wirral, L62 3QF, UK.
† Department of Chemistry, Queen Mary and Westfield College, University of London, Mile End Road, London, E1 4NS, UK. Present address: Department of Chemistry, Imperial College of Science Technology and Medicine, South Kensington, London, SW7 2AY, UK.
‡ Department of Chemistry, University of Minnesota, Minneapolis, MN 55455.

General Comments

The components are all very air-sensitive (most ignite on contact with tissue paper) and all manipulations must be carried out under an inert nitrogen (or argon) atmosphere in standard Schlenk apparatus.[9] If the precursors are to be used for the deposition of high-quality II–VI materials, extreme precautions for the exclusion of contaminants should be taken. The use of a greaseless vacuum line (with Teflon seals) is recommended. Electronic grade dimethylzinc was used throughout as supplied by Epichem Ltd. Bromborough, the Wirral, UK (similar grade material can be obtained from Air Products and Chemicals Inc., Allentown, PA). Dimethylzinc is provided in stainless-steel containers ("bubblers") and must therefore be distilled from the container before weighing. Dimethylzinc is an extremely air-sensitive and pyrophoric compound and is most easily handled frozen during any manipulation. The ligand is added slowly to a 100-mL flask containing a known weight of the zinc alkyl with a 5-mL Segma syringe through a Subaseal rubber septum. Distillation of the adducts is carried out through a glass-transfer tube under dynamic vacuum; the pure product was trapped in a receiver flask immersed in liquid nitrogen (− 196°C). The NMR spectra of the adducts show little or no evidence of impurities such as alkoxides; traces of such compounds are often seen in the spectra of zinc alkyls in the absence of Lewis bases.

■ **Caution.**   *The following compounds are very air-sensitive, have unpleasant odors and are to some degree toxic. Decomposition may lead to the formation of oxide clouds of a dangerous, choking, particle size. Extreme care must be taken in the handling and, in particular, in the disposal of waste material.*

A. DIMETHYL(TRIETHYLAMINE)ZINC

$$Me_2Zn + NEt_3 \longrightarrow Me_2Zn(NEt_3)$$

Triethylamine (2.60 mL, 18.7 mmol, Aldrich) is added from a calibrated syringe via a septum cap to dimethylzinc (1.78 g, 18.7 mmol) at − 196°C and the reaction mixture is allowed to warm to room temperature to give a crude yellow liquid which is then vacuum distilled (40°C, 10^{-2} torr) to give a colorless liquid. The yield on distillation is 3.78 g, 86.4%.

Properties

For $C_5H_{15}NZn$, the ^{1}H NMR spectrum shows peaks (C_6D_6, ppm) at $\delta = - 0.51$ (S, 6H, C$\underline{H}_3$Zn), $\delta = 0.79$ (T, 9H, C$\underline{H}_3$CH$_2$N), and $\delta = 2.39$ (Q, 6H, CH$_3$C$\underline{H}_2$N).

The gas-phase IR spectrum shows bands (cm^{-1}) at 606(s), 621(s), 705(s), 1177(m), 1193(m), 1295(m), 1305(m), 1704(m), 2837(m), 2847(m), 2907(s), and 2925(s). (The first three values refer to Zn-C rocking and asymmetric stretching frequencies[10]). The spectrum is consistent with the almost complete dissociation of the adduct in the vapor phase.[10]

The compound has been used on several occasions to grow extremely high-quality ZnSe by conventional, atmospheric pressure[4] or low-pressure[5] MOCVD.

Note added in proof: Recent studies (1996) indicate that when entrained the eutectic formed by this system contains slightly more NEt$_3$ than Me$_2$Zn.

B. DIMETHYL(*N,N,N',N'*-TETRAMETHYL-1,2-ETHANEDIAMINE-*N,N'*)ZINC

$$Me_2Zn + Me_2N(CH_2)_2NMe_2 \longrightarrow Me_2Zn[Me_2N(CH_2)_2NMe_2]$$

N,N,N',N'-tetramethyl-1,2-ethanediamine (2.9 mL, 19 mmol) is added from a calibrated syringe via a septum cap to dimethylzinc (1.81 g, 19 mmol) at − 196°C and the reaction mixture is allowed to warm to room temperature to give a crude white solid which is further purified by sublimation (55°C, 10^{-2} torr) to give a white crystalline solid. The yield on sublimation is 3.61 g, 89.8%. (mp 60°C).

Properties

For C$_8$H$_{22}$N$_2$Zn, the ^{1}H NMR spectrum shows peaks (C$_6$D$_6$, ppm) at $\delta = -0.50$ (S, 6H, C$\underline{H}_3$Zn), $\delta = 1.73$ (T, 4H, –NC$\underline{H}_2$C$\underline{H}_2$N–), and $\delta = 1.86$ (S, 12H, C$\underline{H}_3$N). The gas-phase IR spectrum shows bands (cm^{-1}) at 600(s), 695(s), 1130(w), 1303(w), 1615(w), 2301(w), 2849(m), 2908(s), 2918(s), 2989(s), and 3039(s). The low-frequency bands are characteristic of the Zn–C interactions discussed in Section 4.A.

C. BIS(HEXAHYDRO-1,3,5-TRIMETHYL-1,3,5-TRIAZINE)-DIMETHYLZINC

$$Me_2Zn + 2(CH_2NMe)_3 \longrightarrow Me_2Zn[(CH_2NMe)_3]_2$$

Hexahydro-1,3,5-trimethyl-1,3,5-triazine (5 mL, 35.6 mmol) is added from a calibrated syringe via a septum cap to dimethylzinc (1.52 g, 15.9 mmol) at − 196°C and the resulting mixture is allowed to warm to room temperature. Any excess triazine is removed *in vacuo*, resulting in a crude white solid which is further purified by sublimation (40°C, 10^{-2} torr) to give a brilliant white crystalline solid, occasionally as massive hexagonal crystals. The yield is 5.76 g, 99.4%. (mp 37.1°C).

Anal. Calcd. C, 46.99%; H, 10.20%; N, 23.08%. Found: C, 47.48%; H, 10.17%; N, 23.74%.

Properties

For $C_{14}H_{36}N_6Zn$, the ^1H NMR spectrum shows peaks (C_6D_6, ppm) at $\delta = -0.39$ (S, 6H, C$\underline{H}_3$Zn), $\delta = 2.04$ (S, 18H, $-CH_2NC\underline{H}_3$), and $\delta = 2.93$ (S, 12H, $-C\underline{H}_2NMe$).

The NMR spectra and the microanalysis often tend to indicate a slight excess of the triazine ligand. Because the material is a crystalline solid, this excess of ligand is presumably occluded in the solid.

The gas-phase IR spectrum shows bands (cm^{-1}) at: 917(w), 1121(m), 1161(s), 1304(m), 1342(m), 1388(m), 1456(w), 2782(m), and 3016(s). The 1:2 stoichiometry of this compound makes observation of the characteristic Me_2Zn frequencies difficult.

The X-ray crystal structures for compounds B and C have been determined.[11, 12]

References

1. K. H. Thiele, *Z. Anorg. Allg. Chem.*, **325**, 156 (1963).
2. For a review, see J. Boersma, in *Comprehensive Organometallic Chemistry*, G. Wilkinson, F. G. A. Stone, and E. W. Abel (eds.), Oxford, New York, 1982, pp. 823–862.
3. P. J. Wright, B. Cockayne, A. C. Jones, E. D. Orrell, P. O'Brien, and O. F. Z. Khan, *J. Cryst. Growth*, **94**, 97 (1988).
4. P. J. Wright, B. Cockayne, P. J. Parbrook, A. C. Jones, P. O'Brien, and J. R. Walsh, *J. Cryst. Growth*, **104**, 601 (1990).
5. (a) K. F. Jensen, A. Annapragada, K. L. Ho, J. S. Huh, S. Patnaik, and S. Salim, *J. de Physique*, **II**, C2–243 (1991); (b) J. S. Huh, S. Patnaik, and K. F. Jensen, *J. Elect. Mat.*, **22**, 509, (1993).
6. (a) P. J. Wright, B. Cockayne, A. C. Jones, and E. D. Orrell, *J. Cryst. Growth*, **91**, 63 (1988); (b) A. C. Jones, S. A. Rushworth, P. O'Brien, J. R. Walsh, and C. Meaton, *J. Cryst. Growth*, **130**, 295 (1993).
7. P. J. Wright, B. Cockayne, P. J. Parbrook, P. E. Oliver, and A. C. Jones, *J. Cryst. Growth*, **108**, 525 (1991).
8. P. J. Wright, B. Cockayne, P. J. Parbrook, K. P. O'Donnell, and B. Henderson, *Semicond. Sci. Technol.* **6**, A29 (1991).
9. D. F. Shriver and M. A. Drezdzon, *The Manipulation of Air-Sensitive Compounds*, 2nd ed., Wiley, New York, 1986.
10. O. F. Z. Khan, P. O'Brien, P. A. Hamilton, J. R. Walsh, and A. C. Jones, *Chemtronics*, **4**, 244 (1989).
11. M. B. Hursthouse, M. Motevalli, P. O'Brien, J. R. Walsh, and A. C. Jones, *Organometallics*, **10**, 3196 (1991).
12. M. B. Hursthouse, M. Motevalli, P. O'Brien, J. R. Walsh, and A. C. Jones, *J. Organomet. Chem.*, **499**, 1 (1993).

5. ARENE CHALCOGENOLATO COMPLEXES OF ZINC AND CADMIUM

SUBMITTED BY MANFRED BOCHMANN,*[‡] GABRIEL BWEMBYA,*
and KEVIN J. WEBB*
CHECKED BY M. A. MALIK,[†] J. R. WALSH,[†] and P. O'BRIEN[†]

As a rule, chalcogenolato complexes of divalent metals, such as zinc and cadmium, exist as coordination polymers with infinite lattices and bridging chalcogenolato ligands.[1] Thermal decomposition leads to polycrystalline metal chalcogenide phases.[2] The solid-state structures of the complexes are sensitive to the steric requirements of the chalcogenolato ligands. With suitably bulky substituents, such as 2,4,6-tri-*tert*-butylphenyl, molecular non-polymeric complexes with low metal coordination numbers are obtained[3, 4] which are volatile and suitable for gas-phase decomposition to give metal chalcogenide films.[5]

Although metal chalcogenolato complexes are often prepared from the metal dihalides (or acetates, nitrates etc.) and the chalcogenolate anion generated in polar solvents, this method was found to be less suitable for the introduction of very bulky ligands, such as those described below, and frequently gave products which proved difficult to purify satisfactorily. The protolysis of metalbis(trimethylsilyl)amides in nonpolar solvents described here is therefore preferable for the synthesis of high-purity materials.

General Procedure

■ **Caution.** *Hydrolysis of metal chalcogenolato complexes may lead to the liberation of chalcogenophenols and their decomposition products H_2S, H_2Se, or H_2Te; these have an unpleasant odor and are toxic. Although the chalcogenolato complexes described here can be handled in air without visible signs of decomposition, they should be stored under inert gas. All preparations should be carried out in a well-ventilated fume hood. Contaminated glassware should be treated with sodium hypochlorite (bleach) solution for several hours and thoroughly rinsed with water and acetone before removal from the fume hood.*

Unless otherwise stated, all operations are carried out in flame-dried glassware under an inert gas atmosphere using standard Schlenk techniques.[6]

* School of Chemical Sciences, University of East Anglia, Norwich, NR4 7TJ, UK.
† Department of Chemistry, Queen Mary and Westfield College, University of London, London, E1 4NS, UK.
Present address: ‡ School of Chemistry, University of Leeds, Leeds LS2 9JT, UK.

Solvents were dried over sodium (light petroleum, bp 40–60°C, toluene) or sodium-benzophenone (diethyl ether, THF). Sodium amide, hexamethyl-disilazane, lithium triethylhydroborate(1 −), and hydrogen tetrafluoro-borate(1 −) diethyl ether complex were obtained from Aldrich and used as supplied.

A. BIS[BIS(TRIMETHYLSILYL)AMIDO] COMPLEXES OF ZINC AND CADMIUM

$$ZnCl_2 + 2NaN(SiMe_3)_2 \longrightarrow Zn[N(SiMe_3)_2]_2 + 2NaCl$$

The procedure for the preparation of $Zn[N(SiMe_3)_2]_2$ is a modification of that described for $Sc[N(SiMe_3)_2]_3$.[7] Into a 250-mL three-necked flask equipped with magnetic stirrer, reflux condenser, and dropping funnel and connected to the inert gas supply of a vacuum line are added sodium amide (5.2 g, 0.133 mol) and toluene (100 mL). 1,1,1,3,3,3-Hexamethyldisilazane (29 mL, 0.137 mol) is added dropwise to the stirred suspension at room temperature over a period of 40 min. The reaction mixture is then refluxed for 6 h and allowed to cool to room temperature; the solvent is then removed in vacuo to give $Na[N(SiMe_3)_2]$ as a white residue which is dried in vacuo for 2 h, yielding 23.8 g (0.130 mol, 98%). This compound is dissolved in diethyl ether (100 mL), cooled to − 78°C, and anhydrous $ZnCl_2$ (7.62 g, 56 mmol) is added. The stirred mixture is allowed to warm to room temperature, refluxed for 1 h, and filtered, and the white residue is washed with diethyl ether (2 × 50 mL). The combined filtrates are evaporated in vacuo and the pale-yellow liquid residue is distilled at 106°C/7 mmHg to give $Zn[N(SiMe_3)_2]_2$ as a colorless liquid (15.1 g, 39 mmol, 70%).[8] ^{1}H NMR (CDCl$_3$): δ 0.078(s).

$Cd[N(SiMe_3)_2]_2$[8] is obtained from CdI_2 (43.7 g, 0.119 mol) and $NaN(SiMe_3)_2$ (43.8 g, 0.239 mol) following an analogous procedure as a pale-yellow liquid, bp 101°C/0.1 mmHg (41.3 g, 0.095 mol, 80%). ^{1}H NMR (CDCl$_3$): δ 0.073(s).

Bis(bis(trimethylsilyl)amido) complexes of zinc and cadmium are thermally stable but highly sensitive to hydrolysis and must be manipulated under inert gas.

B. BIS(2,4,6-TRI-*tert*-BUTYLBENZENETHIOLATO) AND -SELENOLATO COMPLEXES OF ZINC*

$$Zn[N(SiMe_3)_2]_2 + 2Ar''EH \longrightarrow Zn(EAr'')_2 + 2HN(SiMe_3)_2$$

$$E = S, Se; Ar'' = 2,4,6-t-Bu_3C_6H_2$$

* See cautionary note under *General Procedure*.

Into a 100-mL, three-necked flask, equipped with a magnetic stirring bar and connected to the inert gas supply of a vacuum line via a stopcock adaptor, are injected tri-*tert*-butylbenzene thiol[9] (1.0 g, 3.59 mmol) and light petroleum (20 mL). To the stirred solution is added via syringe $Zn[N(SiMe_3)_2]_2$ (0.66 g, 1.71 mmol). The mixture is heated gently with a hair dryer and allowed to stir at room temperature for 1 h. The white precipitate is filtered off, washed with light petroleum (2 × 10 mL), dissolved in warm toluene (ca. 10 mL) and recrystallized at room temperature to give fine, colorless crystals (0.88 g, 1.42 mmol, 82.9%) which sublime > 170°C/0.1 mmHg.

Anal. Calcd. for $C_{36}H_{58}ZnS_2$: C, 69.71; H, 9.42; S, 10.34. Found: C, 69.37; H, 9.67; S, 10.24.

The Se complex, $Zn(SeAr'')_2$, is made by an analogous procedure from 2,4,6-tri-*tert*-bytylbenzeneselenol[9] (1.03 g, 3.16 mmol) and $Zn[N(SiMe_3)_2]_2$ (0.66 g, 1.71 mmol) to give a white precipitate (0.6 g). More product can be recovered from the mother liquor by adding light petroleum (20 mL) and refluxing for 2–3 h followed by filtration. The combined yield is 0.8 g (1.12 mmol, 72% based on Zn).

Anal. Calcd. for $C_{36}H_{58}ZnSe_2$: C, 60.55; H, 8.18; Found: C, 60.52; H, 8.19; mp 210°C (decomp).

Properties

The zinc complexes are colorless solids which are soluble in chloroform, dichloromethane, and warm toluene, but only sparingly in light petroleum. The solids may be handled in air without visible signs of deterioration but should be stored under inert gas. Solutions are sensitive to hydrolysis. The sulfur complex sublimes above 170°C/0.01 mmHg. $Zn(SeAr'')_2$ is thermally more sensitive and melts with decomposition at 210°C. The dimeric nature of the compounds in the solid state was confirmed by the crystal structure of $Zn(SAr'')_2$.[10] The complexes form isolable crystalline three-coordinate adducts with diethyl ether,[4,11] tetrahydrofuran,[4] 2,6-dimethylpyridine, or diphenylmethylphosphine.[12] $Zn(SAr'')_2$ reacts with *N*-methylimidazole to give 1:1 and 1:2 adducts, depending on the concentration of the donor ligand. With ketones or aldehydes, four-coordinate monomeric or dimeric adducts are formed.[13] ^{1}H NMR (CDCl$_3$): $Zn(SAr'')_2$: δ 1.04 (s,9 H), 1.23 (s, 18 H), 6.87(s, 2 H); $Zn(SeAr'')_2$: δ 1.14 (s, 9 H), 1.40 (s, 18 H), 7.20 (s, 2 H).

C. BIS(2,4,6-TRI-*tert*-BUTYLBENZENETHIOLATO) AND -SELENOLATO COMPLEXES OF CADMIUM*

$$Cd[N(SiMe_3)_2]_2 + 2Ar''EH \rightarrow Cd(EAr'')_2 + 2HN(SiMe_3)_2$$

$$E = S, Se; Ar'' = 2,4,6-t-Bu_3C_6H_2$$

The same apparatus and procedure as in Section 5.A are used. To prepare $Cd(SAr'')_2$, 2,4,6-tri-*tert*-butylbenzenethiol[9] (1.5 g, 5.39 mmol) is dissolved in light petroleum (20 mL). $Cd[N(SiMe_3)_2]_2$ (1.08 g, 2.5 mmol) is added to the stirred solution via syringe. After a few seconds, a white solid precipitates. The reaction is warmed on a warm water bath (ca. 40°C) for 1 h. The supernatant liquid is filtered off and the residue recrystallized from toluene at 10°C to give colorless crystals (1.0 g). A second fraction is obtained from the mother liquor; the combined yield is 1.2 g (72% based on cadmium).

Anal. Calcd. for $C_{36}H_{58}CdS_2$: C, 64.79; H, 8.76; S, 9.61. Found: C, 64.83; H, 8.84; S, 9.29, mp 320°C.

$Cd(SeAr'')_2$ is prepared by an analogous procedure. To a stirred solution of 2,4,6-tri-*tert*-butylbenzeneselenol[9] (1.5 g, 4.62 mmol) in light petroleum (40 mL) is added $Cd[N(SiMe_3)_2]_2$ (1.0 g, 2.3 mmol) by syringe. The solution turns bright yellow and the product begins to precipitate after a few minutes. Stirring is continued for 3 h. The solvent is removed in vacuo and the yellow residue is washed with light petroleum (2×20 mL) and dissolved in hot (100–110°C) toluene (90 mL). Bright yellow crystals are obtained on cooling the solution overnight to 10°C; these are filtered off, washed with petroleum, and dried in vacuo. More product is obtained from the filtrate; the total yield is 1.33 g (1.75 mmol, 76% based on cadmium).

Anal. Calcd. for $C_{36}H_{58}CdSe_2$: C, 56.81; H, 7.68. Found: C, 56.89; H, 7.91, mp 300°C (dec).

Properties

$Cd(SAr'')_2$ and $Cd(SeAr'')_2$ are soluble in chloroform, dichloromethane, and toluene and sparingly soluble in light petroleum. Both complexes are dimeric in the solid state.[3,4] $Cd(SAr'')_2$ sublimes above 240°C/0.001 mmHg and the Se analogue sublimes above 220°C; they deposit cadmium chalcogenide films from the vapor phase at 480 and 360°C, respectively.[5] ^{1}H NMR (CD_2Cl_2): $Cd(SAr'')_2$: δ 1.25 (s, 9 H), 1.77 (s, 18 H), 7.20 (s, 2 H); Cd $(SeAr'')_2$: δ 1.28 (s, 9 H), 1.56 (s, 18 H), 7.24 (s, 2 H).

* See cautionary note under *General Procedure*.

D. BIS(2,4,6-TRIMETHYLBENZENETELLUROLATO) COMPLEXES OF ZINC AND CADMIUM*

$$M[N(SiMe_3)_2]_2 + 2ArTeH \rightarrow M(TeAr)_2 + 2HN(SiMe_3)_2$$

$$M = Zn, Cd; Ar = 2,4,6\text{-}Me_3C_6H_2$$

The apparatus and procedure described in Section 5.A is used. To a cold ($-30°C$) solution of 2,4,6-trimethylbenzenetellurol in light petroleum, prepared in situ from $(2,4,6\text{-}Me_3C_6H_2Te)_2$ (1.0 g, 2.03 mmol),[9] is added via syringe $Zn[N(SiMe_3)_2]_2$ (0.67 g, 1.73 mmol). A white to pale beige solid precipitates from the reddish solution. After stirring at $-10°C$ for 4 h, the product is filtered off, washed with warm toluene and light petroleum (2×20 mL), and dried in vacuo to give an off-white to beige microcrystalline solid (0.68 g, 1.22 mmol, 70%).

Anal. Calcd. for $C_{18}H_{22}Te_2Zn$: C, 38.70; H, 4.25. Found: C, 39.14; H, 4.39, mp 224–225°C (decomp).

$Cd(TeC_6H_2Me_3\text{-}2,4,6)_2$ is prepared by an analogous procedure from 2,4,6-$Me_3C_6H_2TeH$ (0.55 g, 2.2 mmol) and $Cd[N(SiMe_3)_2]_2$ (0.47 g, 1.08 mmol) at $-20°C$ to room temperature as a yellow precipitate which is recrystallized from N,N-dimethylformamide to give yellow crystals (0.37 g, 0.54 mmol, 54%).

Anal. Calcd. for $C_{18}H_{22}CdTe_2$: C, 35.35; H, 3.66. Found: C, 35.63; H, 3.68, mp 230–240°C (decomp).

Properties

The zinc and cadmium tellurolato complexes are air-sensitive polymeric solids[14] which should be handled under inert gas. They are sparingly soluble in petroleum and chlorocarbons but readily dissolved by coordinating solvents (N,N-dimethylformamide, dimethylsulfoxide, or pyridine). $Zn(TeAr)_2$ forms a crystalline 1:2 adduct with 2-methyl-1H-imidazole, $Zn(TeAr)_2(imid)_2$. Heating solutions of $Zn(TeAr)_2$ and $Cd(TeAr)_2$ in mesitylene leads to formation of ZnTe and CdTe, respectively. ^1H NMR: $Zn(TeAr)_2$ (DMSO-d_6): δ 2.1 (s, 3H), 2.4 (s, 6H), 6.8 (s, 2H); $Cd(TeAr)_2$ (pyridine-d_5): δ 2.13 (s, 3H), 2.3 (s, 6H), 6.72 (s, 2H).

---

* See cautionary note under *General Procedure*.

References

1. I. G. Dance, *Polyhedron*, **5**, 1037 (1986).
2. J. G. Brennan, T. Siegrist, P. J. Carroll, S. M. Stuczynski, P. Reynders, L. E. Brus, and M. L. Steigerwald, *Chem. Mater.*, **2**, 403 (1990); K. Osakada and T. Yamomoto, *Inorg. Chem.*, **30**, 2328 (1991).
3. M. Bochmann, K. J. Webb, M. Harman, and M. B. Hursthouse, *Angew. Chem.*, **102**, 703 (1990); *Angew. Chem. Int. Ed. Engl.*, **29**, 638 (1990).
4. M. Bochmann, K. J. Webb, M. B. Hursthouse, and M. Mazid, *J. Chem. Soc., Dalton Trans.*, 2317 (1991).
5. M. Bochmann and K. J. Webb, *Mater. Res. Soc. Symp. Proc.*, **204**, 149 (1991); M. Bochmann, K. J. Webb, J. E. Halis, and D. Wolverson, *Eur. J. Solid State Inorg. Chem.*, **29**, 155 (1992).
6. D. F. Shriver and M. A. Drezdzon, *The Manipulation of Air-Sensitive Compounds*, 2nd ed., Wiley, New York, 1986.
7. D. C. Bradley and R. G. Copperthwaite, *Inorg. Synth.*, **18**, 114 (1978).
8. H. Bürger, W. Sawodny, and U. Wannagat, *J. Organomet. Chem.*, **3**, 113 (1965).
9. See preparative procedure in Chapter 2, Section 26.
10. M. Bochmann, G. Bwembya, R. Grinter, J. Lu, K. J. Webb, D. J. Williamson, M. B. Hursthouse, and M. Mazid, *Inorg. Chem.*, **32**, 532 (1993).
11. P. P. Power and S. C. Shoner, *Angew. Chem. Int. Ed. Engl.*, **29**, 1403 (1990).
12. M. Bochmann, G. Bwembya, R. Grinter, A. K. Powell, K. J. Webb, M. B. Hursthouse, K. M. A. Malik, and M. A. Mazid, *Inorg. Chem.*, **33**, 2290 (1994).
13. M. Bochmann, K. J. Webb, M. B. Hursthouse, and M. Mazid, *J. Chem. Soc., Chem. Commun.*, 1735 (1991).
14. M. Bochmann, A. P. Coleman, K. J. Webb, M. B. Hursthouse, and M. Mazid, *Angew. Chem.*, **103**, 973 (1991); *Angew. Chem. Int. Ed. Engl.*, **30**, 973 (1991); M. Bochmann, G. C. Bwembya, A. K. Powell, and X. Song, *Polyhedron* **14**, 3495 (1995).

6. ARENE THIOLATO, SELENOLATO, AND TELLUROLATO COMPLEXES OF MERCURY

SUBMITTED BY MANFRED BOCHMANN*‡ and KEVIN J. WEBB*
CHECKED BY M. A. MALIK and P. O'BRIEN†

Chalcogenolato complexes of mercury can be prepared by a variety of methods. Early preparations involve the reactions of thiols with mercury cyanide,[1] the reaction of mercury salts with alkali chalcogenolates, electrochemical methods,[2] and the oxidative addition of dichalcogenides to metallic mercury.[3] The last method is very convenient for the preparation of complexes with sterically undemanding ligands, but becomes less facile as the

* School of Chemical Sciences, University of East Anglia, Norwich NR4 7TJ, UK.
† Department of Chemistry, Queen Mary and Westfield College, University of London, London, E1 4NS, UK.
‡ Present address: School of Chemistry, University of Leeds, Leeds LS2 9JT, UK.

mercury–chalcogen bond strength decreases in the sequence S > Se > Te. The resulting mercury complexes show great structural diversity; whereas small alkylthiolato complexes $Hg(SR)_2$ (R = Me, Et) exits as essentially monomeric linear molecules in the solid state,[4] related selenolates as well as other alkyl (R = t-Bu) and arene thiolates form infinite polymeric lattices.[5] The compounds decompose thermally either with reductive elimination of organic dichalcogenides[3d] or formation of mercury chalcogenides.[6]

The preparations given below are representative examples of the synthesis of mercury chalcogenolato compounds by several different routes and apply particularly to the preparation of monomeric, sterically highly hindered complexes.

General Procedure

■ **Caution.** *Chalcogenophenols have an unpleasant odor and are toxic.* H_2S, H_2Se, *or* H_2Te *may be liberated on treatment with acid or exposure to the open air. These compounds should therefore be handled under inert gas in a well-ventilated hood. Contaminated glassware should be treated with sodium hypochlorite (bleach) solution for several hours and rinsed thoroughly with water and acetone before removal from the fume hood. Mercury and mercury salts are highly toxic, and skin and eye contact must be avoided. Mercury residues should be disposed of as toxic heavy metal waste.*

Unless otherwise stated, all operations are carried out in flame-dried glassware under an inert gas atmosphere using standard Schlenk techniques.[7] Solvents were dried over sodium (light petroleum, bp 40–60°C) or sodium benzophenone (diethyl ether, THF) or calcium hydride (acetonitrile). $Li[BHEt_3]$ in THF (Superhydride®) and sodium bis(trimethylsilyl)amide were used as purchased (Aldrich).

A. BIS[BIS(TRIMETHYLSILYL)AMIDO]MERCURY

$$HgCl_2 + 2NaN(SiMe_3)_2 \longrightarrow Hg[N(SiMe_3)_2]_2 + 2NaCl$$

Procedure

The procedure follows that described by Bürger et al.[8] A 500-mL three-necked flask equipped with a reflux condenser, a rubber septum on one sidearm, and a magnetic follower and connected to the inert gas supply is charged with $HgCl_2$ (10 g, 36.8 mmol) and 50 mL of diethyl ether. To this is added through the rubber septum via syringe a solution of $NaN(SiMe_3)_2$ in diethyl ether (1.38 M, 53 mL, 73.4 mmol). The mixture is stirred for 0.5 h and the reaction completed by refluxing for 0.5 h. The pale-yellow solution is allowed to cool and filtered under nitrogen into a 500-mL two-necked flask via a stainless-steel filter cannula. The residue is washed with diethyl ether (2×50 mL), and the washings are combined with the filtrate. The solvent is removed in vacuo at room temperature. The residue is vacuum-distilled at 80°C/0.1 mm Hg to give the product as a colorless liquid (10.3 g, 19.8 mmol, 53%). The compound was used without further purification.

B. BIS(2,4,6-TRI-*tert*-BUTYLBENZENETHIOLATO)MERCURY*

$$2Ar''SH + Hg[N(SiMe_3)_2]_2 \longrightarrow Hg(SAr'')_2 + 2HN(SiMe_3)_2$$

$$Ar'' = 2,4,6\text{-}t\text{-}Bu_3C_6H_2$$

Procedure

A 100-mL, three-necked reaction flask equipped with a rubber septum on one of the joints and a magnetic stirring bar and connected via a stopcock adaptor to the inert gas supply of the vacuum line is used. The flask is charged with 2,4,6-tri-*tert*-butylbenzenethiol[9] (0.5 g, 1.8 mmol), and light petroleum (20 mL) is added. A solution of $Hg[N(SiMe_3)_2]_2$ (0.53 g, 0.9 mmol) in light petroleum (5 mL) is added via syringe at room temperature. The mixture is stirred for 2 h and filtered under inert gas through a stainless-steel filter cannula. The solvent is removed in vacuo and the residue recrystallized from toluene (ca. 5 mL) at $-20°C$. The product is obtained as white microcrystals (0.47 g, 0.62 mmol, 69% based on Hg).

Anal. Calcd. for $C_{36}H_{58}HgS_2$: C, 57.23; H, 7.74; S, 8.49. Found: C, 56.90; H, 7.72; S, 8.20.

Properties

$Hg(SAr'')_2$ is a monomeric complex and forms colorless crystals which dissolve readily in chloroform or toluene but are less soluble in light petroleum. The IR spectrum shows a strong Hg-S stretching band at 378 cm^{-1} (ν_{asym}), while ν_{sym} occurs at 332 cm^{-1} in the Raman spectrum. The ^{199}Hg NMR resonance is found at $\delta = -1274$ ppm (relative to neat $HgMe_2$). The compound sublimes above 170°C/0.01 mmHg and decomposes at 270°C.[10] It forms a crystalline T-shaped 1:1 adduct with pyridine.[11] ^{1}H NMR (CDCl$_3$): δ 1.52 (s, 9H), 1.80 (s, 18H), 7.40 (s, 2H).

---

* See cautionary note under *General Procedure*.

C. BIS(2,4,6-TRI-*tert*-BUTYLBENZENESELENOLATO)MERCURY*

$$Ar''_2Se_2 + 2\ LiBHEt_3 \longrightarrow 2Ar''SeLi + 2\ BEt_3 + H_2$$

$$2Ar''SeLi + HgCl_2 \longrightarrow Hg(SeAr'')_2 + 2LiCl$$

$$Ar'' = 2,4,6\text{-}t\text{-}Bu_3C_6H_2$$

The apparatus described in Section 6.B is used. LiSeAr'' is prepared by injecting a solution of Li[BHEt$_3$] in THF (Super-Hydride®, 1 M, 1.6 mL, 1.6 mmol) via syringe into a solution of Ar''$_2$Se$_2$[9] (0.50 g, 0.77 mmol) in THF (20 mL) at room temperature. The mixture is stirred for 30 min, after which the residual Li[BHEt$_3$] is destroyed by injecting a small amount of ethanol (ca. 0.5 mL). To this mixture is added HgCl$_2$ (0.20 g, 0.77 mmol) at room temperature. Stirring is continued for 2 h. The solvent is removed in vacuo and the residue is extracted with toluene (2 × 10 mL). Concentration in vacuo to about 10 mL and cooling to − 20°C overnight gives pale yellow crystals of the product (0.48 g, 0.57 mmol, 74%).

Anal. Calcd. for $C_{36}H_{58}HgSe_2$: C, 50.91; H, 6.78. Found: C, 51.16; H, 6.95.

Properties

Hg(SeAr'')$_2$ is a monomeric complex and forms yellow crystals which are soluble in chloroform and toluene. It should be protected from strong sunlight and is best stored in the dark below 10°C. The complex is not accessible from Ar''$_2$Se$_2$ and Hg. The mercury–selenium stretching frequencies occur at 270 cm^{-1} (ν_{asym}, IR active) and 187 cm^{-1} (ν_{sym}, Raman active). The ^{199}Hg NMR resonance is found at $\delta = -1734$ ppm (relative to neat HgMe$_2$). The compound sublimes above 150°C/0.01 mmHg and decomposes at 330–350°C[10]. ^{1}H NMR (CDCl$_3$): δ 1.40 (s, 9H), 1.67 (s, 18H), 7.55 (s, 2H).

D. BIS(2,4,6-TRIMETHYLBENZENETELLUROLATO)MERCURY*

$$Hg + Ar_2Te_2 \longrightarrow Hg(TeAr)_2$$

$$Ar = 2,4,6\text{-}Me_3C_6H_2$$

The reaction is carried out in a 100-mL, three-necked flask connected via a stopcock adaptor to the inert gas supply of a vacuum line. Mercury metal (0.40 g, 2.0 mmol) is added to a stirred solution of Ar$_2$Te$_2$[12] (1.09 g, 2.2 mmol)

---

* See cautionary note under *General Procedure.*

in toluene (20 mL) at room temperature. The mixture is stirred vigorously until the mercury has disappeared (ca. 24 h). The pale yellow precipitate is filtered off, washed with toluene and diethyl ether, dissolved in hot pyridine (5 mL), and filtered hot to remove mercury metal residues. The filtrate is left to crystallize at $-20°C$ to give yellow crystals of the product. A second fraction can be obtained by the addition of light petroleum to the mother liquor. The combined yield is 0.97 g (1.4 mmol, 70% based on Hg).

Anal. Calcd. for $C_{18}H_{22}HgTe_2$: C, 31.15; H, 3.19. Found: C, 31.14; H, 3.18.

Properties

$Hg(TeAr)_2$ is a polymeric compound; it forms deep yellow crystals which darken on heating above 100°C, sublime with some decomposition at 184°C, and decompose rapidly at 215°C. ^1H NMR (pyridine-d_5): δ 2.20 (s, 3H), 2.62 (s, 6H), 6.84 (s, 2H).

References

1. E. Wertheim, *J. Am. Chem. Soc.*, **51**, 3662 (1929).
2. F. F. Said and D. G. Tuck, *Inorg. Chim. Acta*, **59**, 1 (1982).
3. (a) H. Lecher, *Ber. Dtsch. Chem. Ges.*, **48**, 1425 (1915); (b) D. Spinelli and C. Dell'Erba, *Ann. Chim. (Rome)*, **51**, 45 (1961); (c) E. Kostiner, M. L. N. Reddy, D. S. Urch, and A.G. Massey, *J. Organomet. Chem.*, **15**, 383 (1968); (d) Y. Okamoto and T. Yano, *J. Organomet. Chem.*, **29**, 99 (1971).
4. (a) D. C. Bradley and N. R. Kunchur, *J. Chem. Phys.*, **40**, 2258 (1964); (b) *Can. J. Chem.*, **43**, 2786 (1965).
5. (a) N. R. Kunchur, *Nature (London)*, **204**, 468 (1964); (b) A. P. Arnold, A. J. Canty, B. W. Skelton, and A. H. White, *J. Chem. Soc., Dalton Trans.*, 607 (1982).
6. M. E. Peach, *J. Inorg. Nucl. Chem.*, **35**, 1046 (1973); M. L. Steigerwald and C. R. Sprinkle, *J. Am. Chem. Soc.*, **109**, 7200 (1989).
7. D. F. Shriver and M. A. Drezdzon, *The Manipulation of Air-Sensitive Compounds*, 2nd ed., Wiley, New York, 1986.
8. H. Bürger, W. Sawodny, and U. Wannagat, *J. Organomet. Chem.*, **3**, 113 (1965).
9. The preparation of this compound is described in CHAPTER 2, Section 26.
10. M. Bochmann and K. J. Webb, *J. Chem. Soc., Dalton Trans.*, 2325 (1991).
11. M. Bochmann, K. J. Webb, and A. K. Powell, *Polyhedron*, **11**, 513 (1992).
12. M. Akiba, M. V. Lakshimikantham, K. Y. Jen, and M. P. Cava, *J. Org. Chem.*, **49**, 4819 (1984).

7. ELECTRONIC GRADE ALKYLS OF GROUP 12 AND 13 ELEMENTS

SUBMITTED BY DOUGLAS F. FOSTER* and DAVID J.
COLE-HAMILTON*
CHECKED BY RICHARD A. JONES[†]

Epitaxy is the deposition of a thin, single-crystalline layer of a material onto a single crystal substrate in such a way that the crystal structure of the substrate orders the growth of the thin layer. When the growing film is deposited by the codecomposition of volatile organometallic compounds from the gas phase, the technique is generally known as metal-organic vapor phase epitaxy (MOVPE). From the early studies of Manasevit,[1] the technique has been developed and improved so that it now provides production technology for a variety of different materials, usually semiconductors.[2]

The electrical properties of the grown films are of fundamental importance to their use in devices and these properties are greatly modified by the presence of impurities. It is, therefore, essential to prepare organometallic compounds in a very high state of purity (< 1 ppm of total metallic impurities). Although physical methods such as fractional distillation, fractional crystallization, or zone refining can provide this level of purification, it is usually at the expense of discarding up to 70% of the material. The chemical method, adduct purification, which has been developed by two groups,[3,4] effectively separates the boiling points of desired alkyl from those of any impurities, as shown in Scheme 1. Materials of very high purity[‡] can be produced by this method, often in yields approaching 100%.

The first examples of dissociable adducts of organometallic compounds were used for the removal of coordinating solvents[6,7] (e.g., ethers) or for analysis.[8] However, the technique is now used routinely for the production of electronic grade organometallic compounds. In this section, the preparation and/or purification of the common precursors trimethylindium,[5] trimethylgallium,[5] trimethylaluminum,[5] triethylindium,[5] dimethylcadmium,[9] dimethylzinc,[10] and diethylzinc[10] are fully described. These metal alkyls can be prepared by many different routes;[11-13] the syntheses of unpurified trimethylgallium[14] and dimethylzinc[15] previously appeared in *Inorganic*

---

* School of Chemistry, University of St. Andrews, St. Andrews, Fife KY16 9ST, Scotland, UK.
[†] Department of Chemistry and Biochemistry, The University of Texas at Austin, Austin, TX-78712.
[‡] Impurities are undetectable by inductively coupled plasma emission spectroscopy, sensitivity ~ 50 ppb.[4,5]

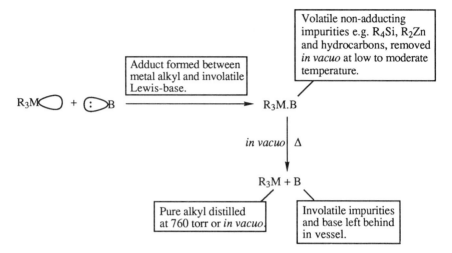

Scheme 1. Principles of adduct purification, taking a Group 13 metal alkyl as the example.

Syntheses. The syntheses presented here offer two major advantages: (1) base-free metal alkyls (pyrophoric) are not present until the final purification stage when they are isolated, and (2) our syntheses involve purification steps that result in products of very high (electronic grade) purity. For example, using inductively coupled plasma emission spectroscopy we have shown that common electronic impurities such as silicon and magnesium can be removed to levels below the detection levels of the instrument (parts per billion).[5] More importantly, photoluminescence spectra and carrier mobilities of, for example, indium phosphide, grown with trimethylindium before and after adduct purification, are significantly improved.[16]

GENERAL COMMENTS

Equipment

The user should be familiar with the Schlenk technique for handling air- and moisture-sensitive compounds. Familiarity with cannula transfer and filter techniques is also required.[17] For the preparation of ultrapure metal alkyls it is essential to avoid the use of any grease both on the vacuum-gas line and on the apparatus, as illustrated in Figs. 1–6. Thus, "grease-free" and vacuum-tight PTFE "Young" piston valves and piston gas valves are employed and all apparatus is assembled with rigid PTFE sleeves in the ground-glass joints to ensure vacuum tight seals.

All glassware is thoroughly cleaned and dried in an oven at $> 100°C$ before assembly. The assembled apparatus is further dried by evacuating the assembly and flaming its outer surfaces with a blue bunsen flame until moisture from the flame no longer condenses on the glass (avoid the ground-glass joints) before filling with purified argon or nitrogen. Glassware used for the preparation of a group 12 electronic grade metal alkyl should not then be used for the preparation of a group 13 electronic grade metal alkyl (and vice versa), thus avoiding cross-contamination.

Magnetic stirring bars should be the large oval type—choose the largest that will fit into the flask. These work well in round-bottomed vessels and can move large amounts of solid.

Owing to the hazardous nature of the products and the toxic nature of several of the solvents and starting materials, all operations are routinely carried out in an efficient fume hood.

Gases

Purified argon or nitrogen are recommended as inert gases. Either gas is dried by passing through two columns (50-cm length, 2.5-cm diameter) filled with 4-Å molecular sieves. These columns are placed on each side of a similar column filled with chromium(II) oxide on silica gel, which acts as an oxygen scavenger. This column is prepared by reduction of chromium(VI) oxide on silica gel with carbon monoxide.[17]

■ **Caution.** *Chromium (VI) oxide is carcinogenic. Inhalation of the fine CrO_3-silica powder must be avoided by working in a well-ventilated fume hood. The reduction of this powder with CO also necessitates the use of an efficient fume hood. The final Cr (II) catalyst is pyrophoric and should not be allowed to come into contact with air.*

Solvents

Solvents are all of Analar grade and should be freshly distilled and degassed prior to use. Diethyl ether, petroleum spirit (40/60°C), and THF are all dried and freed from dissolved molecular oxygen by distillation, under nitrogen, from a solution of the solvent, benzophenone, and sodium. Benzene and toluene are distilled under nitrogen from sodium.

Starting Materials

Trimethylaluminum is available from many commercial sources, including Aldrich (97%), which also supplies a 2.0-molar solution in hexanes which may be substituted for the neat metal alkyl.

If ultrapure metal alkyl products are required, the following are recommended as sources for the metal halides: $GaCl_3$ (Puratronic®, Johnson Matthey; 99.99%, Aldrich), $InCl_3$ (Puratronic®, Johnson Matthey; 99.999%, Aldrich), $CdCl_2$ (Anhydrous 99.999%, Johnson Matthey; 99.99%, Aldrich) and $ZnCl_2$ (Ultra-Dry 99.99%, Johnson Matthey; 99.999%, Aldrich). These materials are very expensive and the following may be used when ultra-pure metal alkyls are not required: $GaCl_3$ (as above), $InCl_3$ (98%, Aldrich), $CdCl_2$ (99 + %, Aldrich) and $ZnCl_2$ (98%, Aldrich).

The Grignard reagents are prepared by standard procedures.[18] If ultrapure metal alkyl products are required, magnesium pieces (Puratronic®, Johnson Matthey) should be used; otherwise standard magnesium turnings (99.9%, Johnson Matthey; 99.95%, Aldrich) will suffice. Methyliodide (99.5%, Aldrich) and ethylbromide (99 + %, Aldrich) are freshly distilled from P_2O_5 prior to use.

Methyllithium complexed with lithium bromide (1.5-molar Et_2O) is preferred over uncomplexed methyllithium owing to its greater stability (Aldrich).

The Lewis bases—4,4'-methylenebis(N,N'-dimethylaniline) (98%, Aldrich), 2,2'-bipyridine (99 + %, Aldrich), and 1,3-bis(4-pyridyl)propane (98%, Aldrich)—are all purified prior to use. They are dissolved in THF, decolorized with activated charcoal, and dried over CaH_2. The resulting clear colorless solutions of the Lewis bases are reduced in volume, in vacuo, and the Lewis bases are crystallized by addition of petroleum spirit and cooling to $-30°C$.

Yields

The preparations reported here are all on a ~ 0.5 molar scale and the yields given are typical, within a few percent, of what might be expected by an experienced worker. The preparations can be scaled up or down, but on a small scale the overall yield of a purified metal alkyl is likely to fall significantly from the figure given in the text.

■ **General caution and safety.** *The metal alkyls trimethylaluminum, trimethylgallium, trimethylindium, triethylindium, dimethylzinc, and diethylzinc all spontaneously inflame in air (pyrophoric) and are violently reactive toward water and carbon dioxide. Dimethylcadmium fumes upon exposure to air but does not inflame, and decomposes, sluggishly, in water. As with cadmium dichloride, and in common with all cadmium compounds, dimethylcadmium is extremely toxic. A dry powder extinguisher, sand bucket, and fire blanket should be at hand at all times.*

It has been reported that base-free trimethylindium can detonate during sublimation; the sublimation should be carried out behind a blast screen.

Face protection and gloves should be worn at all times and all manipulations should be carried out in a well-ventilated fume hood.

The procedures described have been carefully designed so that base-free metal alkyls are not present except in the final purification stages when they are isolated. None of the intermediates is pyrophoric (except methyllithium if it is dried from solvent) so that the risk of fire is minimal if a breakage should occur during the syntheses.

Benzene is carcinogenic and methyliodide and ethylbromide are highly toxic. They should be handled with the same care and precautions as detailed above.

Treatment of fires involving metal alkyls.   *If a flask containing base-free metal alkyl breaks, there will be a fire at the point of breakage. This may extinguish itself as the break seals with metal oxide and hydroxide. Alternatively, it should be extinguished with a powder extinguisher.*

Broken flasks containing metal alkyls in which the break has self-sealed by oxide/hydroxide formation, or powder from an extinguisher used to put out a fire, should be treated as follows: A metal can or wastebasket large enough to accommodate the broken flask is filled to a suitable depth (enough to cover the flask) with a 1:1 mixture of propan-2-ol and xylene. Liquid nitrogen is added to cool the mixture and provide a blanket of nitrogen gas. A skilled chemist wearing full face mask, rubber apron, and gloves then carefully lifts the broken flask and places it in the can. If the oxide/hydroxide plug becomes dislodged during lifting of the flask, a fire will result, so the break should always be pointing away from the operator, and the transfer should be continued. If the flask floats on the propan-2-ol/xylene mixture, it should be weighted down (a wrench or similar device is often appropriate) and a vigorous reaction will occur once the oxide/hydroxide plug is penetrated by the solvent mixture. This procedure has been used successfully without incident for the disposal of Me_3Al (~ 100 g in a broken dropping funnel) and Me_2Zn (~ 30 g in a storage flask from which the neck had broken off). In both cases the break sealed with oxide/hydroxide after a short fire and the plug remained intact during the transfer process.

A. TRIMETHYLINDIUM

(a) $InCl_3 + 3MeMgI \xrightarrow{Et_2O} Me_3In \cdot OEt_2 + 3ClMgI$

or $InCl_3 + 3MeLi \xrightarrow{Et_2O} Me_3In \cdot OEt_2 + 3LiCl$

(b) $Me_3In \cdot OEt_2 \xrightarrow[\text{in vacuo}]{25°C} Me_3In + OEt_2\uparrow$

(c) $2Me_3In\cdot OEt_2$ + $\left(Me_2N-\!\!\left\langle\bigcirc\right\rangle\!\!-_2CH_2\right)$ $\xrightarrow[\text{(ii) petroleum spirit}]{\text{(i) benzene}}$

$\left(Me_3In\cdot Me_2N-\!\!\left\langle\bigcirc\right\rangle\!\!-_2CH_2\right)$ + $2OEt_2\uparrow$

or $2Me_3In$ + $\left(Me_2N-\!\!\left\langle\bigcirc\right\rangle\!\!-_2CH_2\right)$ $\xrightarrow[\quad]{\text{petroleum spirit}}$ $\left(Me_3In\cdot Me_2N-\!\!\left\langle\bigcirc\right\rangle\!\!-_2CH_2\right)$

(d) $\left(Me_3In\cdot Me_2N-\!\!\left\langle\bigcirc\right\rangle\!\!-_2CH_2\right)$ $\xrightarrow[\text{in vacuo}]{80-130°C}$ $\left(Me_2N-\!\!\left\langle\bigcirc\right\rangle\!\!-_2CH_2\right)$ + $2Me_3In\uparrow$

■ **Caution.** *Trimethylindium is pyrophoric. Methyllithium is potentially pyrophoric. The chemicals benzene and methyliodide are highly toxic. See general caution and safety section.*

Procedure

(a) Trimethylindium Etherate. The apparatus as shown in Fig. 1 is assembled and flushed thoroughly with argon. This is most easily achieved by closing valve C, and using valve B successively fully to evacuate and then fill the whole system to atmospheric pressure with argon. With a slight positive pressure of argon applied to the system through valve B, valve C is reopened to vent the argon through the paraffin-oil bubbler.

The argon flow through the system is now increased while the reflux condenser is gently lifted and 110 g of indium trichloride (0.497 mol) together with 200 mL of diethyl ether are placed in the reaction flask. Indium trichloride is hygroscopic and should be handled in air for as short a time as possible. The reflux condenser is rehoused and the rapid purge of argon through valve B is maintained for 10–15 min to ensure that any air which may have entered the system during this process is flushed out. Valves B and A are now closed and a slow argon flow is maintained by passing argon directly over the top of the reflux condenser and out through the paraffin-oil bubbler. The flow rate of argon through the paraffin-oil bubbler is adjusted as necessary during heating or cooling procedures.

The Grignard solution, prepared by the reaction of 110 mL of methyliodide (1.77 mol) with 44 g of magnesium pieces (1.81 mol) in 1000 mL of diethyl ether is added, through the "Suba-Seal," to the dropping funnel using a stainless-steel cannula.[17] *Alternatively*, 1000 mL of a 1.5-molar solution of methyllithium in diethyl ether (1.50 mol of MeLi) is used instead of the Grignard reagent. This is a matter of choice and has no significant effect upon

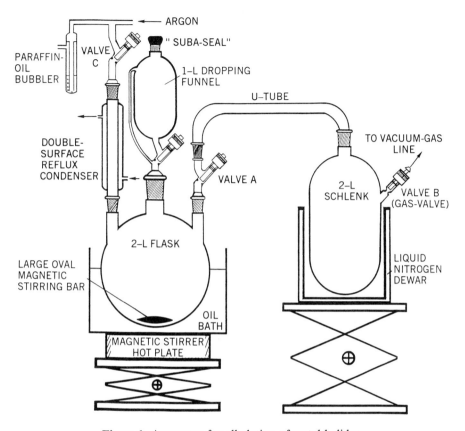

Figure 1. Apparatus for alkylation of metal halides.

the subsequent yield of trimethylindium. The Grignard solution (or MeLi solution) is added dropwise to the stirred indium trichloride suspension so as to maintain a steady reflux (about 2 h). To ensure the complete reaction of the indium trichloride, the resulting dirty grey (or white in the case of methyl-lithium) suspension is externally refluxed, with stirring, for 1 h using the oil bath.

On cooling (check argon flow through the bubbler!), valve A is reopened and both the reflux condenser and dropping funnel are replaced quickly by stoppers while a rapid purge of argon is applied through valve B. Valve A is again closed and the Schlenk flask is fully evacuated through valve B. Valve B is closed and valve A is reopened. This process is repeated several times so that the whole system is fully evacuated of argon without undue loss of volatiles. With valve B closed, the Dewar around the Schlenk flask is now filled slowly with liquid nitrogen while avoiding excessive bumping and frothing in the reaction flask by adjusting valve A. The reaction flask is warmed slowly to 200°C over several hours, maintaining stirring while possible, so that all the trimethylindium and diethyl ether are collected, by trap-trap distillation under a static vacuum, in the Schlenk flask. This process is complete when the grey-white solid residue in the reaction flask is seen to be completely dried out.

■ **Caution.** *Because temperatures reach 200°C, a silicone oil bath is to be used. An ordinary hydrocarbon oil bath may ignite at this temperature.*

Valve A is now closed and both the oil bath and liquid nitrogen Dewar are removed—the reaction flask and Schlenk flask being supported by "lab jacks." On equilibration to room temperature, valve A is reopened and the whole system is slowly filled to atmospheric pressure with argon through valve B. Valve A is again closed and a fast purge of argon is applied through valve B while the Schlenk flask is dislodged from the U tube and stoppered.

■ **Caution.** *Argon should not be allowed to enter a flask immersed in liquid nitrogen as it will condense (boiling point − 186°C) and cause dangerous pressure buildup on warming.*

A stopper is lifted from the reaction flask while 500 mL of toluene are added to cover the solid residues. Propan-2-ol is added gradually to destroy any remaining organometallic compounds (excess Grignard or methyllithium). When this is complete (no further effervescence), water is added cautiously, followed by dilute hydrochloric acid to complete the solvolysis of the magnesium salts.

The apparatus shown in Fig. 2 is now assembled and flushed thoroughly with argon. This is achieved most satisfactorily by closing valves F and G, and using valve E successively to evacuate and then fill the whole system to atmospheric pressure with argon—repeat this process several times to ensure complete removal of air. With a slight positive pressure of argon applied through valve E, valve G to the paraffin-oil bubbler is reopened. Valve E is now closed and a slight argon pressure maintained throughout the system by passing a slow argon stream directly over the top of the reflux condenser and out through the paraffin-oil bubbler. This flow rate is adjusted as necessary during heating or cooling procedures.

The trimethylindium–diethyl ether solution prepared above is now added to the distillation flask using a cannula inserted through the sidearm "Suba-Seal." Finally, with a rapid purge of argon applied through the paraffin-oil bubbler, the "Suba-Seal" is replaced quickly by a stopper and the argon flow is readjusted to its former flow rate.

Diethyl ether is fractionally distilled off from the trimethylindium solution at atmospheric pressure (bp 34–35°C). Distillation is continued until, with an oil bath temperature of ~100°C, the collection of the diethyl ether distillate

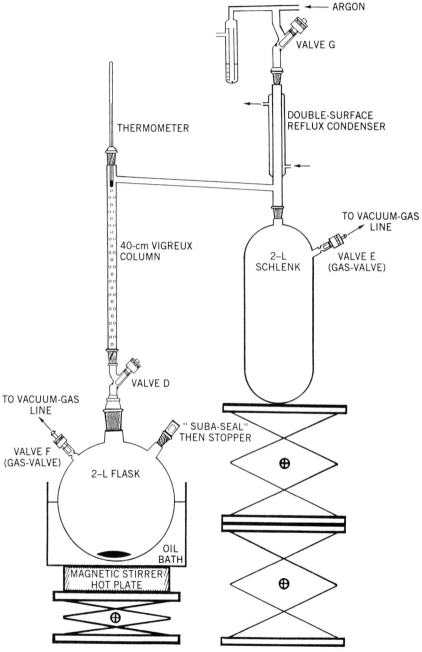

Figure 2. Apparatus for distillation of "free" diethyl ether.

proceeds very slowly—at which point the argon flow through the paraffin-oil bubbler is increased to a rapid purge while the oil bath is lowered.*

On equilibration to room temperature the etherate is *either* dried by repeated distillation at room temperature under a static vacuum to afford unpurified trimethylindium (follow procedure *b*), *or* used directly in the preparation of the Lewis-base adduct with 4,4′-methylenebis(N,N'-dimethylaniline) (follow procedure *c*) which is subsequently thermally dissociated to afford ultrapure trimethylindium.

(b) Solvent-Free Trimethylindium. The trimethylindium etherate prepared above can be dried of diethyl ether (adduct purification not required) in the following way: valve D is closed and the 2-L flask containing the diethyl ether adduct of trimethylindium, now sealed at room temperature under argon, is dislodged from the rest of the apparatus shown in Fig. 2 by carefully disengaging the Vigreux column from the top of the adaptor which houses valve D. This flask is now connected, through a U tube, to a 250-mL storage flask, as illustrated in Fig. 3.

With valve D still closed, the storage flask and glass tubing are fully evacuated through valve H. Valve H is closed and valve D opened. This process is repeated several times so that the whole system is fully evacuated of argon without undue loss of volatiles. When complete, both valves H and D are closed while the Dewar vessel around the storage bulb is filled with liquid nitrogen. Valve D is opened gradually while the etherate is stirred and warmed with an oil bath (30–40°C), causing it to condense, under a static vacuum, into the storage flask. A hot-air gun is used to warm the interconnecting glass tubing where the etherate tends to condense. Occasional warming of the neck of the storage flask is required to prevent the formation of a frozen plug of etherate. When all the etherate is transferred, valve I is closed to seal the etherate under vacuum, and the oil bath and Dewar are removed before cautiously filling the rest of the system (make sure valve I is closed!) with air through valve H. The storage flask is now attached to the drying apparatus, as illustrated in Fig. 4.

Flasks 2 and 3, and all the interconnecting glass tubing, are fully evacuated through one of the sidearm gas valves and the whole system is sealed under vacuum.

---

* Analysis by ^{1}H NMR shows the residue in the distillation flask to have a composition close to that expected for the etherate, $Me_3In \cdot OEt_2$. This etherate has a reported bp of 139°C. However, since the etherate is readily dissociated into its constituent molecules (procedure b) and since pure Me_3In has a bp of 136°C, it seems probable that 139°C refers to a vapor mixture of both uncomplexed Me_3In and its undissociated etherate, $Me_3In \cdot OEt_2$.

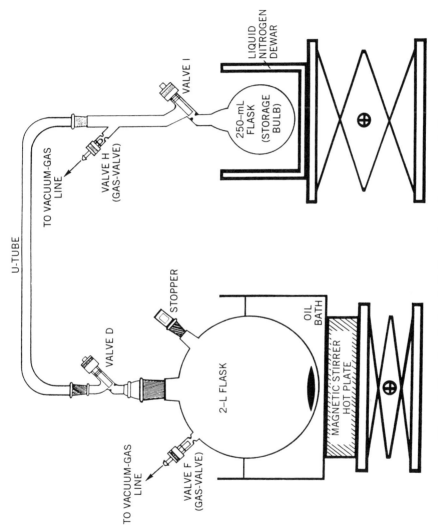

Figure 3. Transfer of trimethylindium etherate to storage bulb.

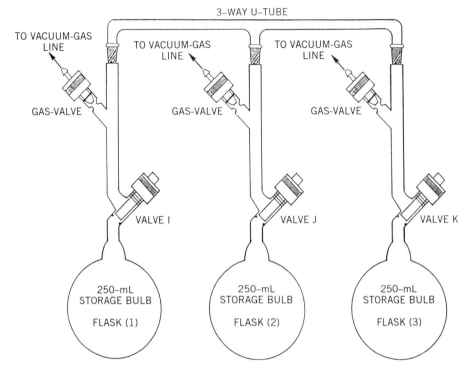

Figure 4. Apparatus for "drying" trimethylindium.

Valve K is closed and a filled liquid nitrogen Dewar vessel is placed around flask 2. Valve I is opened and diethyl ether (containing some trimethyl-indium) is collected in flask 2 by slow trap-trap distillation, under a static vacuum, away from the liquid etherate in flask 1. The etherate in flask 1 is not warmed during this process because gradual cooling facilitates preferential evaporation of a distillate with a higher percentage of diethyl ether as opposed to trimethylindium. Warming the neck of flask 2 with a hot-air gun is required occasionally to prevent the buildup of a frozen plug of distillate. When the trimethylindium residue in flask 1 is virtually dry (appearing as a glassy-white crystalline solid), valve I is closed and valve K is opened. The liquid nitrogen Dewar is transferred from around flask 2 and placed around flask 3. As flask 2 now slowly warms, diethyl ether is trapped away into flask 3 in the same manner as above, thus affording a second crop of crystalline trimethylindium. Valve K is closed and valve I is opened while the liquid nitrogen Dewar is transferred from flask 3 to flask 1, thus retrapping the trimethylindium from flask 2 into the main batch of product within flask 1. Gentle warming with a hot-air gun is useful when trapping crystalline trimethylindium from one flask to another.

Several further trap-trap distillations between flasks 2 and 3 yield successively smaller crops of crystalline trimethylindium which are in each case retrapped into the bulk of the product contained in flask 1.

When the diethyl ether distillate from this trapping process is judged to be virtually trimethylindium free, the flask containing the diethyl ether, 2 or 3, is filled to atmospheric pressure with argon, removed from the assembly, and its contents cautiously treated with propan-2-ol to destroy any remaining traces of trimethylindium dissolved in the diethyl ether.

The cleaned and dried flask is mounted back onto the assembly and fully reevacuated, together with the interconnecting glass tubing. The final traces of diethyl ether are removed from the trimethylindium in flask 1 by trap-trap distillation into flask 2 at room temperature. Crystalline trimethylindium has little tendency to sublime when held at room temperature so that the drying process can be maintained until the product is seen to be fully dry.

Finally, any trimethylindium within the distillate collected in flask 2 is fully dried by trapping any diethyl ether away into flask 3 before resublimation of the small amount of trimethylindium in flask 2 into the main body of dry product in flask 1. This flask is ideal for the long-term storage of trimethylindium in vacuo. Yield: 73 g (92% based on indium trichloride).

(c) Hexamethyl-μ-[4,4′-methylenebis(N,N′-dimethylaniline)]diindium

1. The trimethylindium etherate prepared in procedure a can be used directly in the preparation of the adduct in the following way: When the etherate from procedure a has cooled to room temperature, the stream of argon through the paraffin-oil bubbler (Fig. 2) is increased to a rapid purge while the sidearm stopper in the 2-L flask is replaced quickly by a "Suba-Seal." A solution of 62.3 g of 4,4′-methylenebis(N,N′-dimethylaniline) (0.245 mol) in 300 mL of benzene is now added, through the "Suba-Seal," to the stirred etherate using a cannula. Finally, the "Suba-Seal" is replaced quickly by a stopper before the rapid purge of argon through the bubbler is readjusted to its initially slow flow rate.

The remaining diethyl ether (bp 34–35°C) is removed from the resulting solution by fractional distillation at atmospheric pressure, followed by several milliliters of benzene (bp 80°C), thus ensuring the removal of all diethyl ether which would otherwise interfere with amine adduct formation during the crystallization process.[5,19] The resulting benzene solution of the adduct is allowed to cool to ambient temperature (check argon flow through the paraffin-oil bubbler!) and valve D closed to isolate the distillation flask. The Schlenk flask, containing the collected distillate is replaced by an empty Schlenk flask and the paraffin-oil bubbler is replaced by a stopper.

The system up to valve D is fully evacuated through valve E. Valve E is closed and valve D to the flask is reopened. This process is repeated several times to ensure that the whole system is fully evacuated of argon without undue loss of volatiles. With valves D and E eventually closed, a filled liquid nitrogen Dewar is placed around the Schlenk flask. The adduct is stirred while value D is opened gradually so that the remaining benzene within the adduct is removed by trap-trap distillation under a static vacuum. The flask may be warmed during this process with an oil or water bath at 30–40°C; excessive frothing or bumping is controlled by varying the aperture of valve D. When complete, the flask containing the viscous pale-yellow oil of the adduct is again isolated from the rest of the system by closing valve D, and then filled to atmospheric pressure with argon through valve F. The liquid nitrogen Dewar is removed from around the Schlenk flask and the remaining system (make sure valve D is closed!) is filled with air through valve E. The flask containing the oily adduct can now be dislodged from the rest of the system by carefully disengaging the Vigreux column from the top of the adaptor housing valve D. The top of this adaptor is plugged with a stopper.

A fast purge of argon is maintained through valve F while the sidearm stopper of the flask is once again exchanged for a "Suba-Seal." Petroleum spirit (900 mL) is added by cannula, through the "Suba-Seal," to the stirred viscous oil of the adduct, affording a slightly cloudy but virtually colorless solution of the adduct. This solution is filtered by cannula techniques[17] into a second 2-L flask equipped in an identical manner, including a large magnetic stirring bar. This flask is cooled to − 30°C for several hours (preferably overnight) in a freezer, care being taken to seal all the joints of the flask against the collection of condensation, affording a single, massive crop of small, snow-white, needle-shaped crystals of the adduct.

The solvent is removed from these crystals by rapid cannula filtration, while cooling the flask at − 30°C with a dry ice–acetone bath to prevent the crystals redissolving, and the crystals are washed with 200 mL of cold (− 30°C) petroleum spirit before pumping them dry under vacuum through the sidearm gas valve (valve F). When virtually dry, the adduct can be warmed at 60°C for several hours, while still being pumped on, to drive off the final traces of solvent together with other volatile impurities. The adduct may form large lumps during the drying process and these should be broken up intermittently by gentle crushing with a glass rod while a rapid argon flush is maintained through the flask using the sidearm gas valve. Yield: 131 g (93% based on amine, 92% based on indium trichloride).

2. "Dry" (diethyl ether free) trimethylindium (from procedure b or purchased) can be used for the preparation of the adduct in the following way: A 2-L flask with one sidearm gas valve, one sidearm sealed with a "Suba-Seal," and a central neck containing an adaptor/valve (as with the 2-L flasks in Figs. 2, 3, 5 or 6), is charged with 39 g of 4,4'-methylenebis(N,N'-dimethylaniline) (0.153 mol) and a large magnetic stirring bar. The flask and contents are flushed thoroughly with argon by using the sidearm gas valve successively to evacuate fully and then fill the flask to atmospheric pressure with argon—repeating this process several times to ensure the complete removal of air. Petroleum spirit (600 mL) is added to the flask by cannula addition, through the "Suba-Seal," and the "Suba-Seal" is then replaced with a stopper while purging the flask with argon through the sidearm gas valve.

The flask is now connected through a U tube to a storage flask (bulb) containing 50 g of trimethylindium (0.313 mol) (stored under vacuum)—in a manner identical to the system shown in Fig. 3 (as used in procedure b for the reverse transfer of etherate from the 2-L flask to a storage bulb). Keeping valves D and I closed, the glass tubing connecting the two flasks is fully evacuated of air through valve H. Valve H is then closed while valve D is opened. The 2-L flask and glass tubing are then evacuated slowly of argon through valve H while simultaneously freezing the contents of the 2-L flask (amine suspension in petroleum spirit) by slowly raising a filled liquid nitrogen Dewar so that it eventually freezes the bottom half of the flask. When argon evacuation is complete, valve H is closed to seal the system under vacuum.

Valve I is opened and the trimethylindium is sublimed, under static vacuum, into the 2-L flask (trap-trap distilled). A hot-air gun is used to assist this sublimation process and to ensure that all the trimethylindium is collected on, or near, the surface of the cold, or frozen, amine/petroleum spirit suspension. When all the trimethylindium is trapped into the 2-L flask, valve D is closed to seal the 2-L flask and the rest of the system is filled with air through valve H (make sure valve D is closed!). The liquid nitrogen Dewar is removed and the 2-L flask is disengaged from the rest of the system by careful disconnection between the adaptor housing valve D and the U tube. The top of this adaptor is plugged with a stopper.

The 2-L flask is partially filled with argon through valve F and allowed to thaw—a hot-air gun can be used to speed up this process. On equilibration to room temperature the flask is filled fully to atmospheric pressure with argon. On stirring, the suspension dissolves gradually to form a clear colorless solution of the adduct. If the solution is cloudy, it can be filtered into a second flask as described above.

The flask is cooled to − 30°C for several hours (preferably overnight) in a freezer, care being taking to seal all the joints of the flask against the collection of condensation, to afford a single, massive, crop of small snow-white needle-shaped crystals of the product which is worked up as above. Yield: 84.5 g (96% based on amine, 94% based on trimethylindium).

Anal. Calcd. for $C_{23}H_{40}In_2N_2$: C, 48.1; H, 7.0; N, 4.9. Found: C, 47.8; H, 6.8; N, 5.0.

(d) Thermal "Cracking" of Hexamethyl-μ-[4, 4'-methylenebis (N,N'-dimethylaniline)]diindium—Liberation of High-Purity Trimethylindium. The 2-L flask containing the 131 g of adduct (0.228 mol) from procedure c1 is arranged to form a "cracking" assembly, as illustrated in Fig. 5. The storage flask and glass tubing are fully evacuated through valve L before valve D is opened slowly to evacuate the 2-L flask containing the adduct. Valve L is closed after pumping on the whole system for several minutes, and the Dewar around the collection (storage) flask is filled with liquid nitrogen.

The molten adduct (mp 80–82°C) is "cracked" under a static vacuum, while being stirred and heated with the oil bath at temperatures rising from 80 to 130°C over several hours (usually ~4 h) until no more trimethylindium is seen to be collecting around the neck of the receiving flask. Throughout this process, gentle heating with a hot-air gun is required both to melt any sublimed adduct back into the heated part of the 2-L flask, and to sublime any trimethylindium which may crystallize around the glass tubing into the main body of the collection flask.

On completion of the decomposition, the product is sealed, in vacuo, in the collection flask by closing valve M. The oil bath and liquid nitrogen Dewar are removed, and after cooling of the reaction flask to ambient temperature, the remainder of the system is opened cautiously to air through valve L.

The residue in the flask, a pale brown viscous oil when molten and large platelet crystals when solid, is treated with propan-2-ol to destroy any remaining traces of trimethylindium which may be trapped within the amine.

The product is separated from traces of amine and an unidentified non-volatile oil by resubliming the trimethylindium, under static vacuum, from one storage bulb to another through a U tube with gentle warming from a hot-air gun. Yield: 71 g (97% based on adduct, 89% based on indium trichloride).

Properties

The adduct $\left(Me_3In \cdot Me_2N - \langle\bigcirc\rangle - \right)_2 CH_2$ is a white crystalline solid which

melts at 80–82°C. It is very air-sensitive, but nonpyrophoric, turning yellow on contact with air within seconds. It can be stored indefinitely under an

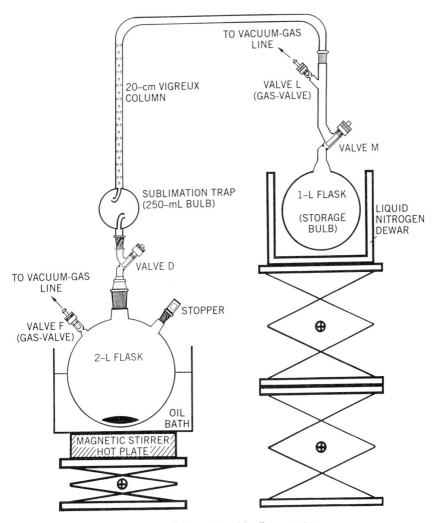

Figure 5. Adduct "cracking" apparatus.

inert atmosphere at room temperature. ^{1}H NMR (C_6D_6, rel. Me$_4$Si, δ ppm): -0.23 (s, 18H, 2InMe$_3$), 2.33 (s, 12H, 2NMe$_2$), 3.67 (s, 2H, CH$_2$), and 6.84 (m, 8H, H Aryl).

Trimethylindium is a glassy-white crystalline solid due to its unique tetrameric nature. In the vapor phase it is monomeric, and has the highest vapor pressure of all the indium alkyls; mp 88°C, bp 135°C. It ignites spontaneously in air (pyrophoric) and decomposes violently in water. It can be stored indefinitely at room temperature under an inert atmosphere, or under its own vapor pressure, in a storage flask such as those illustrated in Fig. 4. It is a strong Lewis acid, forming many stable complexes; its chemical and physical properties have been fully reviewed.[12] ^{1}H NMR (C_6D_6, rel. Me_4Si, δ ppm): $-0.23(s)$.

B. TRIMETHYLGALLIUM

(e) $\quad GaCl_3 + 3MeMgI \xrightarrow{Et_2O} Me_3Ga \cdot OEt_2 + 3ClMgI$

or $\quad GaCl_3 + 3MeLi \xrightarrow{Et_2O} Me_3Ga \cdot OEt_2 + 3LiCl$

(f)

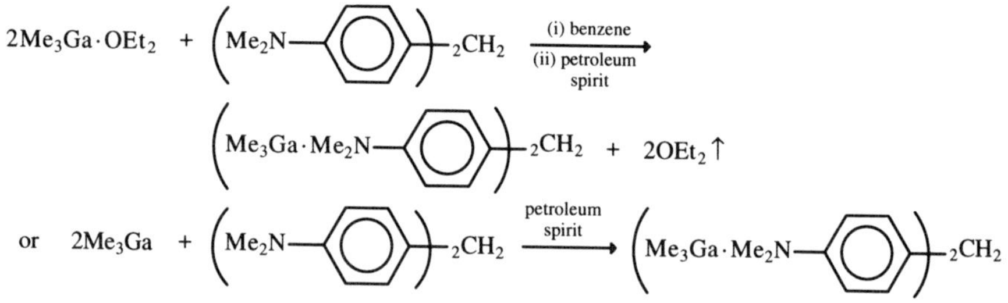

or

(g)

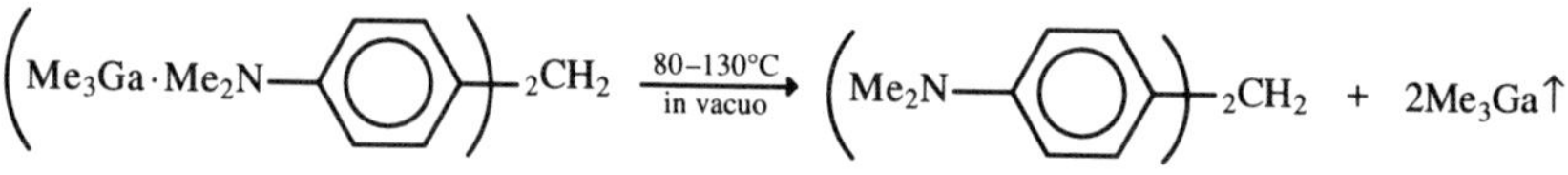

■ **Caution.** *Trimethylgallium is pyrophoric. Methyllithium is potentially pyrophoric. The chemicals benzene and methyliodide are highly toxic. See general caution and safety section.*

Procedure

The procedures and apparatus for the preparation of high-purity trimethylgallium are essentially the same as those employed for the preparation of trimethylindium (as described in Section 7.A, parts a, c, and d), and only the additional pertinent facts relating to the preparation of trimethylgallium are given here.

(e) Trimethylgallium Etherate. The etherate of trimethylgallium is obtained, following the same procedure as for trimethylindium (part a), on addition of the Grignard solution, prepared by the reaction of 125 mL of methyliodide (2.01 mol) with 50 g of magnesium pieces (2.06 mol) in 1000 mL of diethyl

ether, to a stirred solution of 100 g of gallium trichloride* (0.568 mol) in 200 mL of diethyl ether. Methyllithium, 1140 mL of a 1.5 molar solution in diethyl ether (1.71 mol of MeLi) may be used in place of the Grignard reagent with no significant effect upon the subsequent yield of trimethylgallium. Diethylether (bp 34–35°C) is fractionally distilled off from the trimethylgallium solution at atmospheric pressure, until, with an oil bath temperature of ~60°C, collection of the diethyl ether distillate becomes very slow.[†]

(f) Hexamethyl-μ-[4,4'-methylenebis(N,N'-dimethylaniline)]digallium

1. The etherate is treated, following the same procedure as for trimethylindium (part c1), with a solution of 71.2 g of 4,4'-methylenebis(N,N'-dimethylaniline) (0.280 mol) in 300 mL of benzene. After removal of the remaining diethyl ether and benzene, the adduct is recrystallized from 800 mL of petroleum spirit at − 30°C. During the filtration process, the flask containing the adduct must be kept at − 30°C using a dry ice–acetone bath or the crystals will redissolve. The crystals are washed with 200 mL of cold (− 30°C) petroleum spirit and finally dried by pumping under vacuum. The white platelet crystals are heated at 60°C for several hours, while still being pumped on, to drive off the final traces of solvent together with other volatile impurities. As with the trimethylindium adduct, large lumps that form during this drying process must be broken up with a glass rod, thus ensuring that no volatile impurities or solvent are kept trapped within the solid adduct. Yield: 129 g (95% based on amine, 94% based on gallium trichloride).

2. Purchased trimethylgallium ("dry" but unpurified) can be used in the preparation of the adduct by following the same procedure as that described for trimethylindium (part c2). Thus, 50 g of trimethylgallium (0.435 mol) is added, by trap-trap distillation, in vacuo, to a frozen suspension of 55 g of 4,4'-methylenebis(N,N'-dimethylaniline) (0.216 mol) in 600 mL of petroleum spirit. The resulting solution, which forms on warming to room temperature with stirring, is recooled to − 30°C for several hours (preferably overnight) to afford a single, massive crop of white platelet crystals of the adduct. These

---

* Gallium trichloride is an air-sensitive, hygroscopic solid which is usually purchased in sealed glass ampoules. These are most easily handled in the following way: in an inert-atmosphere glove box,[17] the entire ampoule is ground up in a pestle and mortar and transferred to a stoppered Schlenk flask. This Schlenk flask is taken from the glove box and cooled in an ice-water bath while 200 mL of cold diethyl ether is added slowly (EXOTHERMIC!) by cannula to dissolve the gallium trichloride. This solution is then transferred, by cannula filtration, away from the ground glass, to the reaction flask.
† Analysis by ^{1}H NMR shows the residue in the distillation flask to have a composition close to that expected for the etherate, $Me_3Ga \cdot OEt_2$, which has a reported bp of 99°C. This is far higher than that of pure Me_3Ga (bp 56°C).

crystals are worked up as described above. Yield: 102 g (98% based on amine, 97% based on trimethylgallium).

Anal. Calcd. for $C_{23}H_{40}Ga_2N_2$: C, 57.1; H, 8.3; N, 5.8. Found: C, 57.3; H, 8.3; N, 5.6.

(g) Thermal "Cracking" of Hexamethyl-μ-[4,4′-methylenebis(N,N′-dimethylaniline)]digallium—Liberation of High-Purity Trimethylgallium. Following the same procedure as that described for trimethylindium (part d), the molten adduct of trimethylgallium (129 g, 0.267 mol) (mp 83–86°C) is dissociated by heating, in vacuo, at temperatures rising from 80–130°C over a period of several hours (usually ~5 h) until no more trimethylgallium is seen to be collecting around the neck of the receiving flask. The product is separated from traces of amine and an unidentified nonvolatile oil by retrapping the trimethylgallium, at room temperature, in vacuo, from one storage flask to another. Yield: 54 g (88% based on adduct, 83% based on gallium trichloride).

Properties

The adduct $\left(Me_3Ga \cdot Me_2N\text{—}\langle\bigcirc\rangle\text{—}_2CH_2 \right.$ is a white crystalline solid which

melts at 83–86°C. It is very air-sensitive, but nonpyrophoric, turning turquoise on contact with air within seconds. It can be stored indefinitely under an inert atmosphere at room temperature. ^{1}H NMR (C_6D_6, rel. Me$_4$Si, δ ppm): − 0.25 (s, 18H, 2GaMe$_3$), 2.30 (s, 12H, 2NMe$_2$), 3.65 (s, 2H, CH$_2$), and 6.85 (m, 8H, H Aryl).

Trimethylgallium is a colorless liquid; mp − 16°C, bp 56°C. It ignites spontaneously in air (pyrophoric) and decomposes violently in water. It can be stored indefinitely at room temperature under an inert atmosphere, or under its own vapor pressure, in a storage flask such as those illustrated in Fig. 4. It is a strong Lewis acid, forming many stable complexes; its chemical and physical properties have been fully reviewed.[12] ^{1}H NMR (C_6D_6, rel. Me$_4$Si, δ ppm): − 0.12(s).

C. TRIMETHYLALUMINUM

(h) $2Me_3Al + \left(Me_2N\text{—}\langle\bigcirc\rangle\text{—}_2CH_2 \right) \xrightarrow[\text{spirit/toluene}]{\text{petroleum}} \left(Me_3Al \cdot Me_2N\text{—}\langle\bigcirc\rangle\text{—}_2CH_2 \right)$

(i) $\left(Me_3Al \cdot Me_2N\text{—}\langle\bigcirc\rangle\text{—}_2CH_2 \right) \xrightarrow[\text{in vacuo}]{170-220°C} \left(Me_2N\text{—}\langle\bigcirc\rangle\text{—}_2CH_2 \right) + 2Me_3Al\uparrow$

Note: It is more cost effective to purify commercially available, but crude, trimethylaluminum (see starting materials section), than to prepare trimethylaluminum from scratch.

■ **Caution.** *Trimethylaluminum is pyrophoric. See general caution and safety section.*

Procedure

The procedures and apparatus for the purification of trimethylaluminum are essentially the same as those employed for the preparation of trimethylindium (as described in Section 7.A, parts c2 and d), and only the additional pertinent details relating to the purification of trimethylaluminum are given here.

(h) Hexamethyl-μ-[4,4′-methylenebis(N,N′-dimethylaniline)]dialuminum
The adduct is prepared directly from purchased trimethylaluminum by following the same procedure as that described for the preparation of trimethylindium (part c2). Thus, 50 g of trimethylaluminum (0.694 mol) is added, by trap-trap distillation, in vacuo, to a frozen suspension of 88 g of 4,4′-methylenebis(N,N′-dimethylaniline) (0.346 mol) in a solvent mixture of 800 mL of petroleum spirit and 75 mL of toluene. The resulting solution, which forms on warming to room temperature with stirring, is recooled to − 30°C for several hours (preferably overnight) to afford a single, massive crop of small white needle-shaped crystals of the adduct. These crystals are isolated by filtration, while keeping the flask cooled to − 30°C with a dry ice–acetone bath to prevent the crystals from redissolving, washed with 200 mL of cold (− 30°C) petroleum spirit and finally dried by pumping under vacuum. The crystals are heated at 100°C for several hours, while still being pumped on, to drive off the final traces of solvent together with other volatile impurities. As with the trimethylindium adduct, large lumps which tend to form during the drying process must be broken up with a glass rod. Yield: 134 g (97% based on amine or trimethylaluminum).

Anal. Calcd. for $C_{23}H_{40}Al_2N_2$: C, 69.3; H, 10.1; N, 7.0. Found: C, 69.2; H, 10.2; N, 7.1.

(i) Thermal "Cracking" of Hexamethyl-μ-[4, 4′-methylenebis(N,N′-dimethylaniline)]dialuminum—Liberation of High-Purity Trimethylaluminum. Following the same procedure as that described for the preparation of trimethylindium (part d), the molten adduct of trimethylaluminum (134 g, 0.336 mol) (mp 110–115°C) is dissociated by heating, in vacuo, at temperatures rising from 170–220°C for 12 h or until no more trimethylaluminum is

seen to be collecting around the neck of the receiving flask. Although the residue will still contain considerable quantities of trimethylaluminum (as can be detected by ^{1}H NMR spectroscopy), it will not be liberated at 220°C and heating above this temperature will lead to problems associated with the distillation of free Lewis base. The product is separated from traces of amine and an unidentified nonvolatile oil by retrapping the trimethylaluminum, at room temperature, in vacuo, from one storage flask to another. Yield: 31 g (64% based on adduct, 62% based on trimethylaluminum).

Properties

The adduct $\left(Me_3Al \cdot Me_2N \text{—} \hexagon \text{—} _2CH_2 \right)$ is a white crystalline solid which

melts at 110–115°C. It is very air-sensitive, but nonpyrophoric, turning turquoise on contact with air within seconds. It can be stored indefinitely under an inert atmosphere at room temperature. ^{1}H NMR (C_6D_6, rel. Me$_4$Si, δ ppm): -0.55 (s, 18H, 2AlMe$_3$), 2.46 (s, 12H, 2NMe$_2$), 3.78 (s, 2H, CH$_2$) and 6.88 (m, 8H, H Aryl).

Trimethylaluminum is a colorless liquid; mp 15°C, bp 127°C. It ignites spontaneously in air (pyrophoric) and decomposes violently in water. It can be stored indefinitely at room temperature under an inert atmosphere, or under its own vapor pressure, in a storage flask such as those illustrated in Fig. 4. It is a strong Lewis acid, forming many stable complexes; its chemical and physical properties have been fully reviewed.[11] ^{1}H NMR (C_6D_6, rel. Me$_4$Si, δ ppm): -0.34(s) .

D. TRIETHYLINDIUM

(j) $InCl_3 + 3EtMgBr \xrightarrow{Et_2O} Et_3In \cdot OEt_2 + 3ClMgBr$

(k) $Et_3In \cdot OEt_2 \xrightarrow[\text{in vacuo}]{25°C} Et_3In + OEt_2\uparrow$

(l) $2Et_3In \cdot OEt_2 + \left(Me_2N \text{—} \hexagon \text{—} _2CH_2 \right) \xrightarrow[\text{(ii) petroleum spirit}]{\text{(i) benzene}}$

$\left(Et_3In \cdot Me_2N \text{—} \hexagon \text{—} _2CH_2 \right) + 2OEt_2\uparrow$

or $2Et_3In + \left(Me_2N \text{—} \hexagon \text{—} _2CH_2 \right) \xrightarrow{\text{petroleum spirit}} \left(Et_3In \cdot Me_2N \text{—} \hexagon \text{—} _2CH_2 \right)$

(m) $\left(Et_3In \cdot Me_2N \text{—} \hexagon \text{—} _2CH_2 \right) \xrightarrow[\text{in vacuo}]{120-150°C} \left(Me_2N \text{—} \hexagon \text{—} _2CH_2 \right) + 2Et_3In\uparrow$

■ **Caution.** *Triethylindium is pyrophoric. The chemicals benzene and ethyl bromide are highly toxic. See general caution and safety section.*

Procedure

The procedures and apparatus for the preparation of high-purity triethylindium are essentially the same as those employed for the preparation of trimethylindium (as described in Section 7.A, parts a–d), and only the additional pertinent details relating to the preparation of triethylindium are given here.

(j) Triethylindium Etherate. The etherate of triethylindium is obtained, following the same procedure as that described for the preparation of trimethylindium (part a), on addition of the Grignard solution, prepared by the reaction of 140 mL of ethylbromide (1.88 mol) with 48 g of magnesium pieces (1.98 mol) in 1000 mL of diethyl ether, to a stirred suspension of 94 g of indium trichloride (0.425 mol) in 200 mL of diethyl ether. Diethyl ether (bp 34–35°C) is fractionally distilled off from the triethylindium solution at atmospheric pressure, until, with an oil bath temperature of ~ 100°C, collection of the diethyl ether distillate becomes very slow.* This etherate is either "dried" of diethyl ether by repeated distillation at room temperature under a static vacuum (follow procedure k) to afford unpurified triethylindium, or used directly in the preparation of the Lewis base adduct with 4,4′-methylenebis(N,N'-dimethylaniline) (follow procedure 1) which is subsequently thermally dissociated to afford ultrapure triethylindium.

(k) Solvent-Free Triethylindium. Following the same procedure as that described for the preparation of trimethylindium (part b), diethyl ether is separated from the triethylindium by repeated trap-trap distillation, in vacuo, at room temperature. The difficulty here is judging when the liquid product is free of all diethyl ether; however, the relatively low volatility of triethylindium (bp 184°C at 760 torr), even under vacuum (at room temperature), allows for the preferential trapping of any diethyl ether without excessive distillation of triethylindium. The product can be analyzed intermittently by ^{1}H NMR

---

* Analysis by ^{1}H NMR shows the residue in the distillation flask to have a composition close to that expected for the etherate, $Et_3In \cdot OEt_2$. This etherate has a reported bp of 83–84°C at 12 torr. However, since the etherate is readily dissociated into its constituent molecules (procedure k) and since pure Et_3In has a bp of 80°C at 12 torr, it seems probable that the bp of 83–84°C at 12 torr refers to a vapor mixture of both uncomplexed Et_3In and its undissociated etherate, $Et_3In \cdot OEt_2$.

during the trapping process to detect any remaining diethyl ether. Yield: 81 g (94% based on indium trichloride).

(l) Hexaethyl-μ-[4,4'-methylenebis(N,N'-dimethylaniline)]diindium

1. The etherate from j is treated, following the same procedure as that described for trimethylindium (part c1), with a solution of 53 g of 4,4'-methylenebis(N,N'-dimethylaniline) (0.208 mol) in 300 mL of benzene. After removal of the remaining diethyl ether and benzene, the adduct is recrystallized from 300 mL of petroleum spirit by cooling to − 80°C (using a dry ice–acetone bath) for several hours, washed with 100 mL of cold (− 80°C) petroleum spirit and finally dried by pumping under vacuum while keeping the small white platelet crystals cooled to − 80°C rising to − 30°C. On warming to room temperature, these crystals melt slowly to form a viscous, virtually colorless oil. This oil is heated at 60°C for several hours, while still being pumped on, in order to drive off the final traces of solvent together with other volatile impurities. Yield: 127 g (93% based on amine, 91% based on indium trichloride).

2. "Dry" (diethyl ether free) triethylindium (from procedure k or purchased) can be used in the preparation of the adduct by following the same procedure as that described for trimethylindium (part c2). Thus, 50 g of triethylindium (0.248 mol) is added, by trap-trap distillation, in vacuo, to a frozen suspension of 31 g of 4,4'-methylenebis(N,N'-dimethylaniline) (0.122 mol) in 300 mL of petroleum spirit. The resulting solution, which forms on warming to room temperature with stirring, is recooled to − 80°C for several hours to afford a single, massive crop of white platelet crystals of the adduct. These crystals are worked up as described above. Yield: 76 g (95% based on amine, 93% based on triethylindium).·

Anal. Calcd. for $C_{29}H_{52}In_2N_2$: C, 52.9; H, 8.0; N, 4.3. Found: C, 52.5; H, 8.1; N, 4.1.

(m) Thermal "Cracking" of Hexaethyl-μ-[4,4'-methylenebis(N,N'-dimethylaniline)]diindium—Liberation of High-Purity Triethylindium. Following the same procedure as that described for the preparation of trimethylindium (part d), the liquid adduct of triethylindium (127 g, 0.193 mol) is dissociated by heating, in vacuo, at temperatures rising from 120 to 150°C over several hours (usually ∼6 h) until no more triethylindium is seen to be collecting around the neck of the receiving flask. The product is separated from traces of amine and an unidentified nonvolatile oil by retrapping the triethylindium, at ∼40°C in vacuo, from one storage flask to another. Yield: 71 g (91% based on adduct, 83% based on indium trichloride).

Properties

The adduct $\left(Et_3In \cdot Me_2N - \underset{}{\bigcirc\!\!\!\bigcirc} - 2CH_2 \right.$ is a viscous, virtually colorless oil at

room temperature. It is very air-sensitive, but non-pyrophoric, turning a greeny-yellow on contact with air within seconds. It can be stored indefinitely at 0°C in darkness under an inert atmosphere. ^1H NMR (C_6D_6, rel. Me_4Si, δ ppm): 0.43 [qrm, 12H, 2In(CH_2CH_3)$_3$], 1.34 [tm, 18H, 2In (CH_2CH_3)$_3$], 2.97 (s, 12H, 2NMe$_2$), 3.61 (s, 2H, CH$_2$) and 6.82 (m, 8H, H Aryl).

Triethylindium is a colorless liquid; mp − 32°C, bp 184°C. It ignites spontaneously in air (pyrophoric) and decomposes violently in water. It can be stored indefinitely at 0°C in darkness under an inert atmosphere, or under its own vapor pressure, in a storage flask such as those shown in Fig. 4. It is a strong Lewis acid, forming many stable complexes; its chemical and physical properties have been fully reviewed.[12] ^1H NMR (C_6D_6, rel. Me_4Si, δ ppm): 0.54 (qrm) and 1.31 (tm, $^3J_{HH} = 8.0$ Hz).

E. DIMETHYLCADMIUM

(n) $CdCl_2$ + $2MeMgI$ $\xrightarrow{Et_2O}$ $Me_2Cd \cdot (OEt_2)_2$ + $2ClMgI$

or $CdCl_2$ + $2MeLi$ $\xrightarrow{Et_2O}$ $Me_2Cd \cdot (OEt_2)_2$ + $2LiCl$

(o) $Me_2Cd \cdot (OEt_2)_2$ $\xrightarrow{distillation}$ Me_2Cd + $2OEt_2\uparrow$

(p) $Me_2Cd \cdot (OEt_2)_2$ + 2,2′-Bipy $\xrightarrow{Et_2O}$

or Me_2Cd + 2,2′-Bipy $\xrightarrow[\text{spirit}]{\text{petroleum}}$ $Me_2Cd \cdot 2,2′-Bipy$

(q) $Me_2Cd \cdot 2,2′-Bipy$ $\xrightarrow[\text{in vacuo}]{50-100°C}$ 2,2′-Bipy + $Me_2Cd\uparrow$

■ **Caution.** *Dimethylcadmium is not pyrophoric but, as is the case with cadmium dichloride, it is highly toxic. Methyliodide is highly toxic. Methyllithium is potentially pyrophoric. See general caution and safety section.*

Procedure

The procedures and apparatus for the preparation of high-purity dimethyl-cadmium are, in places, essentially the same as those employed for the preparation of trimethylindium (as described in Section 7.A, parts a and d), and only the additional pertinent details relating to the preparation of dimethylcadmium are given here.

(n) Dimethylcadmium Etherate. The etherate of dimethylcadmium is obtained, following the same procedure as that described for the preparation of trimethylindium (part a), by addition of the Grignard solution, prepared by the reaction of 78 mL of methyliodide (1.25 mol) with 33 g of magnesium pieces (1.36 mol) in 500 mL of diethyl ether, to a rapidly stirred suspension of 100 g of cadmium dichloride (0.546 mol) in 500 mL of diethyl ether. The suspension of cadmium dichloride in a large volume of diethyl ether, coupled with rapid stirring, greatly enhances the yield and reproducibility of this preparation. Methyllithium, 735 mL of a 1.5 molar solution in diethyl ether (1.10 mol of MeLi) may be used in place of the Grignard reagent with no significant effect upon the subsequent yield of dimethylcadmium. Trapping of the diethyl ether solution of dimethylcadmium away from the magnesium halide salts seems to take far longer than in the case of any other metal alkyl. It often takes up to 20 h with an oil bath temperature of up to 180°C.

If it is intended to purify the dimethylcadmium, through adduct formation with 2,2'-bipyridine, then there is no need to distill off the free diethyl ether at this point (follow procedure p). However, if unpurified dimethylcadmium is required, then diethyl ether (bp 34–35°C) is fractionally distilled off from the dimethylcadmium solution at atmospheric pressure as in the procedure, part a, for trimethylindium, until, with an oil bath temperature of ∼80°C, the collection of the diethyl ether distillate becomes very slow.* Solvent-free dimethylcadmium is now obtained, by distillation, by following procedure o below.

(o) Solvent-Free Dimethylcadmium. Having followed the same procedure as that described for the preparation of trimethylindium (part a—including the distilling off of the free diethyl ether), the etherate of dimethylcadmium should at this time be sealed under argon in the dislodged 2-L flask illustrated in Fig. 2. With a rapid purge of argon applied through the sidearm gas valve of this flask (valve F), the stopper in the other sidearm is quickly replaced by

---

* Analysis by ^{1}H NMR shows the residue in the distillation flask to be very low in Et$_2$O content. Clearly the etherate, Me$_2$Cd·(OEt$_2$)$_2$, dissociates far below its boiling point, which might well be expected to be higher than that of pure dimethylcadmium (bp 105–106°C).

a "Suba-Seal." The etherate is now transferred, by cannula insertion through the "Suba-Seal," to a second smaller flask (250 mL) equipped in an identical manner (one sidearm gas valve—to be known as valve F, one sidearm housing a "Suba-Seal" and the central neck equipped with an adaptor housing valve D; and the flask equipped with a magnetic stirring bar), and which has been thoroughly flushed with argon. Finally, the "Suba-Seal" in the sidearm of the 250-mL flask is replaced with a stopper while purging with argon.

The 250-mL flask containing the adduct is now housed onto the base of the Vigreux column, through the central adaptor, so that it once again resembles Fig. 2. The 2-L Schlenk containing the distilled diethyl ether from procedure n having been replaced by a smaller, 250-mL, Schlenk flask. Keeping valves D and G closed, the Schlenk flask, Vigreux column, and condenser are thoroughly flushed with argon using the sidearm gas valve (valve E) of the Schlenk flask to successively evacuate and then fill the system with argon several times. With a slight argon pressure applied through valve E, valve G is reopened to vent the argon through the paraffin–oil bubbler. Valve E is now closed and a slight argon pressure is maintained throughout the system by passing argon directly over the reflux condenser and out through the paraffin–oil bubbler. Valve D can now be opened.

The remaining diethyl ether is fractionally distilled off from the dimethylcadmium at atmospheric pressue (bp 34–35°C). Distillation is continued until pure dimethylcadmium has started to cross the still head (bp 105–106°C)—at which point the argon flow through the paraffin–oil bubbler is increased to a rapid purge while the oil bath is lowered quickly to prevent distillation of further product. The system is allowed to stand for several minutes to allow any dimethylcadmium in the Vigreux column to drain back down into the distillation flask and valve D is then closed. A small amount of dimethylcadmium will decompose during this distillation process, as witnessed by the cadmium metal residue formed in the distillation flask, but this is unavoidable and does not significantly affect the subsequent yield of product.

Valve G is closed and a filled liquid nitrogen Dewar is raised around the Schlenk flask as the system (other than the sealed distillation flask) is evacuated slowly of argon through valve E. Any diethyl ether and dimethylcadmium residues within the Vigreux column, sidearm, and reflux condenser are now condensed into the cold Schlenk flask. Warming of the glassware with a hot-air gun greatly assists this process. When all distillate residues are trapped within the main body of the Schlenk flask, the liquid nitrogen Dewar is lowered and the Schlenk flask is filled with argon through valve E. The Schlenk is now quickly dislodged and stoppered.

■ **Caution.** *The frozen Schlenk flask may become dangerously pressurized on thawing. This is avoidable by venting the Schlenk flask through the pres-*

sure-release bubbler attached to the gas line of a standard laboratory vac-uum-gas line.

A standard 250-mL storage flask (as illustrated in Fig. 4) is now attached to the apparatus in place of the dislodged Schlenk flask. With valves D and G still closed, the storage flask and remaining apparatus (other than the distillation flask) are evacuated through the sidearm gas valve of the storage flask. The distillation flask, now cooled to room temperature, is also evacu-ated by slowly opening valve D; very little product is lost during this evacuation if the dimethylcadmium has been allowed to cool to room temperature. When all the argon is evacuated, the sidearm gas valve of the collection flask is closed to seal the system under a static vacuum.

The colorless, liquid product is now distilled, bp 29–30°C at 40 torr, into the collection flask which is ideal for its long-term storage under vacuum (see properties). Yield: 74 g (95% based on cadmium dichloride).

(p) (2, 2'-Bipyridine)dimethylcadmium

1. Having followed the same procedure as that described for the prepara-tion of trimethylindium (part a) (except here the free diethyl ether has not been distilled off), the diethyl ether solution of dimethylcadmium should at this time be sealed under argon in the dislodged 2-L Schlenk flask (Fig. 1).

The apparatus shown in Fig. 6 is now assembled and flushed with argon—valve Q is closed and valve N is used successively fully to evacuate and then fill the whole system to atmospheric pressure with argon. With a slight positive pressure of argon applied through valve N, valve Q is opened to vent the argon through the paraffin–oil bubbler.

The argon flow through the system is now increased while the sidearm stopper of the flask is dislodged and 84.3 g of 2,2'-bipyridine (0.540 mol) (hygroscopic) is added quickly to the flask. The stopper is rehoused and the rapid purge of argon through valve N is maintained for 10–15 min to ensure that any air which may have entered the system during this process is flushed out. Valve N is now closed and a slow argon flow is maintained by passing argon directly over the top of the reflux condenser and out through the paraffin–oil bubbler. Diethyl ether (200 mL) is added to the flask, by cannula addition through the centrally situated "Suba-Seal," and the stirred 2,2'-bi-pyridine suspension brought to a gentle reflux to form a colorless solution.

The gentle reflux is maintained while, using cannula filtration techniques, the dimethylcadmium solution in the 2-L Schlenk is now run into the rapidly stirred 2,2'-bipyridine solution through the central "Suba-Seal," forming a clear bright yellow solution of the adduct. It should be noted that formation of this adduct solution is slightly exothermic, so that the rate of addition of the dimethylcadmium solution needs to be controlled to prevent too violent

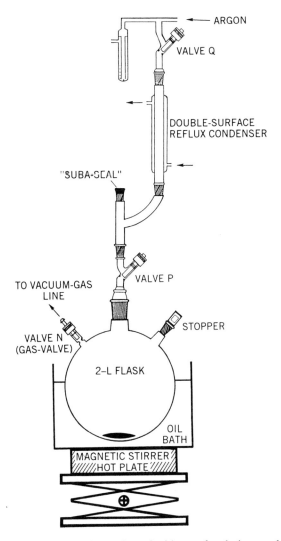

Figure 6. Apparatus for the formation of adducts of cadmium and zinc alkyls.

a reflux (about 20 min). The adduct solution is allowed to cool to room temperature (check argon flow!).

The flask containing the adduct solution is now placed in a large bowl-shaped Dewar and this vessel is half filled with acetone. Dry ice is added gradually to the acetone so that the temperature of the bath is lowered slowly to − 80°C over a period of 1 h (check argon flow!). The bath is maintained at − 80°C for several hours to maximize the yield of the small bright yellow crystals of the adduct which form in the flask—the initially bright yellow solution turns very pale as crystal precipitation nears completion.

When precipitation of the product is judged to be complete, the argon flow through the bubbler is increased to a rapid purge while the sidearm stopper of the flask is replaced quickly with a "Suba-Seal" housing the filtration end of a filter cannula.[17] The other end of the cannula is inserted through a "Suba-Seal" housed in the top of a clean and argon-filled 2-L Schlenk flask. Valve P is closed and the solvent in the flask is transferred to the Schlenk flask by cannula filtration while applying a slight argon pressure through valve N. The crystals are washed with 200 mL of cold ($-80°C$) diethyl ether (added to the flask by cannula addition through the "Suba-Seal" above valve P) and the washings are also filtered into the 2-L Schlenk. Finally, the "Suba-Seal" with its filter cannula is replaced with a stopper while purging rapidly with argon through valve N.

The flask is now lifted from the cold bath and the glassware mounted above the adaptor housing valve P (make sure this valve is closed!) is disassembled. The crystals in the flask (~ 145 g or 90%) are dried at room temperature by applying a vacuum through valve N.

A second batch of crystals may be obtained from the 2-L Schlenk flask. Thus, the diethyl ether in the Schlenk flask is reduced down to about 200 mL by trapping the solvent off through the sidearm gas valve of the Schlenk flask, in batches, directly into the cold-trap of a standard vacuum-gas line. Any product that precipitates during this volume reduction is now redissolved by closing the sidearm gas valve of the Schlenk flask, to seal the latter under vacuum, and using a hot-air gun to gently warm the remaining diethyl ether. Diethyl ether will reflux and condense around the upper part of the Schlenk flask and so wash any solid adduct back down into the base of the Schlenk flask where it will redissolve to form a bright yellow solution. (■ **Caution.** *Great care should be exercised when warming any solvent in a sealed system (even if evacuated).*) The Schlenk flask is now refilled to atmospheric pressure with argon and the contents cooled to $-80°C$ with a dry-ice–acetone bath to afford a second batch of crystals. The solvent is removed by cannula filtration and the crystals (about 13 g or 8%) are washed with 50 mL of cold ($-80°C$) diethyl ether before drying in vacuo. These crystals are added to the main batch of product in the 2-L flask. This is done by applying a rapid purge of argon through valve N while the sidearm stopper of the flask is replaced by a powder funnel. The crystals in the Schlenk flask are now tipped rapidly through the powder funnel into the flask. The stopper is replaced and the flask is evacuated through valve N. The crystalline, free-flowing adduct is pumped on at room temperature for about 2–3 h to drive off the final traces of solvent together with other volatile impurities. The product cannot be warmed during this drying process because a considerable amount of dimethylcadmium will be lost—even at room temperature, the slight vapor pressure of dimethylcadmium above the adduct precludes excessive drying in vacuo. Yield: 158 g (98% based on amine, 97% based on cadmium dichloride).

2. "Dry" (diethyl ether free) dimethylcadmium (from procedure o or purchased) can be used in the preparation of the adduct by using a slightly modified version of the procedure described above. Thus, 50 g of dimethylcadmium (0.351 mol) in 50 mL of petroleum spirit is added, by cannula, to a rapidly stirred and gently refluxing solution of 54.5 g of 2,2'-bipyridine (0.349 mol) in 300 mL of petroleum spirit. The refluxing amine solution instantly turns bright yellow and then, as more dimethylcadmium is added, the yellow microcrystalline adduct begins to precipitate. Precipitation of the crystalline adduct is completed by cooling the flask to $-80°C$ in a dry-ice–acetone bath. These crystals are worked up as described above. Yield: 102 g (98% based on amine, 97% based on dimethylcadmium).

Anal. Calcd. for $C_{12}H_{14}CdN_2$: C, 48.3; H, 4.7; N, 9.4. Found: C, 48.2; H, 4.7; N, 9.5.

(q) Thermal "Cracking" of (2,2'-bipyridine)dimethylcadmium—Liberation of High-Purity Dimethylcadmium. Following the same procedure as that described for the preparation of trimethylindium (part d), the adduct of dimethylcadmium (158 g, 0.529 mol) (mp 75–78°C) is dissociated by heating, in vacuo, at temperatures rising from 50 to 100°C over a period of several hours (usually ~5 h) until no more dimethylcadmium is seen to be collecting around the neck of the receiving flask. Even at low temperatures (50°C), there is a problem associated with adduct sublimation—crystals form around the cooler parts of the 2-L flask and in the interconnecting glass tubing. Gentle heating with a hot-air gun is therefore required throughout this process to melt sublimed adduct back into the heated part of the flask. As the decomposition nears completion it is seen that the sublimate is very pale in color, almost white (almost pure base), as compared with the sublimate at the beginning of the decomposition which is bright yellow (pure adduct).

The product, which at this point is bright yellow due to traces of dissolved base, is retrapped at room temperature, in vacuo, from one storage flask to another. If the product is now not completely colorless, the retrapping process is repeated. Yield: 72 g (96% based on adduct, 93% based on cadmium dichloride).

Properties

The adduct $Me_2Cd·2,2'$-Bipy is a yellow crystalline solid which melts at 75–78°C. It is very air sensitive, but nonpyrophoric, turning white on contact with air within minutes. It can be stored indefinitely at 0°C in darkness under an inert atmosphere. ^{1}H NMR (C_6D_6, rel. Me$_4$Si, δ ppm): -0.35 (s, 6H, CdMe$_2$), 6.7, 7.2, 8.0 and 8.4 (m, 8H, H Aryl).

Dimethylcadmium is a colorless liquid; mp -4.5°C, bp 106°C. It fumes upon exposure to air but does not inflame, and decomposes sluggishly in water. It can be stored indefinitely at 0°C in darkness (light sensitive) under an inert atmosphere, or under its own vapor pressure, in a storage flask such as those in Fig. 4. As a result of the weak acceptor character of dimethylcadmium, only in very few cases are stable coordination complexes formed with donor ligands; its chemical and physical properties have been fully reviewed.[13] ^{1}H NMR (C_6D_6, rel. Me$_4$Si, δ ppm): -0.53(s).

F. DIMETHYLZINC

(r) $ZnCl_2$ + 2MeMgI $\xrightarrow{Et_2O}$ Me$_2$Zn·(OEt$_2$)$_2$ + 2ClMgI

or $ZnCl_2$ + 2MeLi $\xrightarrow{Et_2O}$ Me$_2$Zn·(OEt$_2$)$_2$ + 2LiCl

■ **Caution.** *Dimethylzinc is pyrophoric. Methyllithium is potentially pyrophoric. Methyliodide is highly toxic. See general caution and safety section.*

Procedure

The procedures and apparatus for the preparation of the diethyl ether solution of dimethylzinc and the dissociation of the adduct are essentially the same as those employed for the preparation of trimethylindium (as described in Section 7.A, parts a and d). The procedures and apparatus for the preparation of the adduct of dimethylzinc are essentially the same as those employed for preparation of the adduct of dimethylcadmium (as described in Section 7.E, part p). Only the additional pertinent details relating to the preparation of high-purity dimethylzinc are given here.

(r) Diethylether Solution of Dimethylzinc. The diethyl ether solution of dimethylzinc is obtained, following the same procedure as that described for the preparation of trimethylindium (part a), on addition of the Grignard solution, prepared by the reaction of 78 mL of methyliodide (1.25 mol) with 33 g of magnesium pieces (1.36 mol) in 500 mL of diethyl ether, to a rapidly stirred biphasic solution of 72 g of zinc dichloride (0.528 mol) in 500 mL of diethyl ether. The use of this large volume of diethyl ether with the zinc dichloride, coupled with rapid stirring, greatly enhances the yield and reproducibility of this preparation. Methyllithium, 710 mL of a 1.5 molar solution in diethyl ether (1.07 mol of MeLi), may be used in place of the Grignard reagent with no significant effect upon the subsequent yield of dimethylzinc.

Unlike the situation with trimethylindium, there is no need to distill off the free diethyl ether from the dimethylzinc solution as the adduct of dimethylzinc with 1,3-bis(4-pyridine)propane is readily precipitated from diethyl ether (follow procedure s).

(s) Poly [[1,3-bis(4-pyridyl)propane]dimethylzinc]

1. Following the same procedure as that described for the preparation of the adduct of dimethylcadmium (part p1), the diethyl ether solution of dimethylzinc is added over a period of 30 min, by cannula filtration, to a rapidly stirred and gently refluxing solution of 103 g of 1,3-bis(4-pyridyl)propane (0.520 mol) in 250 mL of diethyl ether. The white microcrystalline adduct begins to deposit after the addition of about one quarter of the dimethylzinc solution and continues to deposit during the addition of the remaining dimethylzinc solution. (■ **Caution.** *The sudden deposition of a mass of microcrystalline solid in the refluxing solution may cause the reflux to become quite vigorous.*) The suspension is allowed to cool to room temperature and left to stand for several hours to maximize the yield of deposited crystals. The solvent is removed by cannula filtration and the crystals are washed with 200 mL of diethyl ether before pumping them dry under vacuum. The crystalline, free-flowing adduct is heated at 60°C for several hours, while still being pumped on, in order to drive off the final traces of solvent together with other volatile impurities. Yield: 150 g (98% based on amine, 97% based on zinc dichloride).

2. Purchased dimethylzinc ("dry" but unpurified) can be used in the preparation of the adduct in exactly the same way as described above. Thus, 50 g of dimethylzinc (0.524 mol) is dissolved in 300 mL of diethyl ether and this solution is added over a period of 30 min, by cannula, to a rapidly stirred and gently refluxing solution of 103 g of 1,3-bis(4-pyridyl)propane (0.520 mol) in 300 mL of diethyl ether. The crystalline product is worked up as above. Yield: 150 g (98% based on amine or dimethylzinc).

Anal. Calcd. for $C_{15}H_{20}N_2Zn$: C, 61.3; H, 6.9; N, 9.5. Found: C, 60.6; H, 6.9; N, 9.5.

(t) Thermal "Cracking" of Poly[[1,3-bis(4-pyridyl)propane]dimethylzinc] —Liberation of High-Purity Dimethylzinc. Following the same procedure as that described for the preparation of trimethylindium (part d), the molten adduct of dimethylzinc (150 g, 0.511 mol) (mp 123–126°C) is dissociated by heating, in vacuo, at temperatures rising from 125 to 140°C over a period of several hours (usually ~5 h) until no more dimethylzinc is seen to be collecting around the neck of the receiving flask. The temperature of the oil bath used in the decomposition is controlled carefully as higher temperatures lead to the formation, by an unknown mechanism, of the adduct (4-methyl-pyridine)dimethylzinc. The product is separated from traces of the amine and the cleavage product, (4-methylpyridine)dimethylzinc, by retrapping the dimethylzinc, at room temperature in vacuo, from one storage flask to another. Yield: 45 g (92% based on adduct, 89% based on zinc dichloride).

Properties

The polymeric adduct is a white crystalline solid which melts at 123–126°C. It is very air sensitive, but nonpyrophoric, turning to a fuming purple liquid on contact with air within seconds. It can be stored indefinitely under an inert atmosphere at room temperature. ^{1}H NMR (THF-d_8, rel. Me$_4$Si, δ ppm): -0.80 (s, 6H, ZnMe$_2$), 1.95 (qu, 2H, $-CH_2CH_2CH_2-$), 2.65 (t, 4H, $-CH_2CH_2CH_2-$), 7.2 and 8.4 (m, 8H, H Aryl).

Dimethylzinc is a colorless liquid; mp -42°C, bp 46°C. It ignites spontaneously in air (pyrophoric) and decomposes violently in water. It can be stored indefinitely at room temperature under an inert atmosphere, or under its own vapor pressure, in a storage flask such as those illustrated in Fig. 4. It is a strong Lewis acid, forming many stable complexes; its chemical and physical properties have been fully reviewed.[13] ^{1}H NMR (C_6D_6, rel. Me$_4$Si, δ ppm): $- 0.63$(s).

G. DIETHYLZINC

(u) $\quad ZnCl_2 \ + \ 2EtMgBr \ \xrightarrow{Et_2O} \ Et_2Zn \cdot (OEt_2)_2 \ + \ 2ClMgBr$

(v) $\quad Et_2Zn \cdot (OEt_2)_2 \ \xrightarrow{\text{distillation}} \ Et_2Zn \ + \ 2OEt_2\uparrow$

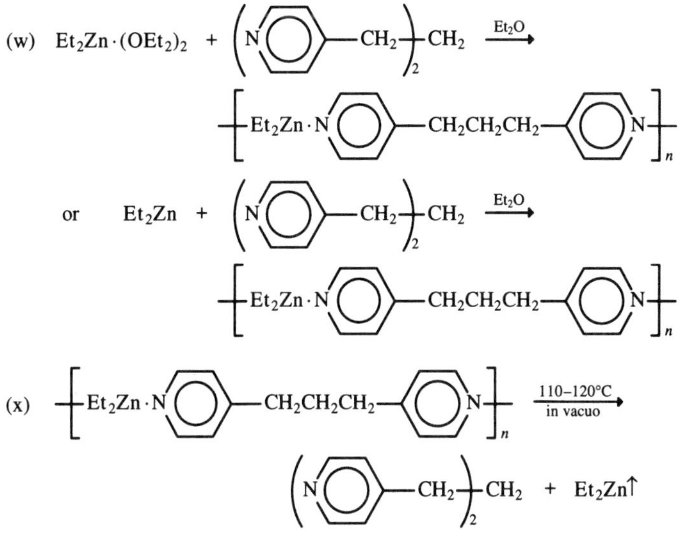

(w)

or

(x)

■ **Caution.** *Diethylzinc is pyrophoric. Ethyl bromide is highly toxic. See general caution and safety section.*

Procedure

The procedures and apparatus for the preparation of the etherate of diethyl-zinc and the dissociation of the adduct are essentially the same as those employed for the preparation of trimethylindium (as described in Section 7.A, parts a and d). The procedures and apparatus for "drying" diethylzinc by distillation and also for the preparation of the adduct of diethylzinc are essentially the same as those employed for the preparation of dimethylcad-mium (as described in Section 7.E, parts o and p). Only the additional pertinent details relating to the preparation of high-purity diethylzinc are given here.

(u) Diethylzinc Etherate. The etherate of diethylzinc is obtained, following the same procedure as that described for the preparation of trimethylindium (part a), on addition of the Grignard solution, prepared by the reaction of 95 mL of ethylbromide (1.27 mol) with 33 g of magnesium pieces (1.36 mol) in 500 mL of diethyl ether, to a rapidly stirred biphasic solution of 72 g of zinc dichloride (0.528 mol) in 500 mL of diethyl ether. The use of this large volume of diethyl ether with the zinc dichloride, coupled with rapid stirring, greatly enhances the yield and reproducibility of this preparation.

If it is intended to purify the diethylzinc, through adduct formation with 1,3-bis(4-pyridine)propane, then there is no need to distill off the free diethyl ether at this point (follow procedure w). However, if unpurified diethylzinc is required, then diethyl ether (bp 34–35°C) is fractionally distilled off from the diethylzinc solution at atmospheric pressure (as in part a for trimethylindium), until, with an oil bath temperature of ~80°C, the collection of the diethyl ether distillate becomes very slow.* Solvent-free diethylzinc is now obtained, by distillation, by the following procedure.

(v) Solvent-Free Diethylzinc. Following the same procedure as that described for the preparation of dimethylcadmium (part o), the diethylzinc etherate is transferred to a smaller (250 mL) flask and the remaining diethyl ether is fractionally distilled off from the diethylzinc at atmospheric pressure (bp 34–35°C); the distillation is continued until pure diethylzinc has started to cross the stillhead (bp 118–119°C) and then is immediately stopped. The product is now collected in a clean Schlenk flask by distillation, in vacuo, bp 41–42°C at 40 torr. Yield: 63 g (97% based on zinc dichloride).

(w) Poly[[1,3-bis(4-pyridyl)propane]diethylzinc]

1. Following the same procedure as that described for the preparation of the adduct of dimethylcadmium (part p1), the diethyl ether solution of diethylzinc is added over a period of 30 min, by cannula filtration, to a rapidly stirred and gently refluxing solution of 103 g of 1,3-bis(4-pyridyl)propane (0.520 mol) in 250 mL of diethyl ether. The pale-yellow crystalline adduct begins to deposit after the addition of about one quarter of the diethylzinc solution and continues to deposit during the addition of the remaining diethylzinc solution. (■ **Caution.** *The sudden deposition of a mass of crystalline solid in the refluxing solution may cause the reflux to become quite vigorous.*) The suspension is allowed to cool to room temperature and left to

---

* Analysis by ^{1}H NMR spectroscopy shows the residue in the distillation flask to be very low in Et_2O content. Clearly the etherate, $Et_2Zn \cdot (OEt_2)_2$, dissociates far below its boiling point, which might well be expected to be higher than that of pure diethylzinc (bp 118–119°C).

stand for several hours to maximize the yield of deposited crystals. The solvent is removed by cannula filtration and the crystals are washed with 200 mL of diethyl ether before pumping them dry under vacuum. The crystalline, free-flowing adduct is heated at 60°C for several hours, while still being pumped on, in order to drive-off the final traces of solvent together with other volatile impurities. Yield: 164 g (98% based on amine, 97% based on zinc dichloride).

2. "Dry" (diethyl ether free) diethylzinc (from procedure v or purchased) can be used in the preparation of the adduct in exactly the same way as described above. Thus, 50 g of diethylzinc (0.405 mole) is dissolved in 300 mL of diethyl ether and this solution is added over a period of 30 min, by cannula, to a rapidly stirred and gently refluxing solution of 80 g of 1,3-bis(4-pyridyl)-propane (0.404 mole) in 200 mL of diethyl ether. The crystalline product is worked up as above. Yield: 127 g (98% based on amine or diethylzinc).

Anal. Calcd. for $C_{17}H_{24}N_2Zn$: C, 63.5; H, 7.5; N, 8.7. Found: C, 63.1; H, 7.4; N, 8.7.

(x) Thermal "Cracking" of Poly[[1,3-bis(4-pyridyl)propane]diethylzinc] *—Liberation of High-Purity Diethylzinc.* Following the same procedure as that described for the preparation of trimethylindium (part d), the molten adduct of diethylzinc (164 g, 0.510 mole) (mp 97–100°C) is dissociated by heating, in vacuo, at temperatures rising from 110 to 120°C over a period of several hours (usually ~5 h) until no more diethylzinc is seen to be collecting around the neck of the receiving flask. The temperature of the oil bath used in the decomposition is controlled carefully as higher temperatures lead to the formation, by an unknown mechanism, of the adduct (4-methyl-pyridine)diethylzinc. The product is separated from traces of the amine and the cleavage product, (4-methylpyridine)diethylzinc, by retrapping the di-ethylzinc, at room temperature in vacuo, from one storage flask to another. Yield: 57 g (91% based on adduct, 87% based on zinc dichloride).

Properties

The polymeric adduct $\left[\text{Et}_2\text{Zn} \cdot \text{N} \bigcirc - \text{CH}_2\text{CH}_2\text{CH}_2 - \bigcirc \text{N} \right]_n$ is a pale-yellow crystalline solid which melts at 97–100°C. It is very air-sensitive, but nonpyrophoric, turning to a fuming purple liquid on contact with air within seconds. It can be stored indefinitely at 0°C in darkness under an inert atmosphere. ^{1}H NMR (THF-d_8, rel. Me$_4$Si, δ ppm): 0.15 [qr, 4H,

Zn($CH_2CH_3)_2$], 1.20 [t, 6H, Zn(CH$_2CH_3)_2$], 2.00 (qu, 2H, –CH$_2CH_2$CH$_2$–), 2.65 (t, 4H, –CH_2CH$_2CH_2$–), 7.2 and 8.4 (m, 8H, H Aryl).

Diethylzinc is a colorless liquid; mp − 28°C, bp 118°C. It ignites spontaneously in air (pyrophoric) and decomposes violently in water. It can be stored indefinitely at 0°C in darkness under an inert atmosphere, or its own vapor pressure, in a storage flask such as those illustrated in Fig. 4. It is a strong Lewis acid, forming many stable complexes; its chemical and physical properties have been fully reviewed.[13] ^{1}H NMR (C$_6$D$_6$, rel. Me$_4$Si, δ ppm): 0.17 (qr) and 1.15 (t, ^{3}J$_{HH}$ = 8.2 Hz).

References

1. (a) H. M. Manasevit, *App. Phys. Letters*, **12**, 156 (1968); (b) H. M. Manasevit and W. I. Simpson, *J. Electrochem. Soc.*, **116**, 1725 (1969).
2. P. Zanella, G. Rossetto, N. Brianese, F. Ossola, M. Porchia, and J. O. Williams, *Chem. Mater.*, **3**, 225 (1991).
3. D. J. Cole-Hamilton, *Chem. Br.*, **26**, 852 (1990) and references cited therein.
4. D. C. Bradley, H. Chudzynska, M. M. Faktor, D. M. Frigo, M. B. Hursthouse, B. Hussain, and L. M. Smith, *Polyhedron*, **7**, 1289 (1988).
5. D. F. Foster, S. A. Rushworth, D. J. Cole-Hamilton, A. C. Jones, and J. P. Stagg, *Chemtronics*, **3**, 38 (1988).
6. G. E. Coates and S. I. E. Green, *J. Chem. Soc.*, 3340 (1962).
7. G. Pajaro, S. Biagini, and D. Fiumani, *Angew. Chem.*, **74**, 901 (1962).
8. G. E. Coates, *J. Chem. Soc.*, 2003 (1951).
9. (a) D. V. Shenai-Khatkhate, E. D. Orrell, J. B. Mullin, D. C. Cupertino, and D. J. Cole-Hamilton, *J. Crystal Growth*, **77**, 27 (1986); (b) P. R. Jacobs, E. D. Orrell, J. B. Mullin, and D. J. Cole-Hamilton, *Chemtronics*, **1**, 15 (1986).
10. D. F. Foster, I. L. J. Patterson, L. D. James, D. J. Cole-Hamilton, D. N. Armitage, H. Yates, A. C. Wright, and J. O. Williams, *Adv. Mat. Opt. Electron.*, **3**, 163 (1994).
11. J. J. Eisch, in *Comprehensive Organometallic Chemistry*; *Aluminum*; Vol. 1, Chapter 6. G. Wilkinson, F. G. A. Stone, and E. W. Abel (eds.), Pergamon Press, Oxford, England, 1982, pp. 555–682 and references cited therein.
12. D. G. Tuck, in *Comprehensive Organometallic Chemistry*; *Gallium and Indium*, Vol. 1, Chapter 7. G. Wilkinson, F. G. A. Stone, and E. W. Abel (eds.), Pergamon Press, Oxford, England, 1982. pp. 683–754 and references cited therein.
13. J. Boersma, in *Comprehensive Organometallic Chemistry*: *Zinc and Cadmium*, Vol. 2, Chapter 16. G. Wilkinson, F. G. A. Stone, and E. W. Abel (eds.), Pergamon Press, Oxford, England, 1982, pp 823–862 and references cited therein.
14. D. F. Gaines, J. Borlin, and E. P. Fody, *Inorg. Synth.*, **15**, 203 (1974).
15. A. L. Galyer and G. Wilkinson, *Inorg. Synth.*, **19**, 253 (1979).
16. N. D. Gerrard, J. D. Nicholas, J. O. Williams, and A. C. Jones, *Chemtronics*, **3**, 17 (1988).
17. D. F. Shriver and M. A. Drezdzon, *The Manipulation of Air-Sensitive Compounds*, 2nd ed., Wiley, New York, 1986.
18. M. S. Kharasch and O. Reinmuth, *Grignard Reactions of Nonmetallic Substances*, Prentice-Hall, New York, 1954.
19. D. F. Foster, S. A. Rushworth, D. J. Cole-Hamilton, R. Cafferty, J. Harrison, and P. Parkes, *J. Chem. Soc., Dalton Trans.*, 7 (1988).

8. TRIMETHYLINDIUM AND TRIMETHYLGALLIUM

SUBMITTED BY D. C. BRADLEY,* H. C. CHUDZYNSKA,*
and I. S. HARDING*
CHECKED BY KAREN S. BREWER,† DANIEL ROSENBLUM,‡
and O. T. BEACHLEY, JR.‡

$$2MCl_3 + 6MeLi + (Ph_2PCH_2)_2 \longrightarrow (Me_3MPh_2PCH_2)_2 + 6LiCl$$

$$(Me_3MPh_2PCH_2)_2 \longrightarrow 2Me_3M + (Ph_2PCH_2)_2$$

$$M = \text{In or Ga}$$

Trimethylgallium and trimethylindium are pyrophoric compounds that are used in the production of semiconductors. Because of the high purity required and difficulty in transportation, commercial sources are correspondingly expensive. For laboratory use it is therefore convenient to generate these alkyls from a compound which is less sensitive to air and moisture and easier to store in standard laboratory glassware. Group III alkyls are Lewis acids and so can form adducts with Lewis bases which are then less susceptible to attack by moisture or air. However, these are correspondingly quite difficult to break apart to regenerate a pure metal trialkyl. Most syntheses have therefore used (1) pure alkylating agents in sealed-ampoule reactions or in noncoordinating solvents, accepting the risks involved with such compounds, or (2) coordinating solvents to reduce the danger, which are then difficult to remove from the product.

Previously published syntheses of trimethylgallium include the lengthy reaction of metallic gallium with dimethylmercury[1] and gallium trichloride with dimethylzinc in sealed ampoules.[2] Trimethylaluminum has also been used as an alkylating agent for gallium trichloride.[3] Trimethylindium has been synthesized previously by a number of different methods. Apart from the lengthy reaction between indium metal and dimethylmercury in a sealed vessel at elevated temperatures,[4,5] these all involve the use of diethyl ether as a solvent[6-8] which has proved difficult to remove.

1,2-Bis(diphenylphosphino)ethane (dppe) forms crystalline adducts with both trimethylindium and trimethylgallium which can be heated in vacuo to release the trimethylmetal vapor[9] in good yield. They are far less reactive to both air and water than the parent trialkyl and so have proved to be useful

* Queen Mary and Westfield College, London E1 4NS, UK.
† Department of Chemistry, Hamilton College, Clinton, NY 13323.
‡ Department of Chemistry, State University of New York at Buffalo, Buffalo, NY 14214.

for long-term storage and safer for handling than the pure trialkyls. The equilibrium dissociation pressures of trimethylindium and trimethylgallium from the adducts have been determined[10] and this pressure applies as long as the adducts remain solid and in equilibrium with the vapor. Even when this is not the case, a large proportion of the trimethylindium or trimethylgallium can be extracted.

The thermal dissociation of these adducts occurs at elevated temperatures and so only impurities that are volatile at these temperatures will distill with the trialkyls. A further distillation of the pure trialkyl at lower temperatures will give effective separation from these impurities, and so the purity of the products is good. The purity can be enhanced by careful recrystallizations of the adduct before thermal dissociation. By avoiding the use of reagents that contain electronic impurities (e.g., Grignard reagents, which contain magnesium), samples are generated which have such low levels of impurities that they can be used in the synthesis of materials for the semiconductor industry. For example, dppe-purified trimethylindium has been used to make indium phosphide that has the lowest carrier concentration and highest mobility measured so far.[11] For this reason, methyllithium is the alkylating agent used in these syntheses.

The following procedure is a modification of the first reported synthesis of the trimethylindium adduct.[9] An important feature in the method is the removal of diethyl ether, which forms no isolable adduct with trimethylindium but is very difficult to remove completely from neat trialkyl. With trimethylgallium this is even more difficult because the adduct bond is stronger and the adduct is isolable. In the present case, it is achieved by displacement of the diethyl ether by dppe and concomitant formation of the isolable dppe adduct. Even here the diethyl ether has to be removed by careful fractionation, to avoid loss of the volatile trialkyl. The original solvent for the preparation was benzene, but this has been replaced successfully with toluene.

Starting Materials

Diethyl ether and toluene (both BDH, GPR grade) are dried and freed of dissolved molecular oxygen by distillation under nitrogen from a solution of the solvent and sodium benzophenone. Anhydrous indium trichloride (Aldrich) and an ethereal solution of methyllithium (Aldrich) are used as supplied; dppe (Fluorochem) is also used as supplied. Anhydrous gallium trichloride (Aldrich) is sublimed before use.

■ **Caution.** *Methyllithium in diethyl ether is both pyrophoric and very moisture-sensitive. Anhydrous trichlorides of indium and gallium are hygroscopic as well as highly toxic[12]. The trimethylindium and trimethylgallium* *adducts formed in this reaction are also air- and moisture-sensitive. These compounds should be handled under an inert atmosphere at all times.*

Procedure

A carefully dried, two-necked, 250-mL flask equipped with a magnetic stirrer bar is evacuated and filled with dry nitrogen. Anhydrous indium trichloride (8.15 g, 36.8 mmol) is added and a stirred suspension formed with dry diethyl ether (35 mL). A pressure-equalizing dropping funnel is attached and charged with an ethereal solution of methyllithium (61 mL of 1.8 M solution, 110 mmol). It is important to use only stoichiometric amounts of methyllithium, since, for example, a 10% excess will lead to formation of lithium tetramethylindate in 30% yield. The flask is cooled in an ice bath and the methyllithium is added dropwise. After addition, the flask is allowed to warm to room temperature and the dropping funnel is replaced with a reflux condenser. To help coagulate the precipitate, the solution is heated under reflux for 1 h then removed from the heating source and allowed to cool. This also drives the reaction with indium trichloride to completion. After the solution has cooled, the reflux condenser is replaced by a nitrogen-filled No. 4 sintered-glass filter which is connected to another 250-mL flask, also connected to the nitrogen supply. By inverting the apparatus, the solution can be filtered. Care must be taken not to allow the solution to flow down the nitrogen inlet. It may be necessary to evacuate the collecting flask slightly, to draw the solvent through the fine sintered-glass filter.

The precipitate is washed with diethyl ether condensed back into the partially evacuated reaction flask from the collection flask. This is done by closing the nitrogen supply to both flasks and then cooling the upper, reaction flask. (A woollen glove stretched over the flask, then soaked in liquid nitrogen, is excellent for this purpose.) The reaction flask is evacuated until the diethyl ether in the collection flask just starts to boil. This is allowed to condense in the cold reaction flask, which is then warmed, and the condensed diethyl ether is allowed to flow down over the precipitate. It is forced through the filter by refilling the apparatus with nitrogen through the reaction flask inlet. It is not necessary to condense large amounts of diethyl ether using this technique. Adding extra diethyl ether to wash the precipitate is not recommended because of the time it takes to remove it by fractionation.

The sinter filter is removed and dppe (7.32 g, 18.4 mmol), toluene (70 mL), and a magnetic stirrer are added to the filtrate. A Perkin triangle with a 12-cm Vigreux column is attached (Fig. 1). The Perkin triangle is a piece of apparatus which is used to remove volatile fractions from a boiling mixture. It includes a fractionating column to separate the most volatile fraction,

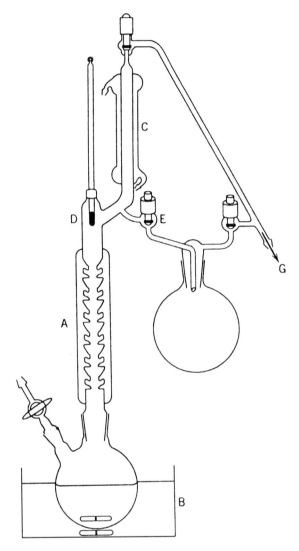

Figure 1. Reaction flask with Perkin triangle: (A) 12-cm vacuum-jacketed Vigreux column, (B) oil bath on hot plate, (C) condenser, (D) thermometer, (E) extraction tap, (F) collection flask, (G) to nitrogen supply and bubbler.

topped by a reflux condenser to maintain a steady flow of cooled liquid down the column, which is essential for its proper functioning. Part of the most volatile fraction at the top of the column can continually be removed through an extraction tap.

The flask is heated in an oil bath on a hot plate (to give a steady heat supply) while it is stirred magnetically. During heating the dppe will dissolve. The solution will start to boil when the temperature of the oil bath is about 110°C, but it may need to become hotter to support reflux in the Perkin triangle. The extraction tap above the fractionating column is kept shut until a steady reflux state is reached. The extraction tap is then opened slightly to remove some of the condensed fractionated vapor; the rate of removal, in drops, should preferably be about 1 per 10 that return to the flask. The temperature of the vapor at the top of the column will initially be at the boiling point of diethyl ether (36°C), but will rise as most of the diethyl ether is removed. When this temperature reaches the boiling point of toluene, the distillation is continued, but at a faster rate of collection. Distillation is stopped when the solution becomes slightly turbid; if this occurs before all the diethyl ether has been removed, the apparatus is cooled, more toluene is added, and the fractionation is continued. At this point it will become apparent whether the ratio of methyllithium to indium trichloride used was exact, because any lithium tetramethylindate formed will not redissolve in the diethyl ether-poor solution. This will reduce the purity of the adduct and the final yield of trimethylindium, but lithium tetramethylindate is also effectively involatile and so will not interfere with generation of trimethyl-indium.

(■ **Caution.** *Lithium tetramethylindate is, however, pyrophoric, and the contaminated adduct should be handled with care.*)

After all the diethyl ether has been removed (the vapor at the top of the column is at 110°C), toluene is distilled from the clear solution until the boiling solution turns turbid. It is allowed to cool slowly overnight to room temperature without stirring to grow crystals of the adduct. The supernatant liquid is decanted with a stainless-steel canula to leave colorless crystals which are finally dried in vacuo for 1 h. Yield 8.62 g (65% based on indium trichloride.) A slight increase in yield is produced by refrigerating the solution before decanting the supernatant liquid. The scale of the reaction can be increased to start with 100 g of indium trichloride using 1-L flasks with no great difficulty.

The reaction to form the trimethylgallium adduct is similar. The only difference in procedure is that, before adding dry diethyl ether (25 mL) to anhydrous gallium trichloride (4.9 g, 27.8 mmol), the flask containing the gallium trichloride is cooled in an ice/water bath or the exothermic reaction between these reagents will lead to substantial smoking in the flask.

An ethereal solution of methyllithum (84 mL of 1.0 M solution, 84 mmol; as close to a 3:1 ratio as possible) is added dropwise, and the solution is heated under reflux, cooled, and filtered as described for the trimethyl indium preparation. Toluene (70 mL), dppe (5.53 g, 13.9 mmol), and a stirrer bar are added to the solution. Diethyl ether is removed by fractionation using the Perkin triangle as previously described, with addition of further toluene as necessary. Cooling of the saturated supernatant liquid in a refrigerator produces colorless crystals which are isolated by decanting the supernatant liquid and are dried in vacuo. Yield 6.25 g (71% based on gallium trichloride).

Trimethylindium is generated from its adduct when required in a trap-to-trap apparatus. A two-necked 100-mL flask with a stopper and gas inlet adaptor is charged with the powdered adduct under an inert atmosphere. This is connected by a short length of vacuum tubing to a two-necked 100-mL flask fitted with two gas inlet adaptors. (It is important that this second trap can be sealed before the apparatus is dismantled, because it will contain the pyrophoric metal alkyl.) The whole apparatus is evacuated through the remaining inlet adaptor. The empty flask is immersed in liquid nitrogen and the adduct is heated to approximately 120°C at 10^{-2} mmHg. This generates appreciable pressures of the trialkyl, which condenses in the liquid nitrogen trap, while the dppe remains relatively nonvolatile.[13] Any small traces of dppe that do sublime are removed from the generated trimethylindium by resubliming it at room temperature from the trap into a storage or reaction vessel. Yield of white solid; about 80%, based on heating for half an hour per gram of powdered adduct. Slightly more can be obtained by prolonged heating.

■ **Caution.** *Storing pure trimethylindium is much less safe than storing the dppe adduct and generating pure trimethylindium when necessary.*

Generation of trimethylgallium from its dppe adduct is identical, except for the thermolysis temperature of the adduct. The powdered gallium adduct needs to be heated to approximately 110°C to generate appreciable pressures of trimethylgallium. The rate of supply may be increased by heating more strongly. However, at temperatures above 130°C it will melt as it generates the trialkyl,[14] causing a slow decrease in the rate of dissociation, but this will not change the final yield from that for the indium analogue. Trimethylgallium is a colorless oil at room temperature.

■ **Caution.** *Trimethylindium and trimethylgallium are both pyrophoric and very moisture-sensitive. These compounds should be handled under an inert atmosphere at all times. Owing to their pyrophoric nature and high volatility, it is important not to allow nitrogen to flow over them when they are at room temperature. Failure to take this precaution will result in a great fire and toxicity hazard wherever the nitrogen is exhausted to air.*

The residue that is left after in vacuo pyrolysis of the dppe adduct can be recycled. The residue is placed in a Soxhlet extractor and washed with refluxing methanol. Dppe will then accumulate in the refluxing methanol as white crystals.

Properties

The dppe adduct of trimethylindium is an air- and moisture-sensitive white solid. The ^{1}H NMR spectrum (C_6D_6, 80 MHz) consists of a singlet at δ 0.13 (9H, Me–In), a doublet at 2.49 (2H, J = 2 Hz, CH_2) and two sets of aromatic peaks in the ranges 6.9–7.1 and 7.25–7.6 ppm (10H, PPh_2). The IR spectrum (Nujol mull) contains the following peaks: 3050(m), 1965(w,br), 1882(w,br), 1808(w), 1755(w), 1581(w), 1473(m,sh), 1455(s,br), 1430(s), 1417(m), 1372(m), 1325(w), 1302(m), 1275(w), 1180(w), 1163(m), 1141(m), 1092(m), 1061(m), 1020(w), 993(m), 913(w), 740(s), 715(s), 682(vs,br), 532(w), 501(s), 479(s,br), 461(s,sh), 443(s), and 328(w) cm^{-1}. The crystals melt cleanly at 150–152°C. Elemental analysis found (calculated); C: 53.31% (53.51%), H: 5.90% (5.89%).

Trimethylindium forms colorless needles or blocks that are pyrophoric in air and extremely moisture sensitive. The ^{1}H NMR spectrum (C_6D_6, 80 MHz) consists of a singlet at δ − 0.19 ppm. The crystals melt sharply 88°C.

The dppe adduct of trimethylgallium is also an air- and moisture-sensitive white solid. The ^{1}H NMR spectrum (C_6D_6, 80 MHz) consists of a singlet at δ 0.15 (9H, Me–Ga), a doublet at 2.52 (2H, J = 4 Hz, CH_2), and two sets of aromatic peaks in the ranges 6.9–7.1 and 7.25–7.6 ppm (10H, PPh_2). The IR spectrum (Nujol mull) contains the following peaks: 3060(m), 1970(w,br), 1896(w,br), 1813(w), 1764(w), 1587(w), 1483(m,sh), 1463(s,br), 1436(s), 1422(m), 1379(m), 1332(w), 1310(m), 1283(w), 1263(vw), 1187(w), 1170(m), 1101(s), 1088(sh), 1070(m), 1026(m), 999(m), 970(w), 920(w), 748(s), 722(s), 695(s), 681(s), 544(s), 508(s), 486(s) 454(s), and, 335(w) cm^{-1}. The crystals melt cleanly at 162–164°C. Elemental analysis found (calculated); C: 61.39% (61.20%), H: 6.73% (6.74%).

Trimethylgallium is a colorless liquid that is pyrophoric in air and extremely moisture sensitive. It boils at 56°C. The ^{1}H NMR spectrum (C_6D_6, 80 MHz) consists of a singlet at δ − 0.12 ppm.

Any hydrolysis of the adducts will be evident in the IR spectrum as an extra broad peak at around 3500 cm^{-1}. Yellowing of these compounds is also diagnostic of slight decomposition.

References

1. G. E. Coates, *J. Chem. Soc.*, **1951**, 2003.
2. E. Wiberg, T. Johannsen, and O. Stecher, *Z. Anorg. Chem.*, **251**, 114 (1943).
3. D. F. Gaines, J. Borlin, and E. P. Fody, *Inorg. Synth.*, **15**, 203 (1974).
4. L. M. Dennis, R. W. Work, E. G. Rochow, and E. M. Chamot, *J. Am. Chem. Soc.*, **56**, 1047 (1934).
5. P. Krommes and J. Lorbeth, *Inorg. Nucl. Chem. Lett.*, **9**, 587 (1973).
6. H. C. Clark and A. L. Pickard, *J. Organomet. Chem.*, **8**, 427 (1967).

7. J. Runge, W. Zimmerman, H. Pfeiffer, and D. Pfeiffer, *Z. Anorg. Allg. Chem.*, **267**, 39 (1951).
8. E. Todt and R. Dotzer, *Z. Anorg. Allg. Chem.*, **321**, 120 (1963).
9. D. C. Bradley, H. Chudzynska, M. M. Faktor, D. M. Frigo, M. B. Hursthouse, B. Hussain, and L. M. Smith, *Polyhedron*, **7**, 1289 (1988).
10. D. C. Bradley, M. M. Faktor, D. M. Frigo, and D. H. Zheng, *Chemtronics*, **3**, 53 (1988).
11. E. J. Thrush, C. G. Cureton, J. M. Trigg, J. P. Stagg, and B. R. Butler, *Chemtronics*, **2**, 62 (1987).
12. N. I. Sax and R. J. Lewis, Sr., *Dangerous Properties of Industrial Materials*, 7th ed., Van Nostrand Reinhold, New York, 1989.
13. D. C. Bradley, M. M. Faktor, and D. M. Frigo, *Chemtronics*, **2**, 22 (1987).
14. D. C. Bradley, M. M. Faktor, D. M. Frigo, and I. S. Harding, *Chemtronics*, **3**, 235 (1988).

9. (*N*,*N*-DIMETHYLETHANAMINE)TRIHYDRIDOALUMINUM

SUBMITTED BY YOSHIHIDE SENZAKI,* DARRELL UHRHAMMER,*
EVERETT C. PHILLIPS,* and WAYNE L. GLADFELTER*
CHECKED BY ROWENA RANOLA,† AMY HARDING,† FREDERICK C.
SAULS,† and LEONARD V. INTERRANTE‡

$$Me_2EtN + HCl \longrightarrow [Me_2EtNH]Cl$$

$$[Me_2EtNH]Cl + Li[AlH_4] \longrightarrow AlH_3(NEtMe_2) + LiCl + H_2$$

The trimethylamine adduct of aluminum hydride (alane) has been of recent interest as a precursor for the chemical vapor deposition (CVD) of aluminum metal[1] and aluminum gallium arsenide thin films.[2] Because of the absence of aluminum–carbon covalent bonds in the precursor, carbon incorporation in the resulting films can be suppressed significantly. In addition, the deposition temperature can be lowered.

We have recently reported an alternative liquid precursor for the CVD of aluminum thin films.[3] The main advantage of (*N*,*N*-dimethylethanamine)-trihydridoaluminum, frequently referred to by its trivial name dimethylethyl-amine alane (DMEAA), over (trimethylamine) trihydridoaluminum is that DMEAA is a liquid at room temperature, which provides stationary pressure conditions for better control of precursor transport. Analogous to the previously reported synthesis of (trimethylamine)trihydridoaluminum,[4] the reaction of lithium tetrahydroaluminate(1 −) with *N*,*N*-dimethyl-ethanaminium chloride in diethyl ether generates the stable liquid precursor DMEAA with high yield.

* Department of Chemistry, University of Minnesota, Minneapolis, MN 55455.
† Department of Chemistry, King's College, Wilkes-Barre, PA 18711.
‡ Department of Chemistry, Rensselaer Polytechnic Institute, Troy, NY 12180.

General Procedure

■ **Caution.** *DMEAA is air-sensitive and inflames upon contact with water. All the reactions are performed under an inert atmosphere. In the synthesis of DMEAA, it is crucial to remove all the aluminum particles from the reaction solution by filtration after the reaction is complete. The aluminum particles remaining from the incomplete filtration can cause autocatalytic and exothermic gas-evolving decomposition of DMEAA which may lead to an explosion. Also, the product should not be overheated during the vacuum distillation. For disposal, DMEAA should be diluted with heptane and allowed to react slowly with isopropyl alcohol with cooling in a fume hood.*

N,N-Dimethylethanamine, hydrogen chloride (anhydrous, 1.0 M solution in diethyl ether), and lithium tetrahydroaluminate $(1 -)$ are used as purchased from Aldrich. The diethyl ether is distilled from sodium benzophenone under nitrogen prior to use.

Preparation of $[Me_2EtNH]Cl$. Me_2EtN (20 mL, 185 mmol) is added via a syringe to a 250-mL, round-bottomed flask containing HCl in Et_2O (185 mL, 185 mmol) and a magnetic stirring bar for 30 min at 0°C with constant stirring. This reaction is exothermic with instantaneous formation of a white precipitate. After the addition is complete, the reaction mixture is allowed to warm to room temperature. The removal of Et_2O under vacuum affords a white solid product 19.5 g (96% yield) that is used without further purification. ^1H NMR (δ, $CDCl_3$); 3.08 (q, 2H, CH_2CH_3), 2.77 [s, 6H, $N(CH_3)$], 1.44 (t, 3H, CH_2CH_3). The checkers suggest using this solution directly (with appropriate adjustment of the quantities) for the next step in the synthesis.

Synthesis of $AlH_3(NEtMe_2)$. In a glove box a 250-mL, three-neck round-bottomed flask is equipped with a septum cap, a glass stopper, a gas inlet adaptor, and a magnetic stirring bar and is charged with $Li[AlH_4]$ (4.6 g, 121 mmol) and $[Me_2EtNH]Cl$ (11.0 g, 100 mmol). An excess of $Li[AlH_4]$ is used to avoid the formation of aluminum chloride derivatives.[4b] The flask is removed from the glove box and attached to a Schlenk line in a fume hood. After the flask is cooled by a liquid nitrogen bath, Et_2O (125 mL) is added by syringe to the reactants. Under a nitrogen gas flow, the mixture is allowed to warm to between $- 40$ and $- 30$°C (best maintained using a slush bath of anisole). At this temperature the reaction begins with the evolution of hydrogen gas. The reaction temperature is maintained at this temperature until the generation of hydrogen gas stops (about 4 h at $- 37$°C), at which time the solution is allowed to warm to room temperature. The reaction mixture is filtered via a cannula and a filter stick, and the solvent is removed from the filtrate under vacuum (30 torr, 25°C). All aluminum particles must be removed from the solution, or they may catalyze decomposition of the product during the following distillation. The end point of Et_2O removal is established by using ^1H NMR spectroscopy. The final product (8.0 g, 77% yield) is collected in a dry ice–isopropanol bath trap by vacuum distillation (0.05 torr, 45°C). This distillation is slow; however, the temperature should not be raised above 45°C or decomposition of DMEAA will begin. Use of a wide-bore transfer tube minimizes the blockage of a transfer tube due to freezing of the product during the distillation.

Properties

The compound $AlH_3(NEtMe_2)$ is a stable, colorless liquid at room temperature. However, it should be sealed under vacuum and kept in a freezer for long-term storage.

Anal. Calcd. for $C_4H_{14}NAl$: C, 46.58; H, 13.68; N, 13.58; Al, 26.16. Found: C, 46.45; H, 13.51; N, 13.68; Al, 25.90. IR (gas phase) : 2984, 2774, 1790 (v_{Al-H}), 1467, 1380, 1055, 766 cm^{-1}. ^1H NMR (δ, d_6-benzene); 3.98 (br s, 3H, AlH_3), 2.31 (q, 2H, CH_2CH₃), 1.95 [s, 6H, N(CH_3)], 0.70 (t, 3H, CH₂CH_3). ^{13}C NMR (δ, d_6-benzene); 53.83 (CH_2CH_3), 44.18 [N(CH_3)], 9.12 (CH_2CH_3)., EI-Mass spectrum m/z: 102 (M$^+$-H). Melting point: 5°C. Vapor pressure: 1.5 torr at 25°C.

In addition to the major parent ion peak due to $AlH_3(NEtMe_2)$, the diamine adduct $AlH_3(NEtMe_2)_2$ and the dimer $[AlH_3(NEtMe_2)]_2$ were detected as minor components in the mass spectrum, whereas these species were not observed by ^1H NMR spectroscopy in a solution. The gas-phase IR spectrum also indicated the presence of a weak peak at 1713 cm^{-1} which may be attributable to the diamine adduct.

References

1. (a) W. L. Galdfelter, D. C. Boyd, and K. F. Jensen, *Chem. Mater.*, 1, 339 (1989). (b) D. B. Beach, S. E. Blum, and F. K. LeGoues, *J. Vac. Sci. Technol. A*, 7, 3117 (1989). (c) T. H. Baum, C. E. Larson, and R. L. Jackson, *Appl. Phys. Lett.*, 55, 1264 (1989). (d) A. T. S. Wee, A. J. Murrell, N. K. Singh, D. O'Hare, and J. S. Foord, *J. Chem. Soc., Chem. Commun.* 11, (1990). (e) L. H. Dubois, B. R. Zegarski, C.-T. Kao, and R. G. Nuzzo, *Surf. Sci.*, 236, 77 (1990). (f) M. E. Gross, K. P. Cheung, C. G. Fleming, J. Kovalchick, and L. A. Heimbrook, *J. Vac. Sci. Technol. A*, 9, 57 (1991).
2. (a) C. R. Abernathy, A. S. Jordan, S. J. Pearton, W. S. Hobson, D. A. Bohling, and G. T. Muhr, *Appl. Phys. Lett.*, 56, 2654 (1990). (b) J. S. Roberts, C. C. Button, J. P. R. David, A. C. Jones, and S. A. Rushworth, *J. Crystal Growth*, 104, 857 (1990). (c) A. C. Jones, S. A. Rushworth, *J. Crystal Growth* 106, 253 (1990). (d) A. C. Jones, S. A. Rushworth, D. A. Bohling, and G. T.

Muhr, *J. Crystal Growth*, **106**, 246 (1990). (e) W. S. Hobson, T. D. Harris, C. R. Abernathy, and S. J. Pearton, *Appl. Phys. Lett.*, **58**, 77 (1991). (f) D. A. Bohling, G. T. Muhr, C. R. Abernathy, A. S. Jordan, S. J. Pearton, and W. S. Hobson, *J. Crystal Growth*, **107**, 1068 (1991).
3. M. G. Simmonds, E. C. Phillips, J.-W. Hwang, and W. L. Gladfelter, *Chem-tronics*, **5**, 155 (1991).
4. (a) J. K. Ruff, and M. F. Hawthorne, *J. Am. Chem. Soc.*, **82**, 2141 (1960). (b) J. K. Ruff, *Inorg. Synth.*, **9**, 30 (1967).

10. TERTIARY AMINE AND PHOSPHINE ADDUCTS OF GALLIUM TRIHYDRIDE

SUBMITTED BY GEORGE A. KOUTSANTONIS* and COLIN L. RASTON‡
CHECKED BY ALAN H. COWLEY† and FRANÇOIS P. GABBAI†

Donor adducts of aluminum and gallium trihydride were the subject of considerable interest in the late 1960s and early 1970s.[1] Thin-film deposition and microelectronic device fabrication has been the driving force for the recent resurgence of synthetic and theoretical interest in these adducts of alane and gallane.[2-4] This is directly attributable to their utility as low-temperature, relatively stable precursors for both conventional and laser-assisted CVD,[5-9] and has resulted in the commercial availability of at least one adduct of alane. The absence of direct metal–carbon bonds in adducts of metal hydrides can minimize the formation of deleterious carbonaceous material during applications of CVD techniques, in contrast to some metal alkyl species.[10, 11]

In this section we present the synthesis of adducts of some gallium trihydride (gallane) compounds which are remarkably thermally stable.[2, 12, 13]

General Considerations

All reactions involving gallium hydrides were conducted under an atmosphere of high-purity argon using standard Schlenk[14] and cannula techniques.[15] Previously sodium-dried diethyl ether is distilled from Na/K alloy, under purified oxygen-free nitrogen, and degassed by three freeze-pump-thaw cycles immediately prior to use. The starting materials, Li[GaH$_4$][16] and

* Department of Chemistry, The University of Western Australia, Nedlands, WA, Australia, 6907.
† Department of Chemistry and Biochemistry, The University of Texas at Austin, Austin, TX, 78712-1167.
‡ Department of Chemistry, Monash University, Clayton, Victoria, Australia, 3168.

$H_3Ga \cdot (NMe_3)$,[17] are described in Volume 15 of *Inorganic Syntheses*. It is particularly important to use pure LiH in the preparation of $Li[GaH_4]$. 1-Azabicyclo[2.2.2]octane (quinuclidine) was purchased from Fluka and used as received.

A. (1-AZABICYCLO[2.2.2]OCTANE) HYDROCHLORIDE

$$C_7H_{13}N + HCl \longrightarrow [C_7H_{13}NH]Cl$$

Procedure

This reaction may be carried out in the air.

1-Azabicyclo[2.2.2]octane (quinuclidine) (5.05 g, 45.4 mmol) is placed in a 100-mL, round-bottomed flask and treated slowly with 20 mL of ice cold (ca. 3.4 M solution) HCl. The resulting solution is evaporated to an oil using a rotary evaporator. The oil is then subjected to a high vacuum (ca. 0.1 torr) and heated (80°C) for 24 h to give a white solid. The solid is scraped from the flask and transferred to a Schlenk flask. Yield 5.94 g (89%).

B. (1-AZABICYCLO[2.2.2]OCTANE)TRIHYDRIDOGALLIUM

$$Li[GaH_4] + [C_7H_{13}NH][Cl] \longrightarrow [H_3Ga \cdot (NC_7H_{13})] + LiCl + H_2$$

Procedure

A freshly prepared 100-mL solution of Li $[GaH_4]$[16] (12.40 mmol) in diethyl ether is added by cannula to a 250-mL Schlenk flask containing a slurry of 2.15 g of quinuclidine hydrochloride in diethyl ether (50 mL) at 0°C. After gas evolution has ceased, the mixture is allowed to warm to ambient temperature and stirred for 16 h. The ethereal solution is decanted by filter-tipped cannula into a 250-mL Schlenk flask, concentrated to half volume in vacuo and cooled to ca. − 30°C for 16 h. Colorless crystals form. The supernatant liquid is decanted by cannula into another Schlenk flask and concentrated and cooled as above to give a second crop of crystals. Yield 2.06 g (90%).

Anal. Calcd. for $C_7H_{13}NGa$ (MW 183.94): C, 45.7; H, 8.8; N, 7.6%. Found: C, 46.1; H, 8.7; N, 7.7%.

Properties

The 1-azabicyclo[2.2.2]octane (quinuclidine) gallane complex is a colorless crystalline solid, highly sensitive to air and moisture. Melting point

100–101°C (dec.) It is soluble in diethyl ether and tetrahydrofuran. Small quantities can be sublimed with decomposition (0.2 torr, 65–70°C). The IR spectrum, in Nujol, shows an absorption at 1810 (v_{GaH}) cm^{-1}. The ^{1}H NMR spectrum (250 MHz, C_6D_6) has the following peaks: δ 0.93 (6H, m, CH_2), 1.12 (1H, m, CH), 2.60 (6H, t, NCH_2), 4.80 (broad s, GaH). The ^{13}C NMR spectrum (62.8 MHz, C_6D_6) has the following peaks: δ 19.3 (s, CH), 25.0 (s, CH$_2$), 48.9 (s, NCH$_2$).

C. TRIHYDRIDO(TRICYCLOHEXYLPHOSPHINE)GALLIUM

$$Li[GaH_4] + P(C_6H_{11})_3 \longrightarrow [H_3Ga \cdot P(C_6H_{11})_3] + LiH$$

Procedure

A solution of tricyclohexylphosphine (4.65 g, 16.58 mmol) dissolved in diethyl ether (120 mL) is added to a freshly prepared solution of Li[GaH$_4$]^{16} (16.30 mmol) in diethyl ether (100 mL), cooled in a liquid nitrogen–chloroform slush bath. The mixture is stirred for 30 min and warmed to ambient temperature as it stirs overnight.

The reaction mixture is filtered by filter-tipped cannula into a 250-mL Schlenk flask and the filtrate is reduced to half volume in vacuo and cooled to about ca. − 30°C overnight. The supernatant liquid is decanted from the small colorless crystals that form and reduced to half volume and again cooled overnight. The preceding procedure is repeated once more. Overall yield 4.73 g (82%).

Anal. Calcd. for $C_{18}H_{36}PGa$ (MW 353.17): C, 61.18; H, 10.19%. Found: C, 60.94; H, 10.15%.

Properties

The tricyclohexylphosphine gallane compound is a colorless crystalline solid, which is stable in dry air for weeks and decomposes slowly in water. Melting point > 130°C (dec.) It is soluble in diethyl ether and tetrahydrofuran. The IR spectrum, in Nujol, shows an absorption at 1800 (v_{GaH}) cm^{-1}. The ^{1}H NMR spectrum (250 MHz, C_6D_6) has the following peaks: δ 1.88, 1.81, 1.67, 1.57, 1.41, 1.10 (2H, 1H, 2H, 1H, 2H, 3H, m, C_6H_{11}), 4.32 (broad s, GaH). The ^{13}C NMR spectrum (62.8 MHz, C_6D_6) has the following peaks: δ 26.0 (s CH$_2$), 27.2 (d, CH$_2$, $^2J_{CP}$ 11 Hz), 29.2 (s, CH$_2$), 31.0 (d, CH, $^1J_{CP}$ 15 Hz). The ^{31}P NMR (121.47, C_6D_6) spectrum has one peak: δ 11.8 (s, PC_6H_{11}).

D. HEXAHYDRIDO[1,2-ETHANEDIYLBIS(DIMETHYLPHOSPHINE)] DIGALLIUM

$$2 \cdot H_3Ga \cdot (NMe_3) + (PMe_2CH_2)_2 \longrightarrow [(H_3Ga)_2 \cdot (PMe_2CH_2)_2] + 2NMe_3$$

■ **Caution.** *1,2-Ethanediylbis(dimethylphosphine) has a noxious odor and is toxic. It should only be used in an efficient fume hood and gloves should be worn while performing the procedure.*

Neat 1,2-ethanediylbis(dimethylphosphine) (0.40 mL, 4.40 mmol) is added by degassed pipette to a 100-mL Schlenk flask containing a solution of $[H_3Ga \cdot (NMe_3)]$ (0.56 g, 4.25 mmol) in diethyl ether (50 mL) cooled with an ice bath. The mixture is stirred for 10 min and warmed to ambient temperature as it stirs overnight. The slightly turbid solution is filtered by filter-tipped cannula into another Schlenk flask and cooled to about $-30°C$ overnight. Colorless crystals form. The supernatant liquid is decanted by cannula, concentrated to about 10 mL in vacuo and cooled as above. Yield 0.49 g (78%).

Anal. Calcd. for $C_6H_{22}P_2Ga_2$ (MW 295.63): C, 24.38; H, 7.50%. Found: C, 23.93; H, 6.93%.

Properties

The 1,2-ethanediylbis(dimethylphosphine) gallane compound is a colorless crystalline solid, highly sensitive to air and moisture. Melting point 107°C (dec.) It is soluble in diethyl ether and tetrahydrofuran. The IR spectrum, in Nujol, shows an absorption at 1829 (ν_{GaH}) cm^{-1}. The ^1H NMR spectrum (250 MHz, C_6D_6) has the following peaks: δ 1.67 (12H, virtual t, PCH$_3$, J 3.4 Hz), 2.28 (4H, s, PCH$_2$), 4.67 (6H, broad s, GaH). The ^{13}C NMR spectrum (62.8 MHz, C_6D_6) has the following peaks: δ 9.6 (t, PCH$_3$, $^1J_{CP}$ 9 Hz), 21.7 (t, PCH$_2$, $^1J_{CP}$ 10 Hz). The ^{31}P NMR (121.47, C_6D_6) spectrum has one peak: δ -29.5 (s, CH$_3$PCH$_2$).

References

1. E. Wiberg and E. Amberger, *Hydrides of the Elements of Main Groups I-IV*, Elsevier, London, 1971.
2. J. L. Atwood, S. G. Bott, F. M. Elms, C. Jones, and C. L. Raston, *Inorg. Chem.*, **30**, 3792 (1991).
3. J. L. Atwood, F. R. Bennett, F. M. Elms, C. Jones, C. L. Raston, and K. D. Robinson, *J. Am. Chem. Soc.*, **113**, 8183 (1991).
4. B. J. Duke, C. Liang, and H. F. Schaefer, *J. Am. Chem. Soc.*, **113**, 2884 (1991).
5. M. Hanabusa and M. Ikeda, *Appl. Organomet. Chem.*, **5**, 289 (1991).

6. L. H. Dubois, B. R. Zegarski, M. E. Gross, and R. G. Nuzzo, *Surf. Sci.*, **244**, 89 (1991).
7. G. S. Higashi, *Appl. Surface Sci.*, **43**, 6 (1989).
8. W. L. Gladfelter, D. C. Boyd, and K. F. Jensen, *Chem. Mater.*, **1**, 339 (1989).
9. T. H. Baum, C. E. Larson, and R. L. Jackson, *Appl. Phys. Lett.*, **55**, 1264 (1989).
10. T. K. Kuech, D. J. Wolford, E. Veuhoff, V. Deline, P. M. Mooney, R. Potemski, and J. Bradley, *J. Appl. Phys.*, **62**, 632 (1987).
11. A. C. Jones, J. S. Roberts, P. J. Wright, P. E. Olive, and B. Cockayne, *Chemtronics*, **3**, 152 (1988).
12. J. L. Atwood, K. D. Robinson, F. R. Bennett, F. M. Elms, G. A. Koutsantonis, C. L. Raston, and D. J. Young, *Inorg. Chem.*, **31**, 2674 (1992).
13. D. O'Hare, J. S. Foord, T. C. Page, and T. J. Whitaker, *J. Chem. Soc., Chem. Commun.*, 1445 (1991).
14. D. F. Shriver and M. A. Drezdon, *The Manipulation of Air Sensitive Compounds*, 2nd ed., Wiley, New York, 1986.
15. J. P. McNally, V. S. Leong, and N. J. Cooper, *Experimental Organometallic* Chemistry. A Practicum in Synthesis and Characterization, A. L. Wayda and M. Y. Darensbourg (eds.), Symposium Series 357, American Chemical Society, Washington, DC, 1987, p. 6.
16. D. F. Shriver and A. E. Shirk, *Inorg. Synth.*, **15**, 45 (1977).
17. D. F. Shriver and A. E. Shirk, *Inorg. Synth.*, **15**, 43 (1977).

11. *trans*-1,1-DI-*tert*-BUTYL-2,3-DIMETHYLSILIRANE AND 2,2-DI-*tert*-BUTYL-1,1,1-TRIETHYLDISILANE

SUBMITTED BY PHILIP BOUDJOUK,* ERIC BLACK, and UPASIRI SAMARAWEERA
CHECKED BY DONALD BERRY[†] and BOK RYUL YOO[†]

The first reported attempt to prepare siliranes by alkali metal reduction of dihalosilanes was by Skell and Goldstein[1] using gas-phase techniques. The first unambiguous preparation of a silirane was by Lambert and Seyferth using a carbon–carbon coupling reaction.[2] Since then many siliranes have been prepared as stable compounds or as reactive species and used as synthetic intermediates. They can be used for the synthesis of heterocycles via ring expansion[3] and for generating silylenes.[4] *trans*-1,1-Di-*tert*-butyl-2,3-dimethylsilirane is an isolable silirane which is an efficient silylene precursor using either thermal (180°C) or photochemical (254 nm) means.[5] This silirane can be prepared in high yields and can be stored indefinitely in an inert atmosphere.

* Center for Main Group Chemistry, Department of Chemistry, North Dakota State University, Fargo, ND 58105-5516.
† Department of Chemistry, University of Pennsylvania, Philadelphia, PA 19104-6323.

A. *trans*-1,1-DI-*tert*-BUTYL-2,3-DIMETHYLSILIRANE

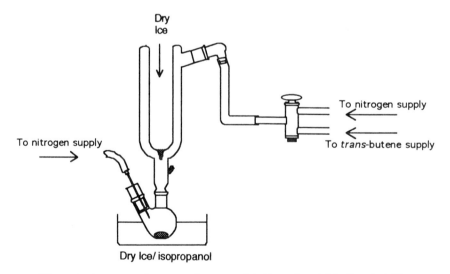

Procedure

Lithium dispersion (0.5% sodium in mineral oil, Aldrich) is washed with hexane to remove the oil before use and dried by passing a stream of nitrogen over it. Di-*tert*-butyldichlorosilane is commercially available (Hüls, bp 191°C) or can be prepared by chlorination of di-*tert*-butylchlorosilane (benzoyl peroxide in refluxing carbon tetrachloride for 8 h, yield > 90%). *trans*-Butene is commercially available (Aldrich) in lecture bottles and is used as is. Tetrahydrofuran is freshly distilled under nitrogen from sodium ketyl benzophenone immediately prior to use.

The apparatus (Fig. 1) consists of a two-necked, 25-mL flask with a magnetic stir bar, in which one neck is equipped with a dry-ice condenser fitted with a gas inlet. This inlet is controlled by a three-way stopcock attached to

Figure 1. Apparatus for preparing *trans*-1,1-di-*tert*-butyl-2,3-dimethylsilirane.

the *trans*-butene supply and to a nitrogen source which has a mineral oil bubbler to release the inert gas pressure. The other neck is closed off with a rubber septum. The glass portions of the apparatus are oven dried and assembled under nitrogen flow, making sure that the system is air and moisture free before starting.

■ **Caution.** *Lithium dispersion is a flammable, moisture-sensitive material. Manipulations should be carried out in a dry box or inert-atmosphere glove bag.*

Tetrahydrofuran (12 mL) is syringed into a 25-mL flask through a rubber septum. Lithium (412 mg, 58.8 mmol) is weighed in a drybox and placed in a capped vial. The septum is removed and the lithium is poured into the reaction vessel from the vial under a positive flow of nitrogen coming from the neck of the flask. The septum is replaced and the system is cooled using a dry-ice/isopropanol bath. Dry ice is added to the condenser and another nitrogen line (also fitted with a mineral oil bubbler) with a needle attached is inserted through the rubber septum to release the pressure during the *trans*-butene addition. The three-way stopcock is rotated to allow the *trans*-butene through and at least 180 drops (5 mL, 56 mmol) are condensed into the reaction flask. The stopcock is rotated back to the nitrogen setting and the pressure release syringe is removed. Di-*tert*-butyldichlorosilane (4.0 g, 18.7 mmol) is syringed in through the septum all at once. The dry ice/isopropanol bath is removed and the reaction mixture is allowed to stir for 12 h or until GC (Hewlett Packard Model 5890, 15 m col; liquid phase DB-1, 100% methylsilicone, film thickness 0.25 μm,) shows the disappearance of di-*tert*-butyldichlorosilane (retention time 2.7 min under these conditions: initial temp 50°C for 2 min, increase temp at 10°C/min to 250°C). The salts in the product mixture are allowed to settle and a distillation apparatus consisting of a 25-mL, one-neck flask, a 20-cm Vigreux column, a short-path distilling head with thermometer, and a 5-mL receiving flask are oven dried and assembled under nitrogen. The product mixture is transferred by syringe to the distillation pot, taking care not to transfer the salts which have settled out. Eicosane, $C_{20}H_{42}$ (Aldrich, bp 220°C/30 torr, 0.2 g, 0.7 mmol) is also added to the distillation pot to be used as a "kicker" during the distillation. Tetrahydrofuran is removed slowly under vacuum (note: a dry ice/isopropanol trap should be placed between the vacuum pump and the distillation apparatus to avoid pulling the THF through the pump) and the silirane (retention time of 4.8 min under the above GC conditions) is then distilled from the brown residue by vacuum distillation. Bp, 45°C/0.5 mm Hg. Yield: 2.96 g (80%). Purity is 95–97% in most cases. Typical impurities are < 2% of the *cis* isomer, resulting from trace amounts of *cis*-2-butene found in *trans*-2-butene and small amounts (usually < 2%) of unreacted di-*tert*-butyldichlorosilane. These products were identified by comparison of their GC and NMR spectra with those of pure samples.

Properties

1,1-Di-*tert*-butyl-2,3-dimethylsilirane is a clear, slightly viscous air- and moisture-sensitive liquid. The NMR data for a solution in benzene are δ 0.80 (m, 2H), 1.23 (s, 18H), 1.63 (d, J = 6.45 Hz, 6H) for ^{1}H NMR; δ 16.66 (CH$_3$), 17.4 (CH), 19.7 [C-(CH$_3$)$_3$], 30.9 [C-(CH$_3$)$_3$] for ^{13}C NMR; and δ $-$ 43.9 for ^{29}Si NMR.

■ **Caution.** *Benzene is a human carcinogen and should be handled in a well-ventilated hood with suitable gloves.*

B. 2,2-DI-*tert*-BUTYL-1,1,1-TRIETHYLDISILANE

Procedure

Triethylsilane is commercially available (Aldrich; bp, 108°C) and is distilled prior to use. Benzene is stirred over sulfuric acid and distilled prior to use. The pyrolysis tubes are vacuum hydrolysis tubes (Kontes Cat. #896860-8910), but any sealable glass tube is sufficient. A sealable glass tube is charged with 9 mL of benzene, *trans*-1,1-di-*tert*-butyl-2,3-dimethylsilirane (450 mg, 2.3 mmol) and triethylsilane (1.8 mL, 11.3 mmol) using syringe and inert atmosphere techniques. The tube is sealed under a nitrogen atmosphere and immersed in a sand or oil bath which has been preheated to 180°C. The tube is pyrolyzed for 2 days. Upon removal from the heating bath, the benzene is stripped away and the disilane is purified by molecular distillation. Bp, 59°C/2 mm Hg. Yield: 416 mg (70%). Typical purity is > 90% by GC and NMR. The impurity is usually unreacted *trans*-1,1-di-*tert*-butyl-2,3-dimethylsilirane (identified by GC and NMR comparisons with an authentic sample).

■ **Caution.** *Pyrolysis tubes may develop enough pressure to explode and cause serious injury. Only high-quality, thick-walled glass should be used. Protective eyewear and a laboratory safety shield should be used at all times.*

The disilane can also be prepared photochemically in 69% yield. The same quantities of silirane and triethylsilane used above are dissolved in 9 mL of degassed hexane in a quartz tube and photolyzed for 4 h in a Rayonet model RPR 1000 reactor fitted with 254-nm bulbs. The disilane is isolated as described above.

Properties

2,2-Di-*tert*-butyl-1,1,1-triethyldisilane is a clear, air- and moisture-stable liquid. The NMR data for a solution in CCl_4 are δ 0.60–1.21 (m, 15H), 1.12 (s, 18H), 3.45 (s, 1H) for ^1H NMR. A sample of analytically pure compound may be obtained by preparative GLC.

Acknowledgments

The financial support of the Air Force Office of Scientific Research through grant no. 91-0197 is gratefully acknowledged. Technical assistance from Ioana Stoenescu is also gratefully acknowledged.

References

1. P. S. Skell and E. J. Goldstein, *J. Am. Chem. Soc.*, **86**, 1442 (1964).
2. R. L. Lambert and D. Seyferth, *J. Am. Chem. Soc.*, **94**, 9246 (1972).
3. D. Seyferth, C. K. Haas, R. L. Lambert, Jr., and D. C. Annarelli, *J. Organomet. Chem.*, **152**, 131 (1978).
4. D. Seyferth, D. C. Annarelli, and D. P. Duncan, *Organometallics*, **1**, 1288 (1982).
5. P. Boudjouk, U. Samaraweera, R. Sooriyakumaran, J. Chrusciel, and K. R. Anderson, *Angew. Chem., Int. Ed. Engl.*, **27**, 1355 (1988). M. Weidenbruch, A. Lesch, and H. Marsmann, *J. Organomet. Chem.*, **385**, C47 (1990). P. Boudjouk, E. Black, and R. Kumarathasan, *Organometallics*, **10**, 2095 (1991).

12. TIN(II) SULFIDE AND TIN(II) SELENIDE

SUBMITTED BY STEVEN R. BAHR and PHILIP BOUDJOUK*
CHECKED BY JOHN ARNOLD[†]

$$3Ph_2SnCl_2 + 3Na_2X \xrightarrow[THF]{} (Ph_2SnX)_3$$

$$(Ph_2SnX)_3 \xrightarrow[He\ atmosphere]{450°C} 2SnX + Ph_4Sn + Ph_2X$$

tetrahydrofuran = THF   X = S or Se   Ph = phenyl

Recently, numerous studies have demonstrated the potential importance of the semiconducting compounds SnS and SnSe in such uses as solar cells, switching devices, and photoelectrochemical electrodes.[1] These compounds are also important for the production of ternary lead–tin chalcogenides potentially useful in photovoltaic devices, photoconductors, photodiodes, and lasers.[2] Reported methods which describe the synthesis of SnS and SnSe include mixing an alkali-metal chalcogenide and $SnCl_2$ in aqueous solution[3] and high-temperature sintering of elemental tin and sulfur or selenium.[4] Also, SnS has been synthesized by combining SnO_2 with molten, anhydrous KSCN at 450°C.[4b] The latter two procedures are anhydrous; however, in the SnO_2/KSCN method, the byproduct K_2S is removed by treatment with water.

The following procedure describes a new low-temperature, completely anhydrous route to SnS and SnSe. These compounds are synthesized by pyrolysis of the phenylated six-membered rings, $(SnSPh_2)_3$ and $(SnSePh_2)_3$, at 450°C in a helium atmosphere. These cyclic compounds are obtained by utilizing anhydrous sodium sulfide[5] or sodium selenide[6] generated from sodium, sulfur, or selenium and a catalytic amount of naphthalene in THF followed by addition of Ph_2SnCl_2. These reactions must be performed under a nitrogen atmosphere and require the use of oven-dried glassware cooled under a flow of dry nitrogen. Tetrahydrofuran must be freed of water and oxygen by distillation from sodium benzophenone immediately before use. Benzene is shaken with concentrated H_2SO_4, dried by distilling off the cloudy benzene/water azeotrope until only clear benzene distills over, and stored in a brown bottle. Hexane is purified by stirring over concentrated H_2SO_4 for 1 d and distilled into a storage bottle.

* Center for Main Group Chemistry, Department of Chemistry, North Dakota State University, Fargo, ND 58105–5516.
† Department of Chemistry, University of California, Berkeley, Berkeley, CA 94720.

Procedure

■ **Caution.** *Benzene is a human carcinogen and the evolution of some hydrogen sulfide or hydrogen selenide during the workup requires that all manipulations be carried out in a well-ventilated fume hood and rubber gloves should be worn. Care should be taken to avoid inhalation of selenium powder which is highly toxic and easily absorbed by the lungs. Sodium reacts violently with water and care should be taken to ensure anhydrous conditions during its handling. The sodium chips used in reactions are prepared by cutting sodium pellets in a glove box.*

A. *CYCLO*-TRIS(DIPHENYL-μ-THIO-TIN)(SnSPh$_2$)$_3$

A Teflon stirbar is placed in a 100-mL, two-necked, round-bottomed flask equipped with a condenser and 50-mL pressure-equalizing addition funnel. A rubber septum is placed on the addition funnel and a nitrogen gas source is attached to the condenser. Sodium chips (1.38 g, 60.0 mmol), sulfur powder (0.97 g, 30.0 mmol), and naphthalene (0.8 g, 0.6 mmol) are added through a neck of the flask while maintaining a counterflow of N$_2$ gas through the reaction vessel. Then 30 mL of THF is added by syringe to the flask through the addition funnel and the mixture is heated to reflux and stirred for 10 h, producing a reaction mixture consisting of a greenish-gray or light yellow suspension. With a positive nitrogen flow, SnCl$_2$Ph$_2$ (Aldrich) (10.3 g, 30.0 mmol) is placed in the addition funnel and 20 mL THF is added to dissolve the SnCl$_2$Ph$_2$. The Na$_2$S is cooled to 0°C, followed by dropwise addition of the SnCl$_2$Ph$_2$ solution over a 30-min period, and then refluxed for 24 h, producing a gray or tan-colored mixture. At this point, the use of inert atmosphere techniques is no longer necessary. The reaction mixture is poured into 75 mL of water saturated with NaCl, and extracted with three 50-mL diethyl ether portions. The portions are combined and dried over MgSO$_4$, and the solvent is removed by rotary evaporator, leaving a light yellow residue. The residue is taken up in 20 mL of benzene and decanted from any undissolved solid, and 25 mL of hexane are added to cause precipitation of the (SnSPh$_2$)$_3$ over 5 h at -10°C. The liquid is decanted and the white crystalline solid dried in vacuo, yielding 5.6 g (61% yield) of product. One additional recrystallization in benzene/hexane will give very pure (SnSPh$_2$)$_3$ for pyrolysis use.

B. *CYCLO*-TRIS(DIPHENYL-μ-SELENO-TIN)(SnSePh$_2$)$_3$

A very similar procedure to that described for (SnSPh$_2$)$_3$ is used for (SnSePh$_2$)$_3$. While maintaining a positive flow of N$_2$ gas, sodium chips

(1.02 g, 44.0 mmol), gray selenium powder (Aldrich, 99.5% +) (1.73 g, 22.0 mmol), naphthalene (0.56 g, 4.4 mmol), and a Teflon stirbar are placed in a 100-mL, two-necked, round-bottomed flask equipped with a 50-mL addition funnel with a rubber septum and a condenser. Then 30 mL of THF is added by syringe, and the reaction mixture is refluxed for 10 h, producing a white or light gray suspension of Na_2Se. (If a purple color persists at this point, small amounts of sodium can be added until a white endpoint is attained. For a green mixture, adding Se powder will eventually give the white suspension.) The $SnCl_2Ph_2$ (7.6 g, 22 mmol) is placed in the addition funnel with 20 mL THF and added dropwise to the 0°C cooled Na_2Se suspension. The reaction is warmed to room temperature and refluxed for 24 h. Workup is performed exactly as described for $(SnSPh_2)_3$, giving 4.5 g (56% yield) of very light yellow crystalline $(SnSePh_2)_3$. One additional recrystallization in benzene/hexane gives clear, colorless crystals for use in pyrolysis reactions.

Properties

$(SnSPh_2)_3$ and $(SnSePh_2)_3$ both exist as moisture-stable, colorless crystalline compounds which recrystallize easily from benzene/hexane solutions. $(SnSPh_2)_3$: mp 180–183°C (lit.[7] mp 183–184°C); ^{119}Sn NMR $(CDCl_3)$: δ 17.6 (lit.[8] δ 16.8). $(SnSePh_2)_3$: mp 176–177°C (lit.[9] mp 176–177°C); ^{119}Sn NMR $(CDCl_3)$: δ − 43.5 (lit.[4] δ − 44); ^{77}Se NMR $(CDCl_3)$: δ − 427.*

C. SnS FROM PYROLYSIS OF $(SnSPh_2)_3$

Pyrolysis Apparatus

The pyrolysis apparatus used is shown in Fig. 1. The oven is a Lindberg tube furnace that is 36 cm long with a 3-cm diameter and a 55 × 2.5 cm fused-quartz tube placed inside. Each end of the tube is fitted with a one-hole Teflon stopper which is wrapped with Teflon tape to ensure a good seal. A short piece of glass tubing is inserted into one of the stoppers and connected to an inert (N_2 or He) gas source by using 1/4 in. Nalgene (polyvinyl chloride) tubing. The inlet to the liquid N_2 trap is connected to the other end. The trap is packed with glass wool to increase its efficiency. A silicone oil bubbler used

---

* The checkers report the following NMR data: $(SnSPh_2)_3$: ^{119}Sn NMR $(CDCl_3)$ δ 16.2 $(^2J_{119Sn117Sn} = 195$ Hz$)$. $(SnSePh_2)_3$: ^{119}Sn NMR $(CDCl_3)$ δ − 44.8 $(J_{119Sn77Se} = 1317$ Hz, $J_{119Sn117Sn} = 239$ Hz$)$. ^{77}Se NMR $(CDCl_3)$ δ − 433 $(J_{119Sn117Se} = 1316$ Hz, $J_{117Sn77Se} = 1259$ Hz$)$.

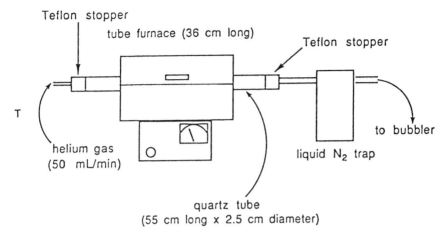

Figure 1. Pyrolysis apparatus.

for monitoring inert gas flow can be installed after the trap. The inert gas flow through the apparatus should be approximately 50 mL/min.*

Procedure

■ **Caution.** *Small amounts of toxic $SnPh_4$, Ph_2S, or Ph_2Se, released as volatile components during the pyrolysis, may not be condensed at the end of the quartz tube or in the liquid N_2 trap. Therefore, the pyrolysis apparatus must be set up in a well-ventilated fume hood or the bubbler outlet vented to a hood.*
$(SnSPh_2)_3$ (0.433 g, 0.473 mmol) is weighed into a ceramic crucible (12 mm height × 10 mm diameter) and placed in a wire holder used to support the sample upright while in the fused-quartz tube. The crucible and holder are placed in the quartz tube so the sample is in the center of the furnace and the apparatus is thoroughly flushed for several minutes with N_2 or He gas. Inert gas flow is set at 50 mL/min and the furnace is heated to 450°C over a 2-h period. The temperature is held there for 1 h before being allowed to cool to room temperature. When the temperature reaches 300°C, a white solid and yellow oil can be seen depositing on the walls of the quartz tube where it exits the furnace and in the liquid N_2 trap. A fine powder of SnS remains in the crucible (145 mg, 33%). At the position where the quartz tube exits the

---

* The checkers found it more convenient to perform the pyrolysis reactions in vacuum-sealed tubes which gave identical results.

furnace, clear needles of $SnPh_4$ deposited along with a light yellow, oily solid. Rinsing the end of the fused-quartz tube and the trap with CH_2Cl_2 followed by removal of the solvent by rotary evaporator left a 260-mg mixture of $SnPh_4$ and Ph_2S (94% total mass recovery).

Anal. Calcd. for SnS: Sn, 78.73; S, 21.27. Found: Sn, 79.33 $\pm$ 0.6; S, 21.31 $\pm$ 0.1.

D. SnSe FROM PYROLYSIS OF $(SnSePh_2)_3$

$(SnSePh_2)_3$ (0.395 g, 0.374 mmol) is pyrolyzed using the same method as that described for $(Ph_2SnS)_3$. A fine, dark gray powder of SnSe (144 mg, 36%) remained in the crucible while a 240-mg mixture of predominantly $SnPh_4$ and Ph_2Se is isolated from the end of the tube and from the liquid N_2 trap (97% total mass recovery).

Anal. Calcd. for SnSe: Sn, 60.05; Se, 39.95. Found: Sn, 60.65 $\pm$ 1.0; Se, 39.40 $\pm$ 0.8.

Properties

Both SnS and SnSe isolated from this method are very fine, dark gray powders. They can be sublimed between 800 and 900°C in the same pyrolysis apparatus producing shiny, black, platelike crystals. Larger crystals of SnS can be grown in comparison to those of SnSe. Combustion analysis shows a lower level of residual carbon impurity for SnS (0.14 $\pm$ 0.1%) compared to SnSe (0.35 $\pm$ 0.1%). Scanning electron micrographs indicate that the SnS particles exist as microcrystalline rosette clusters predominantly in the 10-μm size range with some clusters of 30–50 μm. The SnSe particles are not as uniform as those of SnS but rather exist as 5–100-μm clusters of crystals which are approximately 2–4 $\times$ 0.5 μm in size. X-Ray diffraction data (peak positions and relative intensities) for SnS and SnSe are listed in Table I. The XRD data for SnS compare well with reference patterns (39-354 and 33-1375) in Set 40 of the ICDD Powder Diffraction File (PDF). The peak positions and relative intensities for SnSe are in fair agreement with those of a calculated pattern (PDF 32-1382). However, there are some minor peak shifts in comparison to the calculated pattern. This may be due to a slight stoichiometry range of tin and selenium, such as $SnSe_x$, where x is slightly greater than or less than 1. This may also be true of the SnS made by this method; however, the elemental analyses of both SnS and SnSe support a 1:1 composition. A more extensive discussion of the X-ray diffraction data has been published.[10]

TABLE I. X-Ray Diffraction Data for SnS and SnSe

SnS		SnSe	
d Values	%I	d Values	%I
4.0150	17.6	3.5037	11
3.4111	46.3	3.3703	6
3.2380	50.5	3.0324	19
2.9236	49	2.9307	100
2.8270	100	2.8757	48
2.7924	77	2.3767	42
2.3006	48	2.2099	4
2.1182	12	2.1716	6
2.0214	35	2.0866	23
1.9898	22	2.0388	4
1.8683	27	1.9221	9
1.7851	8	1.8330	28
1.7778	25	1.7893	4
1.7204	15	1.7564	5
1.7114	11	1.7393	6
1.6881	20	1.6881	12
1.6222	18		

References

1. (a) O. P. Singh and V. P. Gupta, *Phys. Status Solidi A*, **136**, 41 (1986). (b) T. Subba Rao, B. K. Samanata Ray, and A. K. Chaudhuri, *Thin Solid Films* **165**, 257 (1988). (c) M. Ristov, G. Sinadinovski, I. Grozdanov, and M. Mitreski, *Thin Solid Films* **173**, 53 (1989). (d) M. Sharon, and K. Basavaswaran, *Indian J. Chem.*, **28A**, 698 (1989). (e) M. T. S. Nair and P. K. Nair, *Semicond. Sci. Technol.*, **6**, 132 (1991).

2. (a) E. E. Matyas and T. E. Borisenko, *Phys. Status Solidi A* **89**, 177 (1985). (b) V. I. Garasim, D. M. Zayachuk, R. D. Ivanchuk, V. B. Lototskii, P. M. Starik, and V. A. Shenderovskii, *Phys. Status Solidi A*, **111**, 99 (1989).

3. A. Korczynski, I. Lubomirska, and T. Sobierajski, *Chem. Stosow.*, **25**, 391 (1981).

4. (a) N. Yellin and L. Ben-Dor, *Mater. Res. Bull.*, **18**, 823 (1983). (b) M. Baudler, *Handbook of Preparative Inorganic Chemistry*, G. Brauer, (ed.), Academic Press, New York, 1963, p. 739.

5. J.-H. So and P. Boudjouk, *Synthesis*, 306 (1989).

6. D. P. Thompson and P. Boudjouk, *J. Org. Chem.*, **53**, 2109 (1988).

7. M. Schmidt, H.-J. Dersin, and H. Schumann, *Chem. Ber.*, **95**, 1428 (1962).

8. K. Furue, T. Kimura, N. Yasuoka, N. Kasai, and M. Kakudo, *Bull. Chem. Soc. Jpn.*, **43** 1661, (1970).

9. H. Schumann, K. F. Thom, and M. Schmidt, *J. Organomet. Chem.*, **2**, 361 (1964).

10. S. R. Bahr, P. Boudjouk, and G. J. McCarthy, *Chem. Mater.*, **4**, 383 (1992).

13. TIN(IV) FLUORIDE (TETRAFLUOROSTANNANE)

SUBMITTED BY DAVID TUDELA* and JUAN A. PATRON*
CHECKED BY DAVID SHERLOCK[†] and ROBERT R. HOLMES[†]

$$2SnF_2 + 2I_2 + 2MeCN \rightarrow SnF_4(MeCN)_2 + SnI_4$$

$$SnF_4(MeCN)_2 \rightarrow SnF_4 + 2MeCN$$

In contrast to SnX_4 (X = Cl, Br, I), which have monomeric tetrahedral structures, SnF_4 has a polymeric fluorine-bridged, two-dimensional sheet structure with hexacoordinated tin.[1] Tetrafluorostannane is a Lewis acid and forms complexes with donor molecules. It also forms graphite intercalation compounds and has been used in the manufacture of fluoride glasses. Tetrafluorostannane has been prepared by several methods involving the use of fluorinating reagents such as anhydrous HF,[2] F_2,[3] BrF_3,[4] ClF_3,[5] SF_4,[6] COF_2,[7] or F_3NO.[8] The manipulation of these fluorinating reagents often requires the use of special equipment not available in every laboratory, and precautions have to be taken against their high reactivity or toxicity. The procedure described below provides a very simple and safe route to tin(IV) fluoride under mild conditions, with the use of standard glassware. Since the title compound is very hygroscopic, all reactions and manipulations should be performed under an atmosphere of dry nitrogen, either in a dry box or using Schlenk techniques.[9]

Procedure

■ **Caution.** *Carbon disulfide and acetonitrile are toxic and flammable. Furthermore, iodine is intensely irritating to the eyes, skin and mucous membranes. Therefore, all manipulations involving these chemicals should be performed in a fume hood.*

A 250-mL, two-necked, round-bottomed Schlenk flask equipped with a magnetic stirring bar and a reflux condenser is charged with 50 mL of dry MeCN (HPLC grade, distilled over CaH_2), 5.50 g (35.1 mmol) of SnF_2 (Aldrich), and 9.34 g (36.80 mmol) of I_2. The suspension is stirred and heated to reflux for 4 h. After cooling to room temperature, 30 mL of freshly distilled dry CS_2 (distilled over phosphorus pentoxide) is added to dissolve most of the SnI_4 formed. The mixture is filtered under nitrogen through a fine porosity

* Departamento de Química Inorgánica, Universidad Autónoma de Madrid, 28049-Madrid, Spain.
† Department of Chemistry, University of Massachusetts, Amherst, MA 01003–4510.

(4–8 μm) frit, and white solid $SnF_4(MeCN)_2$ is washed with three portions of 15 mL of CS_2 and vacuum dried. The solid is transferred to a Schlenk tube and heated under vacuum (ca. 10^{-2} mmHg) to remove the coordinated MeCN. Vigorous gas evolution takes place at about 60°C, and again at about 160°C, due to the loss of MeCN. To avoid projection and spillage of the solid, the vacuum can be conducted through the frit. The temperature is gradually raised to 250°C, yielding 2.90 g (85%) of SnF_4.

Properties

Tetrafluorostannane is an extremely hygroscopic white crystalline solid. The IR spectrum (Nujol mull) shows bands at 720(s) and 558(s) cm^{-1} which agree with those reported in the literature.[10] The Mössbauer parameters at room temperature, IS $= -0.40$, QS $= 1.83$ mm s^{-1}, also agree with those reported previously.[11] The main X-ray powder spacings (Debye-Scherrer, MoK$_\alpha$) are 4.0(w), 3.6(vs), 2.9(w), 2.2(w), 2.0(w), and 1.8(w) Å. This pattern is consistent with that calculated on the basis of the single-crystal X-ray structure determination.[1] Tin(IV) fluoride forms octahedral complexes with a variety of ligands,[10] but they are best prepared by displacing MeCN from $SnF_4(MeCN)_2$.[12]

References

1. R. Hoppe and W. Dähne, *Naturwiss.*, **49**, 254 (1962).
2. O. Ruff and W. Plato, *Chem. Ber.*, **37**, 673 (1904).
3. H. M. Haendler, S. F. Bartram, W. J. Bernard, and D. Kippax, *J. Am. Chem. Soc.*, **76**, 2179 (1954).
4. A. A. Woolf and H. J. Emeleus, *J. Chem. Soc.*, 2865 (1949).
5. A. F. Clifford, H. D. Beachell, and W. M. Jack, *J. Inorg. Nucl. Chem.*, **5**, 57 (1958).
6. A. L. Oppegard, W. C. Smith, E. L. Muetterties, and V. A. Engelhardt, *J. Am. Chem. Soc.*, **82**, 3835 (1960).
7. S. P. Mallela, O. D. Gupta, and J. M. Shreeve, *Inorg. Chem.*, **27**, 208 (1988).
8. O. D. Gupta, R. L. Kirchmeier, and J. M. Shreeve, *Inorg. Chem.*, **29**, 573 (1990).
9. D. F. Shriver and M. A. Drezdzon, *The Manipulation of Air-Sensitive Compounds*, 2nd ed., Wiley, New York, 1986.
10. C. J. Wilkins and H. M. Haendler, *J. Chem. Soc.*, 3174 (1965).
11. L. Fournes, J. Grannec, Y. Potin, and P. Hagenmuller, *Solid State Commun.*, **59**, 833 (1986).
12. D. Tudela and F. Rey, *Z. Anorg. Allgem. Chem.*, **575**, 202 (1989).

14. *N,N,N'*-TRIS(TRIMETHYLSILYL)AMIDINES

SUBMITTED BY RENÉ T. BOERÉ*, ROBIN G. HICKS,[†] and
RICHARD T. OAKLEY[‡]
CHECKED BY TRISTRAM CHIVERS[‡] and KATHERINE McGREGOR[‡]

$$RCN \xrightleftharpoons[\text{2. Me}_3\text{SiCl}]{\text{1. LiN(SiMe}_3)_2 \cdot \text{Et}_2\text{O}} R-\overset{\text{NSiMe}_3}{\underset{\text{N(SiMe}_3)_2}{\Big\langle}} + \text{LiCl}$$

N,N,N'-Tris(trimethylsilyl)amidines have been used recently as precursors for a number of inorganic heterocycles and metallacycles,[1] some of which are being studied in light of their unusual solid state properties[2]. Boeré et al. reported the synthesis of several aryl-substituted persilylated benzamidines and the related compound *N,N,N',N'',N'',N'''*-hexakis(trimethylsilyl)-1,4-benezenedicarboximidamide (hereafter referred to as HBDA);[3] the present syntheses, which are generally based on the same reaction of an aryl-substituted carbonitrile with lithium bis(trimethylsilyl)amide, offer more facile routes to representative mono- and polyfunctional carboximidamides (i.e., amidines) as well as the prototypal derivative *N,N,N'*-tris(trimethylsilyl)formimidamide.[4] As before, the crystalline diethyl ether adduct of lithium bis(trimethylsilyl)amide[5] is favored over the nonsolvated amide in these syntheses; the preparation of the diethyl ether adduct is also described here.

General Experimental Considerations

All reactions are carried out under an atmosphere of dry nitrogen gas unless otherwise stated. Lithium bis(trimethylsilyl)amide (Aldrich), 1,4-dicyanobenzene (Aldrich), chlorotrimethylsilane (Aldrich), and 1,3,5-triazine (Aldrich) are used as received. Benzonitrile (Fisher) is vacuum distilled from P_2O_5 prior to use. Solvents are purified by distillation as follows: hexane, over calcium hydride; diethyl ether, over $LiAlH_4$ (■ **Caution.** *Reacts violently with water*); toluene, over sodium; acetonitrile, over P_2O_5.

* Department of Chemistry, University of Lethbridge, Lethbridge, Alberta, T1K 3M4, Canada.
[†] Guelph-Waterloo Centre for Graduate Work in Chemistry, Department of Chemistry and Biochemistry, University of Guelph, Guelph, Ontario, N1G 2W1, Canada.
[‡] Department of Chemistry, University of Calgary, Alberta, T2N 1N4, Canada.

A. LITHIUM BIS(TRIMETHYLSILYL)AMIDE DIETHYL ETHERATE

Lithium bis(trimethylsilyl)amide (250 g, 1.49 mol) is dissolved in 600 mL of warm (50°C) hexane in a 2-L, round-bottomed flask equipped with a reflux condenser. Diethyl ether (250 mL, 2.4 mol) is added slowly to the amide solution via a pressure-equalizing dropping funnel. The addition should take about 45 min—this corresponds to an addition rate slightly faster than dropwise. (Note: *do not stir or agitate the amide solution during the diethyl ether addition; the resulting crystals are of superior quality using this method.*) The mixture is allowed to cool to room temperature for several hours, during which time large hexagonal crystals of the diethyl ether addition compound begin to precipitate. The flask is then refrigerated overnight. The product is filtered off in air through a Büchner funnel, washed three times with 150-mL portions of hexane, and dried in vacuo. Yields are typically 85–90%, mp 108–110°C.

Anal. Calcd. for $C_{10}H_{28}LiNOSi_2$: C, 49.75; H, 11.69; N, 5.80%. Found: C, 49.58; H, 11.90; N, 6.00%. ^1H NMR (C_6D_6) δ 3.40 (q, CH_3CH_2O), 0.98 (t, CH_3CH_2O), 0.27 (s, $SiCH_3$).

B. *N,N,N′*-TRIS(TRIMETHYLSILYL)-BENZENECARBOXIMIDAMIDE

A solution of benzonitrile (20.6 g, 200 mmol) in 40 mL of toluene is added slowly via a dropping funnel to a slurry of lithium bis(trimethylsilyl)amide diethyl etherate (50 g, 210 mmol) in 200 mL of toluene. The resulting pale yellow solution is warmed to 50°C for 2 h. The mixture is cooled to room temperature and a solution of Me_3SiCl (28 mL, 220 mmol) in toluene (30 mL) is added slowly (dropping funnel). The reaction is then heated at reflux for 2 h, producing a white precipitate (LiCl). After cooling, the solid material is filtered through a Schlenk filter-stick apparatus containing a "D" porosity (10–20 μm) frit. The clear orange filtrate is reduced in volume to 80 mL by distillation. The remaining (now dark) solution is transferred to a 100-mL flask and the remaining toluene is removed under vacuum. Vacuum distillation (Claisen head and air-condenser recommended; ■ **Caution.** *The distillate may solidify prematurely; avoid narrow take-off adaptors and over-cooling of the apparatus*) of the residue (ca. 10^{-2} torr) produces N,N,N′-tris(trimethylsilyl)benzenecarboximidamide as a viscous pale yellow liquid (bp 110–115°C), which solidifies on standing at room temperature. Yield 54.0 g (80%), mp 50–51°C. ^{1}H NMR (CDCl$_3$) δ 7.23 (s, aromatic), 0.22 (s, SiCH$_3$).

Anal. Calcd. for C$_{16}$H$_{32}$N$_2$Si$_3$: C, 57.08; H, 9.58; N, 8.32%. Found: C, 57.27; H, 9.64; N, 8.50%.

C. *N,N,N′,N″,N″,N‴*-HEXAKIS(TRIMETHYLSILYL)-1,4-BENZENEDICARBOXIMIDAMIDE (HBDA)

(1) 2 LiN(SiMe$_3$)$_2$·Et$_2$O   (2) 2 Me$_3$SiCl

Solid 1,4-dicyanobenzene (■ **Caution.** *Highly toxic/irritant—handle with gloves, avoid contact or inhalation.*) (6.4 g, 50 mmol) is added to a slurry of lithium bis(trimethylsilyl)amide diethyl etherate (28 g, 120 mmol) in 250 mL of toluene. The brownish solution is heated to 50°C for 6 h. After cooling to room temperature, a solution of Me$_3$SiCl (16 mL, 130 mmol) in 25 mL of toluene is added slowly (dropping funnel) and the reaction mixture is gently refluxed overnight. A *hot* filtration through a Schlenk filter stick ("D" porosity frit) is performed to remove the precipitated LiCl, and the solution is concentrated to one-third of its original volume by distillation. An equal volume of acetonitrile is then added, causing precipitation of HBDA. The product is filtered under N$_2$, washed with several small portions of acetonitrile, and dried under vacuum. Recrystallization of the crude product from a 2:1 mixture of 120 mL of toluene/acetonitrile affords white needles, yield 23.0 g (83%), mp 162–166°C. ^{1}H NMR (CDCl$_3$) δ 7.27 (s, aromatic), 0.07 (s, SiCH$_3$).

Anal. Calc. for C$_{26}$H$_{54}$N$_4$Si$_6$: C, 52.46; H, 9.82; N, 9.41%. Found: C, 52.28; H, 9.70; N, 9.55%.

D. *N,N,N′*-TRIS(TRIMETHYLSILYL)FORMIMIDAMIDE

(1) 2 LiN(SiMe$_3$)$_2$·Et$_2$O   (2) 2 Me$_3$SiCl

A solution of 1,3,5-triazine (9.38 g, 0.116 mol) in 80 mL of toluene is added slowly (dropping funnel) to a slurry of lithium bis(trimethylsilyl)amide diethyl etherate (57.6 g, 0.24 mol) in 150 mL of toluene. The resulting yellow mixture is heated gently (50°C) for 10 h to give a pale golden brown solution. The solution is cooled to room temperature and a solution of chlorotrimethylsilane (30.0 g, 0.28 mol) in 50 mL of toluene is added (dropping funnel). The reaction is refluxed gently for 16 h. After cooling to room temperature, the mixture is filtered through a Schlenk filter stick ("D" porosity) to remove the lithium chloride precipitate, and the toluene is removed by vacuum distillation. Slow vacuum distillation (Claisen head and water condenser) of the brown residue produces *N,N,N'*-tris(trimethylsilyl)formimidamide as a colorless liquid (bp 34–37°C/10^{-2} torr), yield 24.0 g (80%). ^{1}H NMR (CDCl$_3$) δ 7.96 (s, C*H*), 0.16 (s, SiC*H*$_3$).

Anal. Calc. for C$_{10}$H$_{28}$N$_2$Si$_3$: C, 46.09; H, 10.83; N, 10.75%. Found: C, 46.08; H, 10.65; N, 10.93%.

The distillation residue can be sublimed at 55°/10^{-2} torr to give the side product *N',N',N''*-tris(trimethylsilyl)-1,3,5-triaza-1,4-pentadiene as moisture-sensitive white needles, yield 24.7 g (74%), mp 58–60°C. ^{1}H NMR (C$_6$D$_6$) δ 8.70 (s, H*C*) 0.27, 0.13 (s, SiC*H*$_3$).

Anal. Calc. for C$_{11}$H$_{29}$N$_3$Si$_3$: C, 45.94; H, 10.46; N, 14.61%. Found: C, 45.73; H, 10.47; N, 14.66%.

Properties

The *N*-silylated amidines are easily handled in air, but are best stored under dry nitrogen in a cool, dry place to prevent hydrolysis (as they are slightly moisture-sensitive). This synthetic procedure is applicable to many organic nitriles, and has been used to generate a variety of *N*-silylated amidines.[6]

References

1. (a) R. T. Boeré, A. W. Cordes and R. T. Oakley, *J. Chem. Soc. Chem. Commun.*, 929 (1985); (b) K. Dehnicke, C. Egezinger, E. Hartmann, A. Zinn, and K. Hösler, *J. Organomet. Chem.*, **352**, C1 (1988); (c) H. W. Roesky, B. Meller, M. Noltemeyer, H. G. Schmidt, U. Scholtz, and G. M. Sheldrick, *Chem. Ber.*, **121**, 1403 (1988); (d) E. Hey, C. Ergezinger, and K. Dehnicke, *Z. Naturforsch.*, **44b**, 205 (1989); (e) V. Chandrasekhar, T. Chivers, S. S. Kumaravel, M. Parvez, and M. N. S. Rao, *Inorg. Chem.*, **30**, 4125 (1991); (f) J. K. Buijink, M. Noltemeyer, and F. T. Edelmann, *Z. Naturforsch.*, **46b**, 1328 (1991).
2. (a) P. D. B. Belluz, A. W. Cordes, E. M. Kristof, P. V. Kristof, S. W. Liblong, and R. T. Oakley, *J. Am. Chem. Soc.*, **111**, 9276 (1989); (b) A. W. Cordes, R. C. Haddon, R. T. Oakley,

L. F. Schneemeyer, J. V. Waszczak, K. M. Young, and N. M. Zimmerman, *J. Am. Chem. Soc.*, **113**, 582 (1991); (c) M. P. Andrews, A. W. Cordes, D. C. Douglass, R. M. Fleming, S. H. Glarum, R. C. Haddon, P. Marsh, R. T. Oakley, T. T. M. Palstra, L. F. Schneemeyer, G. W. Trucks, R. Tycko, J. V. Waszczak, K. M. Young, and N. M. Zimmerman, *J. Am. Chem. Soc.*, **113**, 3559 (1991); (d) A. W. Cordes, R. C. Haddon, and R. T. Oakley, in *The Chemistry of Inorganic Ring Systems*, R. Steudel (ed.), Elsevier Science Publications, Amsterdam, 1992, p. 295.

3. R. T. Boeré, R. T. Oakley, and R. W. Reed, *J. Organomet. Chem.*, **331**, 161 (1987).
4. A. W. Cordes, R. C. Haddon, R. G. Hicks, R. T. Oakley, T. T. M. Palstra, L. F. Schneemeyer, and J. V. Waszczak, *J. Am. Chem. Soc.*, **114**, 5000 (1992).
5. U. Wannagat and H. Niederprüm, *Chem. Ber.*, **94**, 1540 (1961).
6. (a) A. W. Cordes, R. C. Haddon, R. G. Hicks, R. T. Oakley, and T. T. M. Palstra, *Inorg. Chem.*, **31**, 1802 (1992); (b) A. W. Cordes, C. M. Chamchoumis, R. G. Hicks, R. T. Oakley, K. M. Young, and R. C. Haddon, *Can. J. Chem.*, **70**, 919 (1992); also ref. 2(c).

15. HOMOLEPTIC BISMUTH AMIDES

SUBMITTED BY CLAIRE J. CARMALT,* NEVILLE A. COMPTON,*
R. JOHN ERRINGTON,* GEORGE A. FISHER,*
ISMUNARYO MOENANDAR,* and NICHOLAS C. NORMAN*
CHECKED BY KENTON H. WHITMIRE[†]

Bismuth is an important element in many of the new high-temperature, oxide superconductors and in a variety of heterogeneous mixed oxide catalysts. Some of the methods employed in the preparation of these materials, namely sol-gel and chemical vapor deposition processes, require bismuth alkoxides as precursors and a number of papers on these compounds have recently been published.[1] One synthetic route to bismuth alkoxides, which avoids the more commonly used trihalide starting materials and the often troublesome separation of alkali metal halides, involves the reaction between a bismuth amide and an alcohol according to the following equation:

$$Bi(NR_2)_3 + 3R'OH \rightarrow Bi(OR')_3 + 3HNR_2$$

Since relatively little has been published on the synthesis and properties of homoleptic bismuth amides,[2,3] details on the preparation of two such compounds, $Bi(NMe_2)_3$ and $Bi\{N(SiMe_3)_2\}_3$, are described herein.

* Department of Chemistry, The University of Newcastle, Newcastle upon Tyne, NE1 7RU, UK.
[†] Department of Chemistry, Rice University, P.O. Box 1892, Houston, TX, 77251, USA.

A. TRIS(DIMETHYLAMINO) BISMUTH Bi(NMe₂)₃

$$BiCl_3 + 3Li[NMe_2] \rightarrow Bi(NMe_2)_3 + 3LiCl$$

Procedure

■ **Caution.** *Li[NMe₂] and n-BuLi can ignite upon contact with water or air. These reagents should be handled only under an inert atmosphere.*
All solvents must be appropriately dried and deoxygenated, all glassware should be dried in an oven and all manipulations should be carried out under an inert atmosphere.

Li[NMe₂] was prepared by condensing an excess of dimethylamine onto frozen *n*-BuLi and allowing the mixture to warm to room temperature with shaking following the procedure described by Chisholm et al.[4] with the slight modification that immersion of the reaction flask in dry ice–propan-2-ol was omitted.

A *n*-BuLi solution in hexanes (Aldrich) (1.77 M, 56.5 mL, 100 mmol) was transfered to a 500-mL, round-bottomed flask and frozen in liquid nitrogen. The flask was then evacuated and connected to a calibrated vacuum manifold. An excess of dimethylamine (HNMe₂) was then condensed into the flask. The tap was closed and the mixture was allowed to warm slowly to room temperature while the flask was agitated (CARE!). Solid Li[NMe₂] precipitates as the reaction proceeds. The volatiles are removed by vacuum and the white solid is pumped dry.

Solid Li[NMe₂] (0.728 g, 14.27 mmol) and BiCl₃ (Aldrich 99.999%) (1.50 g, 4.76 mmol) are placed in a nitrogen filled Schlenk flask, fitted with a second side arm, which is connected, through a ground glass joint, to a second Schlenk flask. The first flask is cooled to 0°C using an external ice-bath. A mixture of Et₂O (diethyl ether) (20 mL) and THF (tetrahydrofuran) (5 mL) is then added and the resulting suspension is stirred using a magnetic stirrer bead. An immediate colour change from white to yellow occurs on addition of the solvent and the mixture becomes yellow-green after stirring for one hour. After this time, the solvents are removed by vacuum at 0°C resulting in a yellow-green oily solid. The second flask is then cooled using a liquid nitrogen filled Dewar flask and the Bi(NMe₂)₃ is sublimed from the first flask into the second using a warm water bath (approx temp, 60°C) and a hot air blower if necessary. Bi(NMe₂)₃ is obtained as bright yellow crystals (1.323 g, 81.5% yield). Due to the extreme moisture sensitivity of Bi(NMe₂)₃, all glassware should be baked out in an oven prior to use and due to the photosensitivity of this compound, it is advisable to cover the second Schlenk flask into which the compound is sublimed, with aluminium foil.

This preparation is a modification of that reported by Ando et al.[5] the main difference being that we did not find it necessary to reflux the reaction mixture or purify by distillation.

Anal. Calcd. for $C_6H_{18}N_3Bi$: Bi, 61.3. Found: Bi, 60.4%.

Properties

$Bi(NMe_2)_3$ is an extremely air-sensitive, volatile, low-melting, yellow crystalline compound that is soluble in hydrocarbon and ether solvents. It is rather photosensitive, turning black on exposure to bright sunlight, although it can be handled under an inert atmosphere in normal laboratory lighting conditions for short periods without substantial decomposition. In the dark and in a freezer at $-30°C$, it is stable for several months. ^1H and $^{13}C\{^1H\}$ NMR spectra at room temperature in d^8-toluene show signals at δ 3.72 and 48.4 respectively. The crystal structure has been reported[3] and the compound is monomeric.

B. TRIS[BIS(TRIMETHYLSILYL)AMINO] BISMUTH $Bi\{N(SiMe_3)_2\}_3$

$$BiCl_3 + 3Na[N(SiMe_3)_2] \rightarrow Bi\{N(SiMe_3)_2\}_3 + 3NaCl$$

Procedure

$NH(SiMe_3)_2$ (Aldrich) (2.12 mL, 1.622 g, 10.05 mmol) is syringed into a nitrogen-filled 100-mL Schlenk flask and cooled to 0°C using an external ice bath. *n*-BuLi (Aldrich) (6.28 mL of a 1.6-M solution in hexane, 10.05 mmol) is then added slowly, which results in a clear solution with a small amount of white precipitate. After stirring for 1 h, the flask is warmed with an external water bath and all volatiles are removed by vacuum, affording white solid $Li[N(SiMe_3)_2]$. This compound is then redissolved in Et_2O (diethyl ether) (20 mL) and cooled to 0°C using an external ice bath. $BiCl_3$ (Aldrich 99.999%) (1.057 g, 3.35 mmol) dissolved in a mixture of Et_2O (20 mL) and THF (5 mL) is then added slowly, resulting in a yellow solution. This mixture is then stirred for 1 h and allowed to warm to room temperature, during which time the solution becomes yellow-green. All volatiles are then removed by vacuum which leaves a yellow-green oil. The crude product is dissolved in pentane (30 mL) and filtered through a medium-porosity glass frit into a second 100-mL Schlenk flask, resulting in a yellow solution. Removal of all volatiles by vacuum affords an oily yellow solid (1.68 g, 70%) which is pure enough for further use. Recrystallization from a minimum of pentane at $-30°C$ gives a pale yellow solid.

Bi{N(SiMe$_3$)$_2$}$_3$ can also be prepared in similar yields from Na[N(SiMe$_3$)$_2$] and BiCl$_3$ in toluene. After removal of all volatiles under vacuum, the product is extracted into hexanes and filtered to remove NaCl. A yellow solid is obtained after removal of the solvent from the filtrate. Na[N(SiMe$_3$)$_2$] is typically prepared as follows: NaH (60% dispersion in oil, 0.72 g, 18.0 mmol) is washed with hexanes (3×5 mL), dried under vacuum and suspended in toluene. HN(SiMe$_3$)$_2$ (3.8 mL, 2.91 g, 18.0 mmol) is added and the mixture heated under reflux for 18 h. The hot solution is filtered through a sintered glass filter and the solvent removed under vacuum from the colourless filtrate to give a colourless solid (2.36 g, 72%).

Properties

Bi{N(SiMe$_3$)$_2$}$_3$[6] is an extremely air-sensitive, pale yellow microcrystalline compound that is very soluble in hydrocarbon and ether solvents. It is much less photosensitive than Bi(NMe$_2$)$_3$. ^{1}H and ^{13}C{^{1}H} NMR spectra at room temperature in d^8-toluene show signals at δ 0.55 and 7.51 respectively.

References

1. R. C. Mehrotra and A. K. Rai, *Ind. J. Chem.*, **4**, 537 (1966); W. J. Evans, J. H. Hain, and J. W. Ziller, *J. Chem. Soc., Chem. Commun.*, 1628 (1989); M. -C. Massiani, R. Papiernik, L. G. Hubert-Pfalzgraf, and J. -C. Daran, *J. Chem. Soc., Chem. Commun.*, 301 (1990); M. A. Matchett, M. Y. Chiang, and W. E. Buhro, *Inorg. Chem.*, **29**, 358 (1990); M. -C. Massiani, R. Papiernik, L. G. Hubert-Pfalzgraf, and J. -C. Daran, *Polyhedron*, **10**, 437 (1991); C. M. Jones, M. D. Burkart, and K. H. Whitmire, *Angew. Chem., Int. Ed. Engl.*, **31**, 451 (1992); K. H. Whitmire, J. C. Hutchison, A. L. McKnight, and C. M. Jones, *J. Chem. Soc., Chem. Commun.*, 1021 (1992).
2. W. Clegg, N. A. Compton, R. J. Errington, N. C. Norman, and N. Wishart, *Polyhedron*, **8**, 1579 (1989) and references therein.
3. W. Clegg, N. A. Compton, R. J. Errington, G. A. Fisher, M. E. Green, D. C. R. Hockless, and N. C. Norman, *Inorg. Chem.*, **30**, 4680 (1991).
4. M. H. Chisholm, D. A. Haitko, and C. A. Murillo, *Inorg. Synth.*, **21**, 51 (1982).
5. F. Ando, T. Hayashi, K. Ohashi, and J. Koketsu, *J. Inorg. Nucl. Chem.*, **37**, 2011 (1975).
6. M. J. S. Gynane, A. Hudson, M. F. Lappert, P. P. Power, and H. Goldwhite, *J. Chem. Soc., Dalton Trans.*, 2428 (1980). See also M. F. Lappert and P. P. Power, unpublished results (1977) taken from *Metal and Metalloid Amides*, M. F. Lappert, A. R. Sanger, R. C. Srivastava, and P. P. Power, Ellis Horwood, London, 1979, p. 459.

16. *CYCLO*-TETRASULFUR(2 +) BIS[HEXAFLUOROARSENATE(1 −)], *CYCLO*-TETRASULFUR(2 +) BIS[UNDECAFLUORODIANTIMONATE(1 −)], *CYCLO*-TETRASELENIUM(2 +) BIS[HEXAFLUOROARSENATE(1 −)] AND *CYCLO*-TETRASELENIUM(2 +) BIS[UNDECAFLUORODIANTIMONATE(1 −)]

SUBMITTED BY MICHAEL P. MURCHIE,* RAMESH KAPOOR,* JACK PASSMORE,* GABRIELE SCHATTE*, and TODD WAY
CHECKED BY DARRYL D. DESMARTEAU†

$$4M + 3AsF_5 \xrightarrow[\text{traces of } X_2; \text{ rt}]{SO_2} [M_4][AsF_6]_2 + AsF_3 \tag{1}$$

$$1/2S_8 + 6SbF_5 \xrightarrow[\text{traces of } X_2; \text{ rt}]{SO_2 \text{ or } AsF_3} [S_4][Sb_2F_{11}]_2 + SbF_3 \cdot SbF_5 \tag{2}$$

$$24Se + 35SbF_5 \xrightarrow[\text{traces of } X_2; \text{ rt}]{SO_2} 6[Se_4][Sb_2F_{11}]_2 + 6SbF_3 \cdot 5SbF_5 \tag{3}$$

$$M = S, Se \qquad X = Br, Cl$$

Sulfur is oxidized by an excess of arsenic or antimony pentafluoride to $[S_8][AsF_6]_2$ and $[S_8][Sb_2F_{11}]_2$, respectively.[1a, b] However, in the presence of a trace of halogen ($X_2 = Cl_2$, Br_2, I_2), the effective oxidizing power of AsF_5 (or SbF_5) is considerably enhanced, and $[S_4][AsF_6]_2$ (or $[S_4][Sb_2F_{11}]_2$) are readily produced quantitatively in a few minutes[2, 3] as described in eqs. (1) and (2). Selenium can be oxidized by AsF_5 (or SbF_5) to $[Se_4][AsF_6]_2$ (or $[Se_4][Sb_2F_{11}]_2$) on heating for several days,[1b] but the reaction is far more conveniently carried out with traces of halogen[2, 3] giving the products at room temperature within minutes, as described in eqs. (1) and (3). The presence of more AsF_5 or SbF_5 than required by eqs. (1), (2), or (3) does not lead to any further oxidation of the $[S_4]^{2+}$ and $[Se_4]^{2+}$ species. A comparative study of the effect of solvent, oxidizing agent (AsF_5, SbF_5), and facilitating reagent (halogen) on the course of reaction is given in reference 3. Other polyatomic cations $[S_8]^{2+}$, $[Se_8]^{2+}$, and $[Se_{10}]^{2+}$ are also very conveniently prepared as their $[AsF_6]^-$ salts in the presence of traces of

* Department of Chemistry, University of New Brunswick, Fredericton, New Brunswick, Canada, E3B 6E2.
† Department of Chemistry, Clemson University, Clemson, SC 29634.

halogen, provided stoichiometric quantities of sulfur (or selenium) and AsF_5 are used. The compounds are of interest as they contain cations rather than anions of the electronegative elements sulfur and selenium. In addition, they contain novel bonding arrangements and have extensive chemistries. The structures, bonding, and energetics of the homopolyatomic cations of Groups 16 and 17 have recently been reviewed.[4a] Some chemical reactions of $[S_4]^{2+}$ and $[Se_4]^{2+}$ that have been investigated are given in references 4b and 4c.

General Procedure

■ **Caution.** *All reagents, particularly AsF_5 and SbF_5, should be handled in a very efficient fume hood with appropriate precautions. Selenium is hazardous through inhalation as dust and through skin absorption from solutions. Arsenic- and antimony-pentafluoride are very poisonous and hydrolyze readily to form HF.*

The arsenic pentafluoride (Ozark-Mahoning) [bp: − 53°C; p (298 K): 214 ± 10 psi (ca. 1475 kPa)] is stored as a compressed gas in a lecture bottle (diameter × length: 5 × 30 cm, 100–500 g of AsF_5). A manual control valve (Matheson [61], stainless steel) has to be attached to the valve outlet connection [CGA 660* (1.030 in. − 14 R.H. Ext.)] at the valve of the cylinder (Fig. 1). A Teflon washer between the valve outlet connection and the manual control valve ensures a gas-tight seal. The AsF_5 cylinder is then connected through the manual control valve via a 1/4 in. Monel tubing and Swagelok compression fittings to the metal (Monel) vacuum line (Fig. 1). The connection was evacuated and leak tested. The manual control** valve allows the removal of gaseous AsF_5 directly from the compressed gas cylinder by adjusting the flow of AsF_5. For further information on safe handling of compressed gases, see reference 5. The aresenic trifluoride (Ozark-Mahoning) (b.p. + 63°C at 752 Torr. m.p. − 8.5°C), is condensed from the cylinder into a 100 ml. Pyrex glass vessel equipped with a Whitey® valve and filled with ca. 5g of vacuum dried NaF to remove HF.

The liquid sulfur dioxide solutions described in the preparations have a vapor pressure of about 3.3 atm at 21°C. Therefore, well-constructed glass vessels and a glass (or metal) vacuum line must be employed to prevent pressure bursts. Thick leather gloves, safety goggles, a face shield, and a rubber apron should be worn and the experiments have to be conducted behind a safety shield or explosion-proof glass in a fume hood to prevent possible contact with the reaction mixtures as well as with AsF_5 and SbF_5.

* Valve outlet connection standardized by the Compressed Gas Association (CGA) (see ref. 5).
** Corrosion of the manual control valve is indicated by sticking on opening and closing, in which case the valve is taken apart and cleaned.

The starting materials AsF_5 and SbF_5 and the target compounds $[M_4][AsF_6]_2$ and $[M_4][Sb_2F_{11}]_2$ are extremely moisture sensitive, especially $[S_4]^{2+}$ salts which react very rapidly with traces of moisture to give a blue-colored material indicative of $[S_8]^{2+}$ (although actually due to S_5^+; ref. 4a), and should be handled in a rigorously moisture-free environment. Prior to use, the reaction vessel $[(g)$ in (Fig. 1)] is evacuated for at least 1 h while heated gently with a Bunsen flame or hot air gun. In addition, the vacuum line is gently heated to remove traces of water and prefluorinated with sulfur tetrafluoride (Matheson) or arsenic pentafluoride.

Sulfur (Fisher Scientific U.S.P.; precipitated) and selenium (BDH, 99.9%) were vacuum dried prior to use. Chlorine (Matheson) and bromine (Fisher Scientific) were distilled onto and stored over P_4O_{10}.

The liquid sulfur dioxide (Canadian Liquid Air Ltd.) is stored over CaH_2 in a 100-mL glass vessel equipped with a sintered-glass filter disk (coarse; o.d., 25 mm) and a Whitey valve (1KS4) joined to the flask by Swagelok Teflon compression fittings. (The checker used a 100-mL glass bulb fitted with a Kontes glass-Teflon valve and recommended cooling the bulb containing the SO_2 to 0°C before opening. Appropriate amounts of SO_2 can be removed by vacuum transfer using PVT measurement.)

Antimony pentafluoride (Ozark-Mahoning) is twice distilled in vacuo in a rigorously dried all-glass apparatus which is identical to the apparatus used for the transfer of SbF_5 as described in Fig. 2. Liquid SbF_5 is poured in a dry-nitrogen filled glove-bag through a standard funnel into a Pyrex round-bottomed flask ($V = 100$ mL) fitted with a 90° Teflon-stemmed Pyrex glass valve (diameter, 10 mm; PTT/10/RA, J. Young, UK). The bulb is then attached via (III) to the transfer vessel (I) (see Fig. 2) and the contents are distilled under a static vacuum into bulb (A) of the transfer vessel (I). The distillation bulb is replaced by the storage bulb (Fig. 2, II) and the SbF_5 is distilled from bulb A into the storage vessel (II) [Pyrex round-bottomed flask ($V = 100$ mL) fitted with a Teflon-stemmed Pyrex glass valve (SPTT/5, J. Young, UK)]. In the following, the procedure and apparatus are described for the transfer of SbF_5 to obtain the amount required for the reactions (Fig. 2).*** The transfer vessel (I) (preweighed) is connected to the storage vessel (II) containing the purified SbF_5 via a stainless-steel Swagelok fitting with a 45-mm piece of FEP®**** tubing inside III [o.d., 4.2 mm; i.d.,

---

*** According to the checker the SbF_5 should be placed by vacuum transfer through a short path in the approximate amount desired into a tared glass vessel fitted with a glass-Teflon valve. The amounts of sulfur or selenium can then be adjusted appropriately and the entire contents of the tared container vacuum-transferred onto the SO_2-sulfur (or selenium) mixture through a short path connection. This procedure should give easy control of the stoichiometry.

**** FEP®: tetraethylene/hexafluoropropylene copolymer. *Supplier:* Fluorocarbon-Bunnel Plastics Division, Penntube Products, Mickleton, NJ.

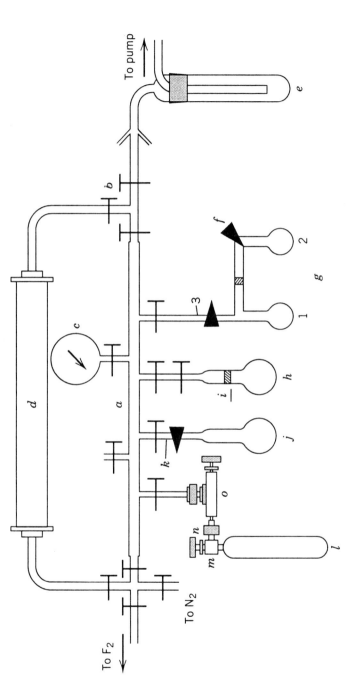

Figure 1. Metal (Monel) vacuum line. (*a*) 1/4-in. Monel tubing silver soldered to 1/2-in. Monel tubing to form the manifold; (*b*) stainless-steel Whitey (1KS4) valve joined to the manifold by Swagelok Teflon compression fittings (Crawford Fitting) using nylon back and Teflon front ferrules; (*c*) stainless-steel pressure gauge (30-in. Hg vacuum to 60 psi) (Dresser Industries); (*d*) copper soda lime trap; (*e*) Pyrex glass cold trap; (*f*) Teflon-stemmed Pyrex glass valves (PTT/5/RA, SPTT/5; J. Young, UK); (*g*) two-bulbed glass reaction vessel: 1,2, thick-walled glass bulbs (*V* = 25 mL); 3, 1/4-in. o.d. glass tubing attached directly to the reaction vessel; (*h*) SO$_2$ reservoir (thick-walled Pyrex glass bulb) (*V* = 100 mL) joined to the vacuum line through a 1/4-in. o.d. glass tubing attached to the vessel, and a Swagelok Teflon compression fitting; (*i*) coarse sintered glass frit; (*j*) halogen (Br$_2$ or Cl$_2$) reservoir (thick-walled Pyrex glass bulb) (*V* = 50 mL) joined to the vacuum line through a 1/4-in. o.d. glass tubing, directly attached to the vessel, and a Swagelok Teflon compression fitting; (*k*) 1/4-in. glass tubing (*V* = 1.1 mL) for measuring traces of halogen; (*l*) lecture bottle of AsF$_5$ attached via a 1/4-in. o.d. Monel tubing and Swagelok Teflon compression fittings to the vacuum line; (*m*) cylinder valve of the AsF$_5$ cylinder with valve handwheel; (*n*) cylinder valve outlet connection [CGA 660 (1.030 in. -14 R.H. Ext.)]; (*o*) manual control valve (Matheson [61], stainless steel).

105

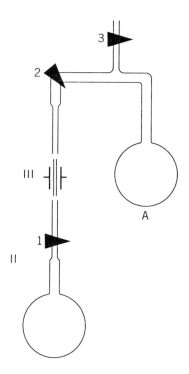

Figure 2. Transfer of SbF$_5$. (I) Transfer vessel; (II) storage vessel containing purified SbF$_5$; (III) stainless-steel Swagelok (SS-400-6 1/4″) fitting with FEP® tubing inside (l = 45 mm; o.d. = 4.2 mm; i.d. = 2.8 mm); (A) transfer bulb (V = 25 mL); (1–3) Teflon-stemmed Pyrex glass valves [SPTT/5 (1,3), PTT/5/RA (2); J. Young, UK].

2.8 mm]. The FEP® tubing is long enough to fit inside the 1/4-in. glass tubing on both sides. The whole apparatus is then attached to the metal (Monel) vacuum line (see Fig. 1) and the valves (2 and 3) are opened to evacuate and leak test this area. The bulb (A) (V = 25 mL) of the transfer vessel is cooled to − 196°C using liquid nitrogen and valve 3 is closed. Upon opening valve **1**, the SbF$_5$ is transferred into bulb A under a static vacuum (p = 10^{-3} mbar). The transfer of SbF$_5$ (bp: 141°C) can be facilitated by heating the storage vessel (**II**) with a heat gun. With moderate heating, the transfer rate of SbF$_5$ is approximately 500 mg/min. After approximating the amount of SbF$_5$ transferred, closing valve 1 and evacuating the transfer vessel through valve 3, all valves are closed, and the apparatus is taken apart at the Swagelok fitting (III). The SbF$_5$ is then thawed and weighed, and if a sufficient amount had been obtained, the transfer vessel is connected to the reaction vessel (Fig. 1g) via a stainless-steel Swagelok fitting (III), and the process is repeated in reverse. If a sufficient amount is not obtained, the transfer vessel must be reconnected to the SbF$_5$ storage vessel and the transfer process has to be continued.

Arsenic pentafluoride is used without further purification. The following procedure is used when working with AsF$_5$ (see also safety section above): The manual control valve (Fig. 1*o*) is opened slowly (to prevent a high pressure in case the main valve is leaking) and evacuated for 5 min. After closing the manual control valve and the valve between vacuum line and glass cold trap (Fig. 1*e*), the cylinder valve (Fig. 1*m*) of the AsF$_5$ cylinder is opened (1/2 turn). The gaseous AsF$_5$ is expanded slowly into the vacuum line by opening the valve to the vacuum line and then opening the manual control valve and monitoring the pressure with the pressure gauge (Fig. 1*c*). The cylinder manual control valve and the valve to the vacuum line are closed and the AsF$_5$ is condensed into the reaction vessel. The AsF$_5$ is transferred in aliquots of not more than 4 mmol ($p = 1$ atm; volume of the vacuum line, 97 mL). If more AsF$_5$ is needed, the procedure is repeated. Before disconnecting the AsF$_5$ cylinder from the vacuum line, the area between the cylinder and the manual control valve is evacuated for 20 min to reduce corrosion.

Reactions are carried out in a one-piece glass apparatus consisting of two thick-walled round-bottomed flasks ($V = 25$ mL) linked by a glass tube (o.d., 10 mm) incorporating a sintered-glass filter disk (coarse). Both bulbs are fitted with Pyrex glass valves with PTFE pistons {SPTT/5 (bulb 1) and STT/5/RA (bulb 2), J. Young, London, UK} (Fig. 1*g*).***** Volatile materials are handled by using a metal (Monel) vacuum line (10^{-3} mbar) as illustrated in Fig. 1. Solids are manipulated in a Vacuum Atmospheres Dri-Lab equipped with a Dri-Train (HE-493) and an internal circulating drying unit containing 1 kg of 3-Å molecular sieves.

The IR spectra of the solids are obtained as finely ground powder between AgCl plates using the Perkin-Elmer IR spectrometers model 457 and 683 (Note: [S$_4$][AsF$_6$]$_2$ reacts with Nujol). Solid samples for Raman spectroscopy are mounted in glass melting-point tubes in the Dri-Lab and sealed under dry nitrogen atmosphere. Raman spectra are recorded using a Spex-Ramalab spectrometer and a Spectra Physics 164-2 W argon ion laser (5145 Å) at room temperature (slit width: 4 cm^{-1}; laser power: 190 mW).

***** The checker used a reaction vessel made from two 25-mL round-bottomed flasks. On each bulb a short 3/8-in. piece of glass tubing was attached, so as not to interfere with the manipulation of the reactor and was connected to the main body of the vessel by a 3/8-in. stainless-steel Swagelock connector using nylon rear and Teflon front ferrules. The product was stored by replacing the connector with a 3/8-in. Swagelock stainless-steel plug. According to the checker, this allowed easy removal of solids after the reactions were complete and the vessel could be reused for each reaction after cleaning.

FT-Raman spectra are measured at room temperature on a FT-IR spectrometer (Bruker IFS66) equipped with an FT-Raman accessory (Bruker FRA106) using a Nd-YAG laser (emission wavelength: 1064 nm). The data are collected in the backscattering mode (180° excitation; resolution: 4 cm^{-1}; 256 scans; 19 mW).

The total weight of the solid product is evidence for the given formulations for the $[AsF_6]^-$ salts as the reduced product AsF_3 is volatile. $[M_4][AsF_6]_2$ salts are slightly soluble in liquid SO_2, and more soluble in liquid AsF_3. $[M_4][Sb_2F_{11}]_2$ salts are more soluble than the $[AsF_6]^-$ salts. The solubility of $[M_4][AsF_6]_2$ and $[M_4][Sb_2F_{11}]_2$ compounds increases upon addition of AsF_5 and SbF_5, respectively.

A. $[S_4][AsF_6]_2 \cdot xSO_2$ $(x \leqslant 1)$

Arsenic pentafluoride (2.45 g, 14.41 mmol), in an excess relative to the amount indicated by eq. (1), is condensed onto a mixture of sulfur (0.51 g, 15.94 mmol of S atoms) and liquid SO_2 (6.38 g) into bulb 1 of the reaction vessel (Fig. 1g) at $-196°C$. A dark-blue solution is obtained upon warming the contents to room temperature. A trace quantity of Cl_2 (0.10 mmol)****** is now condensed onto the solution at $-196°C$. The reaction is allowed to stand without stirring and a colorless crystalline precipitate is formed in 15 min after the solution is warmed to room temperature. The color of the solution over the precipitate changes from dark to light blue within this time, suggesting that the reaction is complete. The solution is transferred through the sintered glass frit into bulb 2. The white precipitate is washed 2–3 times with about 2 mL of the solvent by condensing the SO_2 back into bulb 1 at $-78°C$ and pouring it through the frit into bulb 2. The volatile materials are removed under a dynamic vacuum and a white compound (2.10 g) is obtained in quantitative yield. The calculated weight of $[S_4][AsF_6]_2 \cdot xSO_2$ is 2.27 g $(x = 1)$ and 2.02 g $(x = 0)$. The white crystalline solid collapses to a powder on subjecting it to vacuum for less than 10 min on losing SO_2 molecules of crystallization.

Anal. (Calcd. for $[S_4][AsF_6]_2 \cdot 0.6SO_2$). Found: S, 27.20 (27.09); As, 28.35 (27.52) and F, 41.50 (41.88). Similar experiments using traces of other halogens (Br_2 or I_2) give white $[S_4][AsF_6]_2 \cdot xSO_2$ $(x \leq 1)$ in quantitative yield (Note: The reaction with I_2 proceeds slower than those with Cl_2 or Br_2).

****** Small traces of X_2 [Cl_2 or Br_2] are added by expanding the X_2 gas in a small precalibrated volume (1.1 mL; 1/4 in. glass tubing that is part of the storage vessel for the halogen, Fig. 1j) at room temperature at a known pressure and then condensing it into bulb 1 at $-196°C$.

The compound $[S_4][AsF_6]_2 \cdot xSO_2(x \leq 1)$ is best identified by its Raman spectrum (cm^{-1}):[3, 6, 7] 1146 (0.5) $[v_3(SO_2)]$, 728 (0.5) $[v([AsF_6]^-)]$, 707 (0.5) $[v_3([AsF_6]^-)]$, 673 (2.0) $[v_1([AsF_6]^-)]$, 605 (5.5) $[v_3([S_4]^{2+})]$, 585 (10.0) $[v_1([S_4]^{2+})]$, {572 (sh), 545 (2.0)} $[v_2([AsF_6]^-)]$, 520 (0.5) $[v_2(SO_2)]$, 396 (0.5) $[v_2([S_4]^{2+})]$, 368 (1.5) $[v_5([AsF_6]^-)]$ and its IR spectrum (cm^{-1}):[3, 7] 1330 w, 1305 w $[v_3(SO_2)]$, 1148 w $[v_1(SO_2)]$, 699 vs $[v_3([AsF_6]^-)]$, 575 m $[v_2([AsF_6]^-)]$, 552 ms $[v_5([S_4]^{2+})]$, 539 sh, 522 vw, and 390 s $[v_4([AsF_6]^-)]$. If the compound is not subjected to vacuum for a long time and is still crystalline, then SO_2 molecules of crystallization are easily detected in the IR spectrum of the powder. The appropriate weight changes also aid in compound identification. The pure compound is colorless. Even the slightest hydrolysis or lack of complete reaction will lead to the presence of a blue color indicative of $[S_8]^{2+}$.[4a]

B. $[S_4][Sb_2F_{11}]_2$

Antimony pentafluoride (6.92 g, 31.89 mmol), in an excess relative to the amount indicated by eq. (2), is condensed onto a mixture of sulfur (0.54 g, 16.88 mmol of S atoms) and AsF_3 (12.19 g) in bulb 1 of the reaction vessel (Fig. 1*g*) at $-196°C$. A thick red-brown solution is formed on warming the contents to room temperature. A trace quantity of Br_2 (0.11 mmol) is now condensed onto the solution at $-196°C$.* A white precipitate under a dark blue solution is formed in less than one day at room temperature. The color of the solution changes from dark blue to transparent pale pink-red on standing at room temperature for 2 weeks. The solution is transferred through the sintered frit into bulb 2. The insoluble white solid (reduced antimony fluoride product) is washed three times with about 2 mL of the solvent. The volatile materials are then removed under dynamic vacuum, leaving 4.41 g of a beige-white soluble solid in bulb 2 (calculated weight for $[S_4][Sb_2F_{11}]_2$ based on sulfur: 4.40 g) and 2.24 g of a white insoluble solid in bulb 1 (calculated weight for $SbF_3 \cdot SbF_5$ is 1.68 g). The product in bulb 1 is identified as $SbF_3 \cdot SbF_5$ by its Raman spectrum.[8] A similar reaction, but using SO_2 as a solvent was complete after one week.

Anal. (Calcd. for $[S_4][Sb_2F_{11}]_2$). Found: S, 11.68 (12.38), Sb, 48.47 (47.20), and F, 39.90 (40.43).

The compound $[S_4][Sb_2F_{11}]_2$ is best identified by its Raman spectrum:[3, 6, 7] 694 (w) 682 (1.0), 669 (2.5), 648 (1.0), 603* (3.0), 580* (10.0) and 367* (0.5) cm^{-1} (* identifies vibrations due to the $[S_4]^{2+}$ cation). The $[Sb_2F_{11}]^-$ anion is also identified by its ^{19}F NMR spectrum obtained in liquid SO_2 (ppm):[9] -90.0 (1), -108.9 (8), and -130.9 (2) relative to external CCl_2F (numbers in parentheses refer to peak areas). The product weights and colorless appearance of the product also give evidence for the formation of the compound.

C. $[Se_4][AsF_6]_2$

Arsenic pentafluoride (2.91 g, 17.12 mmol), in an excess relative to the amount indicated by eq. (1), is condensed onto a mixture of selenium (1.27 g, 16.08 mmol) and SO_2 (5.12 g) at $-196°C$ into bulb 1 of the reaction vessel (Fig. 1g) and traces of Br_2 (0.04 mmol) are added directly onto the frozen mixture. On warming to room temperature a bright-yellow precipitate appears under an intense yellow-brown solution in about 5 min. A large amount of precipitate is observed after 2 h and its amount does not change on further standing ($\simeq$ 4 days). The solution is transferred to bulb 2 and the precipitate in bulb 1 is washed about three times with 1–2 mL of the solvent. The volatile materials are removed under vacuum, giving 0.33 g of a soluble yellow-green solid (bulb 2) and 2.50 g of the bright-yellow solid (total weight: 2.83 g; calculated weight for $[Se_4][AsF_6]_2$: 2.79 g; yield: 87%). The soluble green-yellow solid (bulb 2) contains $[Se_4][AsF_6]_2$ with a trace impurity of $[Se_8][AsF_6]_2$, as indicated by its intense green color.

The compound $[Se_4][AsF_6]_2$ is identified by its yellow color, appropriate weight changes, FT Raman spectrum {(cm^{-1}): 718 (0.2) $[v([AsF_6]^-)]$, {683 (2.5), 670 (0.5, sh)} $[v_1([AsF_6]^-)]$, {583 (0.2), 556 (0.5)} $[v_2([AsF_6]^-)]$, 371 (0.5) $[v_5([AsF_6]^-)]$, 324 (10) $v_1([Se_4]^{2+})]$, 304 (0.2) $[v_5([Se_4]^{2+})]$, 181 (2.5) $[v_2([Se_4]^{2+})]$}, Laser. Raman spectrum {324 cm^{-1} (10), $v_1([Se_4]^{2+})$} and IR spectrum {(cm^{-1}): 698 vs, 552 m and 390 vs cm^{-1} due to $[AsF_6]^-$} (see also refs. 6 and 10).

D. $[Se_4][Sb_2F_{11}]_2$

Antimony pentafluoride (3.59 g, 16.53 mmol) and a trace quantity of Br_2 (0.02 mmol) are condensed onto a mixture of selenium (0.73 g, 9.25 mmol) and SO_2 (10.68 g) at $-196°C$. A greenish-yellow solution over a white precipitate is obtained on stirring the contents at room temperature for 4 days. The solution is transferred into bulb 2 and the white solid is washed 3–4 times with about 1–2 mL of SO_2. The volatile materials are removed, under dynamic vacuum leaving 2.96 g of a bright-yellow solid in bulb 2 (calculated weight for $[Se_4][Sb_2F_{11}]_2$: 2.81 g) and 0.74 g of an insoluble white solid (bulb 1) (calculated yield for $6SbF_3 \cdot 5SbF_5$: 0.83 g). The latter compound is identified by its Raman spectrum to be (β) $6SbF_3 \cdot 5SbF_5$.[8]

The compound $[Se_4][Sb_2F_{11}]_2$ is identified by its Raman spectrum[6] (326 (10) cm^{-1} due to $[Se_4]^{2+}$). The ^{19}F NMR chemical shifts (ppm) of the solid in SO_2 at $-70°C$ are -90.0 (1), -108.9 (8), and -130.9 (2) relative to

CCl_3F (external reference) and indicate the presence of the $[Sb_2F_{11}]^-$ anion (relative intensities in the parentheses).

Note: The nature of the antimony(III) fluoride–antimony(V) fluoride adduct may vary depending upon the exact amount of SbF_5 used and the nature of the reaction. Two adducts [A] and [B] with very different Raman spectra were assigned the formula $SbF_3 \cdot SbF_5$ on the basis of elemental analysis.[11] Our subsequent work[8] has shown that one adduct (A) has the composition $SbF_3 \cdot SbF_5$ [Raman spectrum (cm^{-1}):[11] 719(vw), 700(m), 685(m), 678(m), 668(ms), 642(s), 605(ms), 555(mw), 495(mw, br), 456(m, vbr), 298(m), 276(mw), and 206(w)] and the other adduct (B) is in fact (β) $6SbF_3 \cdot 5SbF_5$ [Raman spectrum (cm^{-1}):[8] 681 (0.6), 659 (5.7), 652 (10.0), 622 (0.4), 577 (3.7), 566 (3.0), 294 (0.4), and 282 (1.6)].

Acknowledgments

We thank the National Science and Engineering Research Council (Canada) for financial support of this work, for an International Scientific Exchange Award (R. K.), and for an International Fellowship (G. S.). The assistance of Mr. Dale Wood is gratefully acknowledged.

References

1a. R. J. Gillespie, J. Passmore, P. K. Ummat and O. C. Vaidya. *Inorg. Chem.* **10**, 1327 (1971).

1b. P. A. W. Dean, R. J. Gillespie, and P. K. Ummat, *Inorg. Synth.*, **15**, 213 (1974).

2. J. Passmore, G. W. Sutherland, and P. S. White, *J. Chem. Soc., Chem. Commun.*, 330, (1980).

3. M. P. Murchie, J. Passmore, G. Sutherland, and R. Kapoor, *J. Chem. Soc., Dalton Trans.*, 503 (1992).

4. (a) N. Burford, J. Passmore, and J. C. P. Sanders, in *From Atoms to Polymers: Isoelectronic Analogies*, J. F. Liebman and A. Greensberg (eds.), VCH Verlagsgesellschaft mbH, D-6940 Weinheim (Federal Republic of Germany). VCH; 1989, p. 53 and references therein. (b) Reactions of $[Se_4][AsF_6]_2$ with C_2F_4 (ref. 11), I_2 (ref. 12), Br_2 (ref. 13), and Se_4N_4 (ref. 14). (c) Reactions of $[S_4][AsF_6]_2$ with I_2 (ref. 15), and S_4N_4 (ref. 16).

5. (a) The Matheson Company, Inc. (ed.), *Matheson Gas Data Book*, 4th ed., 1966. (b) Matheson Division, Searle Medical Products, USA, Inc. (ed.), *Guide to Safe Handling of Compressed Gases*, 1982.

6. R. C. Burns and R. J. Gillespie, *Inorg. Chem.*, **21**, 3877 (1982).

7. R. Minkwitz, J. Nowicki, W. Sawodny, and K. Härtner, *Spectrochim. Acta*, **47A**, 151 (1991).

8. W. A. S. Nandana, J. Passmore, and P. S. White, *J. Chem. Soc., Dalton Trans.*, 1623 (1985).

9. J. Bacon, P. A. W. Dean, and R. J. Gillespie, *Can. J. Chem.*, **47**, 1655 (1969).

10. G. M. Begun and A. C. Rutenberg, *Inorg. Chem.*, **6**, 2212 (1967).

11. T. Birchall, P. A. W. Dean, B. Della Valle, and R. J. Gillespie, *Can. J. Chem.*, **51**, 667 (1973).

12. C. D. Desjardins and J. Passmore, *Can. J. Chem.*, **55**, 3136 (1977).

13. M. M. Carnell, F. Grein, M. P. Murchie, J. Passmore, and C. M. Wong, *J. Chem. Soc., Chem. Commun.*, 225 (1986).

14. M. P. Murchie, J. Passmore, and P. S. White, *Can. J. Chem.*, **65**, 1584 (1987).
15. E. G. Awere, J. Passmore, and P. S. White, *J. Chem. Soc., Dalton Trans.*, 1267 (1992).
16. J. Passmore, G. W. Sutherland, T. K. Whidden, and P. S. White, *J. Chem. Soc., Chem. Commun.*, 289 (1980).
17. E. G. Awere and J. Passmore, *J. Chem. Soc., Dalton Trans.*, 1343 (1992).

17. $Fe_2(S_2)(CO)_6$ AND $Fe_3Te_2(CO)_{9,10}$

SUBMITTED BY PAUL F. BRANDT,* DAVID A. LESCH,*
PHILIP R. STAFFORD,* and THOMAS B. RAUCHFUSS*
CHECKED BY JOSEPH W. KOLIS† and LISA C. ROOF†

Iron carbonyl chalcogenides were first prepared by Hieber and Geisenberger in their investigations on the effect of sulfur on the carbonylation of iron powder. They prepared $Fe_3S_2(CO)_9$ and $Fe_3Se_2(CO)_9$ by the reaction of H_2S or Se_8 with iron powder under 200 atm of CO at 70°C.[1] Hieber and Gruber later developed low-pressure syntheses of these same compounds by the reaction of $HFe(CO)_4^-$ solutions with the oxyanions EO_3^{2-} where E = S, Se, and Te.[2] Of this series, the compound $Fe_3Te_2(CO)_9$ has found the widest use. Its preparation is easy and requires only inexpensive reagents. It has been shown that Hieber's synthesis affords three products, $Fe_3Te_2(CO)_{10}$, $Fe_3Te_2(CO)_9$, and small amounts of $Fe_2(Te_2)(CO)_6$.[3] Hieber and Gruber also found that the reaction of $HFe(CO)_4^-$ with polysulfide anions gives $Fe_2(S_2)(CO)_6$ in addition to $Fe_3S_2(CO)_9$. The diiron compound has found wide use in organometallic and inorganic synthesis. Our modification of the original procedure using elemental sulfur is simpler and affords a purer product.

A. HEXACARBONYLDISULFIDODIIRON

$$Fe(CO)_5 + 2OH^- \rightarrow HFe(CO)_4^- + CO_2(OH)^-$$

$$2HFe(CO)_4^- + 3/8S_8/H^+ \rightarrow Fe_2(S_2)(CO)_6 + \cdots$$

Procedure

■ **Caution.** *Metal carbonyl compounds are toxic and the reactions herein should be done in a well-ventilated fume hood.*

* School of Chemical Sciences, University of Illinois, Urbana, IL 61801.
† Department of Chemistry, Clemson University, Clemson, SC 29634.

A 3-L, three-necked flask is fitted with a mechanical stirrer, a nitrogen gas inlet, and a rubber septum. The system is purged with N_2. The flask is charged with 100 mL degassed MeOH, 40 mL degassed 50% KOH (aq), and 20 mL of $Fe(CO)_5$ (0.152 mol, Aldrich). The mixture is stirred vigorously for 15 min. The orange-red solution is cooled to 0°C with an ice bath and 46 g (1.44 mol) sulfur is added in one portion through a sidearm. (■ **Caution.** *Gas is expelled and is vented through a syringe needle in the septum.*) The mixture immediately turns dark brown. After 15 min of vigorous stirring, 250 mL degassed H_2O and 500 mL degassed pentane are added. At this time the flask is fitted with a pressure-equalizing addition funnel which is charged with 200 mL 3 M degassed HCl. The acid is added cautiously to the iron carbonyl solution over the course of 20 min. (■ **Caution.** *Large quantities of toxic H₂S gas are evolved causing the solution to foam extensively if the acid is added too quickly. The addition funnel should not be stoppered during this part of the procedure.*) After acidification, the ice bath is removed and the mixture is stirred for 1 h and filtered. The black solid is extracted with three 40-mL portions of pentane. The red pentane layer is separated from the aqueous layer which can be discarded. The pentane solution is washed with three 100-mL aliquots of H_2O, dried over Na_2SO_4, and evaporated to dryness ($< 25°C$). The crude solid/oil is dissolved in a minimum of hexanes (~ 10 mL). Thin layer chromatographic analysis of the hexane solution shows two mobile reddish bands as well as substantial amounts of polar material at the origin. Purification of the product was accomplished by chromatography on a 40×4 cm column of silica gel, eluting with hexanes. The product elutes as the first intensely colored band, closely followed by a reddish band of $Fe_3S_2(CO)_9$. The isolated yield of red crystalline $Fe_2(S_2)(CO)_6$ is 8.0–9.2 g (0.024–0.026 mol, 30–35%).

The IR spectrum of the complex in cyclohexane solution is characterized by the three strong bands in the carbonyl region at 2083(m), 2042(s), and 2006(s) cm^{-1}. The most common impurities are elemental sulfur and $Fe_3S_2(CO)_9$ [v_{co}(hexanes) = 2062(s), 2043(s), 2024(m) cm^{-1}]. The melting point is 46°C. The compound is thermally unstable above room temperature. It can be purified by vacuum sublimation at room temperature.

Anal. Calcd. for $C_6Fe_2O_6S_2$: C, 20.96; Fe, 32.48; S, 18.65. Found: C, 20.95; Fe, 32.19; S, 18.42.

Properties

$Fe_2(S_2)(CO)_6$ is obtained as red-brown crystals which are stable in air for several weeks. The compound dissolves in nonpolar organic solvents to give red solutions which are slightly air sensitive. Like $Fe_3Te_2(CO)_9$,

$Fe_2(S_2)(CO)_6$ has been widely used in the preparation of mixed metal cluster compounds. Low-valent metal centers add across the S–S bond to give Fe_2S_2M clusters.[4,5,6] Of bioinorganic interest, $Fe_2(S_2)(CO)_6$ has been used to prepare a number of Fe–S and Mo–Fe–S clusters.[7] The S–S bond in $Fe_2(S_2)(CO)_6$ can be reduced with hydride reagents to give $Fe_2S_2(CO)_6^{2-}$, which reacts at sulfur with a variety of electrophiles.[8] Photolysis of solutions of $Fe_2(S_2)(CO)_6$ affords $Fe_4S_4(CO)_{12}$.[5]

B. DECACARBONYL AND NONACARBONYL TRIIRONDITELLURIDE

$$Fe(CO)_5 + 2OH^- \rightarrow HFe(CO)_4^- + CO_2(OH)^-$$

$$HFe(CO)_4^- + [TeO_3]^{2-}/H^+ \rightarrow Fe_3Te_2(CO)_{10}$$

$$Fe_3Te_2(CO)_{10} \rightarrow Fe_3Te_2(CO)_9 + CO$$

■ **Caution.** *Metal carbonyl compounds are toxic and the reactions described herein should be done in a well-ventilated fume hood.*

A 2-L, three-necked flask is fitted with a rubber septum, an addition funnel, and a N_2 bubbler. The system is purged with N_2 and is charged with 60 mL degassed MeOH and 22 mL degassed 50% KOH (aq). To the magnetically stirred mixture, 10 mL $Fe(CO)_5$ is added via syringe. After a further 15 min, 400 mL degassed CH_2Cl_2 is added to the orange-red solution, followed by a fresh solution of 13 g K_2TeO_3 (0.015 mol) in 100 mL of H_2O via the addition funnel. (■ **Caution.** *Large amounts of CO gas are expelled.*) Nitrogen is passed over the black solution while it is diluted with 150 mL H_2O. This solution is cooled to 0°C and is acidified slowly with 75 mL of 6 M HCl via the addition funnel. (■ **Caution.** *Large quantities of gas are evolved causing the solution to foam extensively if the acid is added too quickly.*) After acidification, $Fe_3Te_2(CO)_{10}$ is obtained by filtration through a coarse-fritted glass filter. The product is washed with three 50-mL aliquots of H_2O, three 50-mL aliquots of 95% EtOH, and three 50-mL aliquots of hexanes. If the final hexane extract is not colorless, additional hexane washings may be necessary to remove all $Fe_3Te_2(CO)_9$. Additional $Fe_3Te_2(CO)_{10}$ can be obtained by filtration of the organic layers; however, the presence of this decacarbonyl does not interfere with the subsequent step. The yield of the greenish-black microcrystalline $Fe_3Te_2(CO)_{10}$ is typically 8–10 g (0.011–0.014 mol, 48–56%).

The filtrate from the above extraction is evaporated to dryness and treated with 100 mL of hexane to give ~ 5.5 g of a black residue. The resulting slurry is heated under reflux for 30 min, followed by cooling and filtration. The yield of $Fe_3Te_2(CO)_9$ is 5.2 g (7.7 mol, 30%). The previously isolated $Fe_3Te_2(CO)_{10}$

can be converted quantitatively to the purplish-black $Fe_3Te_2(CO)_9$ by slurrying in 200 mL refluxing hexanes for 30 min.

The IR spectrum of $Fe_3Te_2(CO)_{10}$ in cyclohexane solution is characterized by the bands in the carbonyl region at 2102(w), 2052(s), 2044(s), 2032(w), 2014(m), 1989(m), 1978(w), and 1976(w) cm^{-1}.

Anal. Calcd. for $C_{10}Fe_3O_{10}Te_2$: C, 17.09; Fe, 23.84; Te, 36.31. Found: C, 16.64; Fe, 23.73; Te, 35.73.

The IR spectrum of a cyclohexane solution of $Fe_3Te_2(CO)_9$ is characterized by the three bands in the carbonyl region at 2045(vs), 2025(vs), and 2005(s) cm^{-1}. A shoulder can be seen at 1998 cm^{-1}. $Fe_3Te_2(CO)_9$ exhibits a single peak in the ^{125}Te NMR spectrum at 1123 ppm downfield of Me_2Te.

Anal. Calcd. for $C_9Fe_3O_9Te_2$: C, 16.01; Fe, 24.83; Te, 37.82. Found: C, 15.83; Fe, 24.81; Te, 37.76

Properties

$Fe_3Te_2(CO)_{10}$ is a greenish-black compound which dissolves in acetone to give orange solutions which decompose quickly to $Fe_3Te_2(CO)_9$ and unidentified oxides. Solid samples of the decacarbonyl undergo decarbonylation over a period of weeks at room temperature while solid samples of $Fe_3Te_2(CO)_9$, which are black-brown, are moderately air stable. $Fe_3Te_2(CO)_9$ decomposes slowly in nonpolar solvents such as benzene and hexane.

Several reactions have been identified for $Fe_3Te_2(CO)_9$, the most prominent being its ability to form adducts with Lewis bases.[9] Compounds of the type $Fe_3Te_2(CO)_9L$ are characteristically orange and can be isolated for a variety of organophosphorus ligands. These adducts can be decarbonylated to give species of the type $Fe_3Te_2(CO)_8L$. $Fe_3Te_2(CO)_9$ can be converted to $Fe_2(Te_2)(CO)_6$ which, like $Fe_2(S_2)(CO)_6$, can be used as a precursor to mixed-metal clusters. There are numerous examples of transmetalation reactions involving $Fe_3Te_2(CO)_9$ and other metal carbonyls. This method can be used for the synthesis of many metal carbonyl tellurides, including those that do not contain iron.[4,5,10] Variations in this synthesis have been fruitfully pursued.[11]

References

1. W. Hieber and O. Geisenberger, *Z. Anorg. Allg. Chem.*, **262**, 15 (1950).
2. W. Hieber and J. Gruber, *Z. Anorg. Allg. Chem.*, **296**, 91 (1958).
3. D. A. Lesch and T. B. Rauchfuss, *Inorg. Chem.*, **20**, 3583 (1981).
4. L. E. Bogan, Jr., T. B. Rauchfuss, and A. L. Rheingold, *J. Am. Chem. Soc.*, **107**, 3843 (1985).

5. L. E. Bogan, Jr., D. A. Lesch, and T. B. Rauchfuss, *J. Organometal. Chem.*, **250**, 429 (1983).
6. D. Seyferth, G. B. Womack, R. S. Henderson, M. Cowie, and B. W. Hames, *Organometallics*, **5**, 1568 (1986). V. W. Day, D. A. Lesch, and T. B. Rauchfuss, *J. Am. Chem. Soc.*, **104**, 1290 (1982). P. Braunstein, J. D. Jud, A. Tiripicchio, M. Tiripicchio-Camellini, and M. Sappa, *Angew. Chem. Int. Ed. Engl.*, **21**, 307 (1982). P. D. Williams, M. D. Curtis, D. N. Duffy, and W. M. Butler, *Organometallics*, **2**, 165 (1983). I. L. Eremenko, A. A. Pasynskii, A. S. Abdulaev, A. S. Aliev, B. Orazsakhatov, S. A. Sleptsova, A. I. Nekhaev, V. E. Shklover, and Yu. T. Struchkov, *J. Organometal. Chem.*, **365**, 297 (1989).
7. P. A. Eldredge, R. F. Bryan, and B. A. Avrill, *J. Am. Chem. Soc.*, **110**, 5573 (1988). J. A. Kovacs, J. K. Bashkin, and R. H. Holm, *J. Am. Chem. Soc.*, **107**, 1784 (1985).
8. D. Seyferth and G. B. Womack, *Organometallics*, **5**, 2360 (1986). D. Seyferth, R. S. Henderson, L. Cheng Song, and G. B. Womack, *J. Organometal. Chem.*, **292**, 9 (1985). D. Seyferth, R. S. Henderson, and L. C. Song, *Organometallics*, **1**, 125 (1982).
9. D. A. Lesch and T. B. Rauchfuss, *Organometallics*, **1**, 499 (1982). D. A. Lesch and T. B. Rauchfuss, *Inorg. Chem.*, **22**, 1854 (1983).
10. P. Mathur, B. H. S. Thimmappa, and A. L. Rheingold, *Inorg. Chem.*, **29**, 4658 (1990). P. Mathur, I. J. Mavunkal, V. Rugimini, and M. F. Mahon, *Inorg. Chem.*, **29**, 4838 (1990). P. Mathur, I. J. Mavunkal, and V. Rugimini, *Inorg. Chem.*, **28**, 3616 (1989). P. Mathur, I. J. Mavunkal, and A. L. Rheingold, *J. Chem. Soc. Chem. Commun.*, 382 (1989).
11. R. E. Bachman and K. H. Whitmire, *Inorg. Chem.*, **33**, 2527 (1994). L. C. Roof, W. T. Pennington, and J. W. Kolis, *Angew. Chem. Int. Ed. Engl.*, **31**, 913 (1992).

Chapter Two

LIGANDS AND REAGENTS

18. 5,10,15,20-TETRAKIS(2,6-DIHYDROXYPHENYL)-21*H*,23*H*-PORPHINE*

SUBMITTED BY JAMES P. COLLMAN,[†] VIRGIL J. LEE,[†]
and XUMU ZHANG[†]
CHECKED BY HONG PING YUAN[‡] and L. KEITH WOO[‡]

Synthetic metalloporphyrins serve as excellent starting points in the preparation of a variety of biomimetic systems. For example, several porphyrin-based structures have been prepared which mimic either the dioxygen-binding ability of hemoglobin and myoglobin[1] or the oxidative reactivity of cytochrome P-450.[2] The peripheral substituents on the porphyrin macrocycle play an important role in these synthetic systems. These substituents can sterically prevent the dimerization of the model system[3] and can interact with substrates to produce selectivity in oxygenation reactions.[4] The location and type of these substituents govern the degree of selectivity that can be achieved.

The most common methods of attaching substituents to porphyrins utilize *ortho*-phenyl-substituted derivatives of 5,10,15,20-tetraphenyl-21*H*,23*H*-porphine (TPP). While (2-aminophenyl) TPP derivatives have been extensively used in the preparation of biomimetic systems,[5] the corresponding (2-hydroxyphenyl) TPP derivatives have only recently been utilized.[6] One of the limitations to the use of hydroxy-substituted systems has been the synthesis of 5,10,15,20-tetrakis(2,6-dihydroxyphenyl)-21*H*,23*H*-porphine **3**.

Herein we describe a facile, three-step procedure for the synthesis of **3**. The first step involves the synthesis of 2,6-dimethoxybenzaldehyde **1**. Although

* Systematic name: 2,2′,2″,2‴ (21*H*, 23*H*-porphine-5,10,15,20-tetrayl)-tetrakis[1,3-benzenediol].
† Department of Chemistry, Stanford University, Stanford, CA 94305.
‡ Department of Chemistry, Iowa State University, Ames, IA, 50011.

161

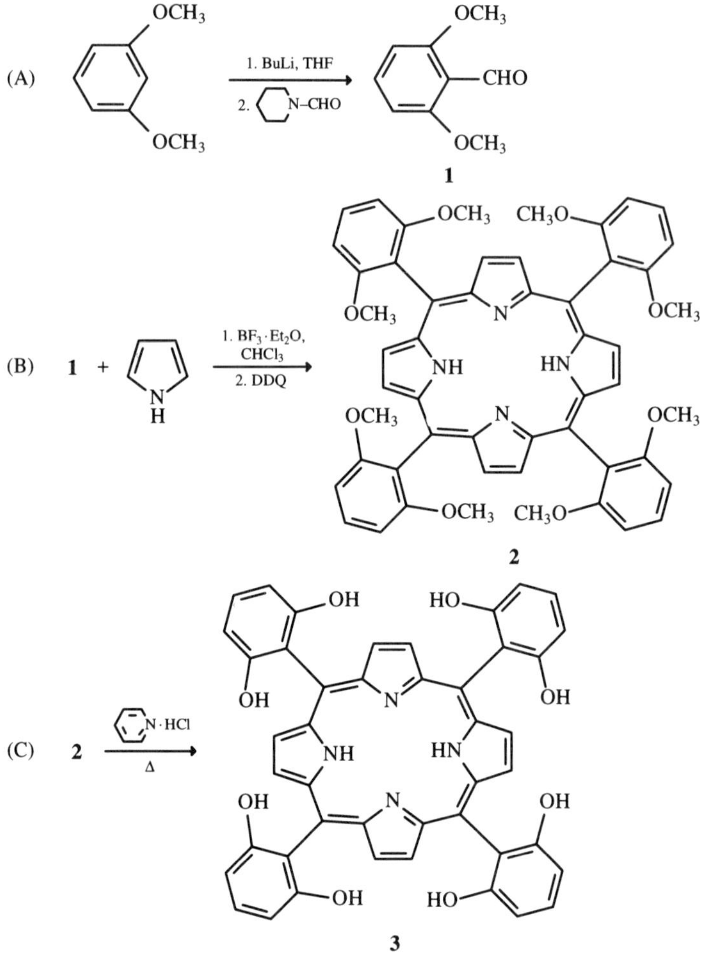

(A)

1. BuLi, THF
2. ⟨N–CHO⟩

1

(B)   **1** + pyrrole

1. BF₃·Et₂O, CHCl₃
2. DDQ

2

(C)   **2**

⟨N·HCl⟩
Δ

3

Scheme 1

1 is a commercially available compound,[7] it is expensive. The modified *ortho*-lithiation procedure of Wittig[8a] described here is a one-step, inexpensive preparation requiring minimal product purification. The condensation of **1** with pyrrole is accomplished under modified Lindsey conditions[9] to give **2** in 20% yield.[10] The conversion of **2** to **3** is achieved in a molten salt reaction under conditions similar to those used in the preparation of the corresponding tetrahydroxy-TPP system.[11] Metallation of this system can be accomplished under standard conditions either before[6b] or after[6c] the desired substituents are attached to the hydroxy units.

Starting Materials

The following chemicals were obtained from Aldrich Chemical Company, Inc. and used without further purification: 1,3-dimethoxybenzene (99%); butyl lithium (1.6 M in hexanes); 1-formylpiperdine (99%); boron trifluoride-diethyl ether (purified, redistilled); 2,3-dichloro-5,6-dicyano-1,4-benzoquinone (98%), and pyridine hydrochloride (98%). All solvents were reagent grade and were obtained from Fisher Scientific and used without further purification except where noted. Silica gel (230–400 mesh) was obtained from EM Scientific. Chloroform was stored over activated, 4-Å molecular sieves for at least 24 h prior to use. Tetrahydrofuran (optima grade) was distilled from sodium benzophenone. Pyrrole (99%) was obtained from Aldrich Chemical Company, Inc. and distilled from calcium hydride.

■ **Caution.** *The toxic nature of the reagents and products requires that these reactions be performed in a well-ventilated fume hood, and gloves should be worn at all times.*

A. 2,6-DIMETHOXYBENZALDEHYDE (1)

Procedure

An oven-dried, 500 mL, three-necked, round-bottomed flask equipped with a magnetic stirring bar, a dropping funnel, and a nitrogen inlet is charged with 1,3-dimethoxybenzene (21.0 mL, 0.16 mol) and dry tetrahydrofuran (40 mL) under nitrogen. To this solution is added a 1.6 M butyllithium solution in hexanes (100 mL, 0.16 mol) (■ **Caution.** *Butyllithium is extremely air and moisture sensitive and must be handled in an inert atmosphere.*) over a 30-min period. This resulting mixture is allowed to stir at room temperature (3 h). The solution is cooled to 0°C and 1-formylpiperdine (17.8 mL, 0.16 mol) (■ **Caution.** *1-Formylpiperdine is toxic.*) is added dropwise over a 30-min period. Stirring is continued at 0°C (1 h) and then at room temperature overnight. The mixture is diluted with dichloromethane (300 mL) and washed with water (3×250 mL). The final organic layer is dried over magnesium sulfate and filtered. The yellow filtrate is diluted with heptane (200 mL) and the methylene chloride is removed on a rotary evaporator. The crystalline precipitate is collected, washed with pentane, and dried under vacuum (10 mm) to yield **1** (15.7 g, 59%).

Properties

Compound **1** is a white crystalline material, mp 96–98°C (uncorrected) (lit. 98–99°C[8]). ^{1}H NMR (CDCl$_3$, 400 MHz) δ: 10.52 (s, 1H), 7.45

(t, J = 8.5 Hz, 1H), 6.58 (d, J = 8.5 Hz, 2H), 3.90 (s, 6H). The solid material appears to be indefinitely stable when stored in an amber vial.

B. 5,10,15,20-TETRAKIS(2,6-DIMETHOXYPHENYL)-21*H*,23*H*-PORPHINE (2)

Procedure

An oven dried, 3-necked, 5-L, round-bottomed flask equipped with a mechanical stirrer and a gas-dispersion tube is charged with **1** (6.72 g, 40 mmol), pyrrole (2.78 mL, 40 mmol), and chloroform (4 L) (■ **Caution.** *Chloroform is a suspected carcinogen.*). Nitrogen is bubbled gently through the stirring solution (10 min) after which the dispersion tube is replaced by a nitrogen inlet. Boron trifluoride-diethyl ether (1.22 mL, 10 mmol) is added to the mixture via syringe and the flask is wrapped with aluminum foil as a cautionary procedure to protect the mixture from light. After stirring under a nitrogen atmosphere at room temperature (1 h), 2,3-dichloro-5,6-dicyano-1,4-benzoquinone (6.81 g, 30 mmol) is added and the mixture is allowed to stir for an additional hour under nitrogen. The solution is concentrated to a dark solid on a rotary evaporator and the product is separated by column chromatography (column size 8×8 cm) on silica gel (500 g). The product fraction is diluted with methanol (200 mL) and concentrated on a rotary evaporator to about 100 mL. The precipitate is collected by filtration and dried at 120°C under vacuum (10 mm) to yield **2** (1.71 g, 20%).

Properties

Compound **2** is a purple amorphous powder. ^{1}H NMR (CDCl$_3$, 400 MHz) δ: 8.64 (s, 8H), 7.66 (t, J = 8.4 Hz, 4H), 6.95 (d, 8.4 Hz, 8H), 3.47 (s, 24H), -2.44 (s, 2H, D$_2$O exchangeable); UV-vis (CH$_2$Cl$_2$) λ_{max}: 418 (ε 388,000 cm^{-1} M^{-1}), 480 (3360), 512 (17,600), 544 (4860), 588 (6400), 642 (2060) nm. The material has good solubility in chlorinated solvents, ethereal solvents, and aromatic solvents. The solid also appears to be indefinitely stable in air.

C. 5,10,15,20-TETRAKIS(2,6-DIHYDROXYPHENYL)-21*H*,23*H*-PORPHINE (3)

Procedure

A 250-mL, round-bottomed flask equipped with a magnetic stirring bar is charged with **2** (1.71 g, 2.0 mmol) and pyridine hydrochloride (100 g) (■ **Caution.** *Pyridine hydrochloride is toxic with a disagreeable order.*). The flask is fitted with a short, Vigreux condenser and a heating mantle is used to bring the mixture to a gentle reflux (222–224°C, 3 h). The heating source is removed and the hot solution is poured into water (1 L). The aqueous mixture is extracted with ethyl acetate (2×350 mL) and the combined organic layers are washed with 1% HCl (2×500 mL), dried over sodium sulfate, and filtered. The filtrate is concentrated to a purple solid which is dried at 120°C under vacuum (10 mm) to give **3** (1.30 g, 88%).

Properties

Compound **3** is a purple amorphous powder. ^{1}H NMR (d^8THF at 60°C, 400 MHz)* δ: 8.85 (s, 8H), 7.40 (t, J = 8.1 Hz, 4H), 7.31 (bs, 8H, D_2O exchangeable), 6.75 (d, J = 8.1 Hz, 8H), -2.37 (s, 2H, D_2O exchangeable); UV-vis (MeOH) λ_{max}: 414 (ε 306,000 cm^{-1} M^{-1}), 479 (2890), 512 (20,700), 544 (3970), 586 (5710), 640 (1260) nm.

Anal. Calcd. for **3**·1/2 H_2O: C, 70.30; H, 4.16; N, 7.45. Found: C, 70.42; H 4.22; N, 7.26. The material has good solubility in alcohols, ethyl acetate, and *N,N*-dimethylformamide. The solid appears to be indefinitely stable in air.

References

1. For a review of hemoglobin and myoglobin model systems see R. D. Jones, D. A. Summerville, and F. B. Basolo, *Chem. Rev.*, **79**, 139 (1979).
2. For a review of cytochrome P-450 model systems see M. J. Gunter and P. Turner, *Coord. Chem. Rev.*, **108**, 15 (1991).
3. J. P. Collman, *Acc. Chem. Res.*, **10**, 265 (1977).
4. For examples of porphyrin-catalyzed regioselective oxidations see (a) B. R. Cook, T. J. Reinert, and K. S. Suslick, *J. Am. Chem. Soc.*, **108**, 7281 (1986) and (b) J. P. Collman, X. Zhang, R. H. Hembre, and J. I. Brauman, *J. Am. Chem. Soc.*, **112**, 5356 (1990). For examples of porphyrin-catalyzed enantioselective oxidations see (c) J. T. Groves and R. S. Myers, *J. Am. Chem. Soc.*, **105**, 5791 (1983) and (d) J. P. Collman, V. J. Lee, X. Zhang, J. Ibers, and J. I. Brauman, *J. Am. Chem. Soc.*, **115**, 3834 (1993).
5. For a review of biomimetic systems that are derived from 5,10,15,20-tetrakis(2-aminophenyl)-21H,23H-porphine see (a) B. Morgan and D. Dolphin, *Structure and Bonding*, **64**, 115 (1987). For biomimetic systems that are derived from 5,10,15,20-tetrakis(2,6-diaminophenyl)-21H,23H-porphine see (b) C. A. Quintana, R. A. Assink, and J. A. Shelnutt, *Inorg. Chem.*, **28**, 3421 (1989) and (c) C. M. Drain and B. B. Corden, *Inorg. Chem.*, **28**, 4374 (1989).

* For compound **3**, the checkers suggest the use of CD_3OD at 22°C as the NMR solvent. This is perhaps more convenient than recording the spectrum in THF-d_8 at 60°C. The drawback is that the hydroxyl protons are not observed in CD_3OD. The resonances in CD_3OD appear at 8.88 (bs, 8H, β-H), 7.47 (t, 4H, *p*-aryl), 6.80 (d, 8H, *m*-aryl), -2.71 (bs, 2H, NH).

6. (a) M. Momenteau, B. Loock, C. Huel, and J. L'hoste, *J. Chem. Soc., Perkin Trans. 1*, 283 (1988) and references therein; (b) E. Tsuchida, E. Hasegawa, T. Komatsu, T. Nakata, K. Nagao, and H. Nishida, *Bull. Chem. Soc. Jpn.*, **64**, 888 (1991) and references therein; (c) Y. Naruta, F. Tani, N. Ishihara, and K. Maruyama, *J. Am. Chem. Soc.*, **113**, 6865 (1991) and references therein.

7. Available from Aldrich Chemical Company, Inc.

8. (a) G. Wittig, in *Newer Methods of Preparative Organic Chemistry*, Vol. 1, Interscience, New York, 1948, p. 579; (b) R. G. F. Giles and M. V. Sargant, *J. Chem. Soc. Perkin Trans. 1*, 2447 (1974) and (c) M. E. Snook, P. F. Mason, and R. F. Arrendale, *J. Chromatogr.*, **324**, 141 (1985) and references therein.

9. J. S. Lindsey and R. W. Wagner, *J. Org. Chem.*, **54**, 828 (1989).

10. Tsuchida et al. have prepared **3** from **1** in 9.3% overall yield by a different procedure. See E. Tsuchida, T. Komatsu, E. Hasegawa, and H. Nishide, *J. Chem. Soc., Dalton Trans.*, 2713 (1990).

11. M. Momenteau, J. Mispelter, B. Loock, and E. Bisagni, *J. Chem. Soc., Perkin Trans. 1*, **189** (1983).

19. TRIBENZOCYCLYNE* (TBC) AND TETRABENZOCYCLYNE† (QBC)

SUBMITTED BY DON SOLOOKI,‡ JOSEPH D. FERRARA,§ DENNIS MALABA,‡ JOHN D. BRADSHAW,‡ CLAIRE A. TESSIER,‡ and WILEY J. YOUNGS‡
CHECKED BY JENS A. JOHN and JAMES M. TOUR§§

Cyclyne ligands are macrocyclic polyalkynes[1] which have available carbon–carbon triple bonds for coordination to transition metal atoms or ions. Cyclynes with three alkynes in a 12-membered ring (cyclotriynes) such as tribenzocyclyne (TBC) are conjugated, antiaromatic, planar molecules and have cavities which are large enough to fit low-oxidation-state, first-row transition metals. The distance from the center of the cavity to the center of the alkynes is about 2.1 Å.[2] Tribenzocyclyne has been shown to be a versatile ligand for coordinating transition metals. Several modes of metal coordination to TBC have been firmly established: a metal can coordinate to the three alkynes in the cavity of TBC to give a planar metallocyclotriyne;[3] two molecules of TBC can coordinate to a metal through the alkynes to yield

* 5, 6, 11, 12, 17, 18-Hexadehydrotribenzo[*a,e,i*]cyclododecene.

† 5, 6, 11, 12, 17, 18, 23, 24-Octadehydrotetrabenzo[*a,e,i,m*]cyclohexadecene.

‡ Department of Chemistry, The University of Akron, Akron, OH 44325-3601.

§ Present address: Molecular Structure Corp., 3200 Research Forest Drive, The Woodlands, TX 77381-4238.

§§ Department of Chemistry, The University of South Carolina, Columbia, SC 29208.

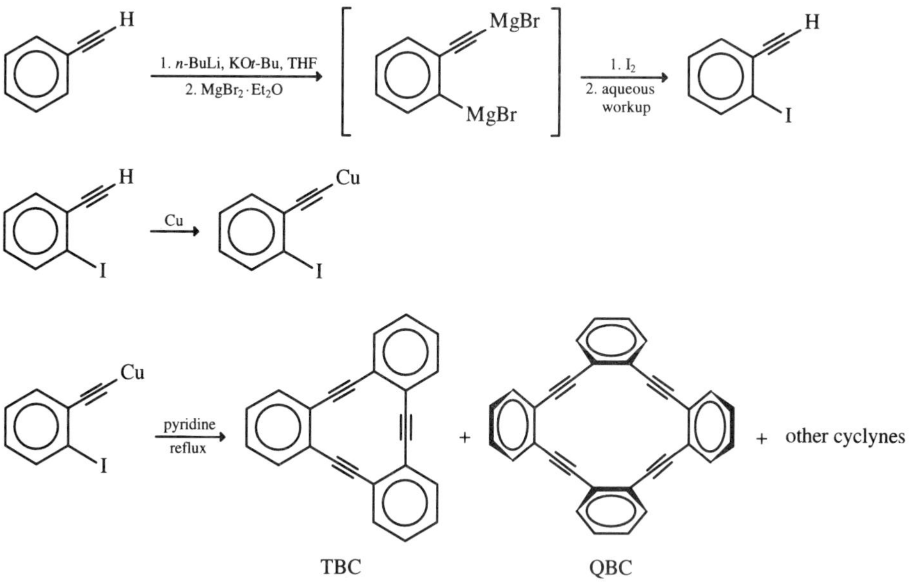

a sandwich complex;[4] and three metals can coordinate to the three alkynes so that the metals are above the plane of the TBC molecule.[5] Tribenzocyclyne can be prepared by Stephens–Castro coupling,[6] a bromination/dehydrobromination route,[7] or through palladium-catalyzed coupling.[8] Here we report a detailed procedure used in our laboratory for the synthesis of TBC by Stephens–Castro coupling of copper(I) (2-iodophenyl)acetylide. In addition to TBC, tetrabenzocyclyne (QBC),[6a, 9] a nonplanar cyclyne with four alkyne units, is also formed in low yield. The synthesis of the precursor (2-iodophenyl)acetylene is as described by Brandsma et al.[10] with some modifications. In addition to the benzocyclynes reported here, thiophene[11] and methoxy-substituted benzene[12] derivatives have also been prepared by this method. This procedure requires approximately 7 days.

General Procedure

Unless otherwise stated, all manipulations are carried out under an atmosphere of argon using standard Schlenk techniques.[13] The glassware used in the preparation of (2-iodophenyl)acetylene and the Stephens–Castro coupling of copper acetylide is dried in an oven at 140°C overnight. Potassium *tert*-butoxide (Aldrich)[a]* is sublimed at 150°C on a diffusion-pump vacuum

---

* Superscript letters refer to "notes added by checkers" at the end of this section.

line (10^{-6} torr) and stored in the dry box (lower yields of (2-iodo-phenyl)acetylene are obtained when commercially available potassium *t*-butoxide is used without purification). *n*-Butyllithium (2.5 M in hexanes, Aldrich)[b] is standardized with diphenylacetic acid before use.[14] Phenylacety-lene (Aldrich) is vacuum distilled from anhydrous sodium sulfate prior to use. Iodine, copper(II) sulfate pentahydrate (Fisher), hydroxylamine hydrochlo-ride, and magnesium bromide diethyl ether (Aldrich)[c] are used as received. Tetrahydrofuran (Aldrich) is freshly distilled under argon from sodium ben-zophenone. Pyridine (Fisher) is distilled under argon from barium oxide. Flash chromatography[15] is carried out using 40 μm silica gel (Baker).

A. (2-iodophenyl)acetylene[d]

Procedure

The reaction vessel consists of a four-necked, 2-L flask fitted with a powder addition tube (300 mL), gas inlet, septum, and mechanical stirrer. In a dry box, the reaction flask is charged with KO-*t*-Bu (21.72 g, 0.1935 mol) and *n*-BuLi (155 mL, 2.5 M in hexanes, 0.388 mol); the powder addition tube is charged with $MgBr_2 \cdot OEt_2$ (100 g, 0.44 mol) and attached to the reaction flask. The flask is equipped with a mechanical stirrer and a septum before it is removed from the dry box. A mineral-oil bubbler is attached to the gas inlet of the flask. The flask is then cooled to $-78°C$ as is 450 mL of THF in a separate flask. Throughout the remainder of the procedure, the reaction mixture is stirred and maintained at $-78°C$. Tetrahydrofuran is transferred slowly over the course of ~ 10 min to the reaction mixture via cannula (an exothermic reaction ensues). The mixture is then allowed to stir until a homogeneous red-brown solution results (~ 1 h). Phenylacetylene (19.4 mL, 0.176 mol) is added via syringe, producing a deep green solution. After 3 h, $MgBr_2 \cdot OEt_2$ is added slowly to the reaction mixture. The reaction turns cream colored after 15 min and is allowed to stir for another 45 min. In a separate flask, iodine (49.14 g, 0.1936 mol) is dissolved in 250 mL of THF, cooled to $-40°C$, and transferred via cannula to the reaction mixture. A more pale, cream color persists. At this point, the reaction mixture is allowed to warm to room temperature over 16 h while stirring.

Workup, in air, is begun by adding 250 mL of water and acidification to pH 5 with 1 M HCl. The reaction mixture is transferred to a 2-L separatory funnel with the assistance of 250 mL of diethyl ether. The aqueous phase is drawn off and set aside. The organic layer is washed twice with 300-mL portions of saturated aqueous $Na_2S_2O_3$ and saturated aqueous NaCl. This is followed by a wash with two 200-mL portions of water. The extract is dried over $MgSO_4$, and concentrated in vacuo. Analysis by gas chromatogram indicates that 60% of the phenylacetylene is converted to (2-iodo-phenyl)acetylene. Considerable decomposition of the product may result from heating the product; therefore, unreacted phenylacetylene is removed from the crude product by trap-to-trap fractionation[13] at 21°C on a diffu-sion-pump vacuum line (10^{-6} torr) through traps at -20 and $-196°C$. The small fraction collected at $-20°C$ generally contains (2-iodophenyl)acety-lene. The fractionation is carried out for approximately 12 h, whereupon the residues of the distillation pot and contents of the $-20°C$ trap are combined and collectively run through a plug of silica gel (10 cm diameter by 10 cm tall column, elution with hexanes). (2-Iodophenyl)acetylene (20.5 g) is isolated in 51% yield based upon phenylacetylene and is deemed to be pure by GC and NMR.

B. COPPER(I) (2-IODOPHENYL)ACETYLIDE[e]

Procedure

Following the procedure of Castro et al.[16] $CuSO_4 \cdot 5H_2O$ (22.6 g, 0.0903 mol) and ammonium hydroxide (28%, 150 mL) are placed in a 2-L Erlenmeyer flask under an argon purge. The deep blue solution is stirred for 15 min before adding 400 mL of water and $NH_2OH \cdot HCl$ (12.55 g, 0.18 mol). Addi-tion of (2-iodophenyl)acetylene (20.6 g in 500 mL of ethanol) gives a canary yellow copper(I) (2-iodophenyl)acetylide precipitate. Water is added as needed to allow stirring. The reaction mixture is filtered. The filter cake is washed three times with 200 mL each of water, ethanol, and ether so that it will not dry completely. The copper(I) (2-iodophenyl)acetylide is allowed to air-dry for 24 h [25.5 g, 97% based upon (2-iodophenyl)acetylene].

■ **Caution.** *When dry, copper acetylides are known to be shock and thermally sensitive. Copper(I) (2-iodophenyl)acetylide should therefore be handled with care.*

C. 5,6,11,12,17,18-HEXADEHYDROTRIBENZO[*a, e, i*] CYCLODODECENE (TBC)[f] AND 5,6,11,12,17,18,23,24-OCTADEHYDROTETRABENZO[*a, e, i, m*]CYCLOHEXADECENE (QBC)

Procedure

A 1-L flask is charged with the copper(I) (2-iodophenyl)acetylide (17.9 g, 0.0616 mol) and placed under an active vacuum to remove remaining vol-atiles and effect constant weight (ca. 18 h at 10^{-4} torr). About 500 mL of pyridine is added to give an approximate 0.1 M suspension of the copper acetylide. The mixture is refluxed for 6.5 h and pyridine is removed in vacuo. The crude product is extracted with about 1000 mL of either acetone or diethyl ether, and the insolubles are removed by filtration before the solvent is removed on a rotavapor. Tribenzocyclyne is sublimed at 160°C on a diffusion-pump vacuum line (10^{-6} torr) for approximately 48 h [yield 2.9 g, 47.7% based upon copper(I) (2-iodophenyl)acetylide]. Recrystallization from boiling toluene (ca. 5 mL), followed by washing with cold hexanes, provides spectroscopically pure TBC. Tetrabenzocyclyne[9] is isolated by extraction of the residue left in the sublimator with acetone followed by column chromatography (elution with a 1:4 mixture of dichloromethane/hexanes, 0.50 g isolated, 8% yield). A trace amount of the hexabenzocyclyne (HBC) is also isolated.[9]

Properties

(2-Iodophenyl)acetylene should be protected from light and stored over copper metal. The ^{1}H NMR spectrum (CDCl$_3$) exhibits doublets at 7.81 (1H) and 7.48 (1H), triplets at 7.27 (1H) and 6.99 (1H), and a singlet at 3.3 δ (1H). The ^{13}C NMR spectrum (CDCl$_3$) exhibits peaks at 138.7, 133.4, 129.9, 128.6, 127.7, 100.4, 85.1, and 81.0 δ. A very weak C≡C stretch is observed at 2130 cm^{-1} in the IR spectrum.

Copper(I) (2-iodophenyl)acetylide exhibits a stronger C≡C stretch at 1935 cm^{-1} in the IR spectrum.

Tribenzocyclyne is a bright yellow solid which decomposes at 215–220°C. The ^{1}H NMR spectrum (CDCl$_3$) exhibits multiplets centered at 7.32 (6H) and 7.17 δ (6H) typical of an AA'BB' spin system. The ^{13}C NMR spectrum (CDCl$_3$) exhibits peaks at 132.0, 128.6, 126.7, and 92.9 δ. A very weak C≡C stretch is observed at 2217 cm^{-1} in the IR spectrum. Mass spectroscopic evidence indicates a parent ion at $m/e = 300$.

Tetrabenzocyclyne is a colorless solid which decomposes at 220°C. An AA'BB' spin system is again observed in the ^{1}H NMR spectrum (CDCl$_3$) where QBC exhibits multiplets centered at 7.53 (8H) and 7.26 δ (8H). The ^{13}C NMR (CDCl$_3$) exhibits peaks at 132.3, 128.0, 125.6, and 91.1 δ. A very weak C≡C stretch is observed at 2214 cm^{-1} in the IR spectrum. Mass spectroscopic evidence for QBC includes an M^{+} ion at $m/z = 400$, a peak corresponding to the loss of acetylene at $m/z = 374$, and a doubly charged M^{++} ion at $m/z = 200$.

Hexabenzocyclyne has been characterized by mass spectroscopy (EI) giving $m/z = 600$ (M^{+}), 574 (loss of HC≡CH), and 300 (M^{++}). ^{1}H NMR (CDCl$_3$) shows multiplets centered at 7.33 (12H) and 7.05 δ (12H).

Acknowledgments

We would like to thank Dr. Michael J. Taschner for helpful discussions. C. A. T. and W. J. Y. would like to thank the Petroleum Research Fund, Office of Naval Research, State of Ohio through the Edison Center for Sensors and Technology at CWRU, and the National Science Foundation for supporting different aspects of this research.

Notes Added by Checkers

(a) Commercially available potassium *t*-butoxide (Aldrich) was not purified further.

(b) *n*-Butyllithium (1.6 M in hexane, Lithco) was titrated with a standard solution of *t*-butanol in xylenes using 1,10-phenanthroline as an indicator.

(c) Magnesium bromide diethyl ether is freshly prepared as follows: Magnesium (2.00 g, 0.0823 mol) and ether (80 mL) are combined. 1,2-Dibromoethane (5.50 mL, 0.0638 mol) is added dropwise so that the reaction is refluxing vigorously. After the addition is complete, the reaction mixture is stirred for 0.5 h. The product is then removed via cannula and used directly.

(d) The procedure was modified as follows. A three-necked flask (250 mL) is equipped with a mechanical stirrer. The flask is charged with KO-*t*-Bu (3.10 g, 0.0275 mol) and *n*-BuLi (35.0 mL, 0.0560 mol) under nitrogen. After cooling to $-78°C$, precooled THF (65 mL) is added slowly. The mixture is stirred under nitrogen at $-78°C$ for 0.5 h. Phenylacetylene (2.75 mL, 0.0250 mol) is added via syringe and the resulting green mixture is stirred for 2 h. Then freshly prepared $MgBr_2 \cdot OEt_2$ (0.0638 mol, see note c) is added slowly and the reaction mixture is stirred for 1 h at $-78°C$. A solution of iodine (7.00 g, 0.0276 mol) in THF (36 mL) is added and the reaction is allowed to come to room temperature overnight. The mixture is worked up as described in the experiment and then purified by chromatography (silica gel, hexane) to yield 2.18 g (38%) of the title compound as a yellow oil.

(e) The procedure was adhered to using the following amounts: $CuSO_4 \cdot 5H_2O$ (2.38 g, 0.00952 mol), NH_4OH (28%, 16.0 mL), H_2O (40 mL), $NH_2OH \cdot HCl$ (1.32 g, 0.0190 mol) and (2-iodophenyl)acetylene (2.18 g, 0.00954 mol) in ethanol (50 mL). The described workup yielded 2.52 g (91%) of the title compound as a yellow solid.

(f) The following modifications were used: Copper(I) (2-iodophenyl) acetylene (2.52 g, 0.00868 mol) is placed under vacuum (1 mmHg, 12 h) to remove volatiles. Pyridine (87 mL) is added and the suspension is heated to reflux for 6 h. The solvent is removed in vacuo. The solid residue is dissolved in diethyl ether and the insolubles are removed by filtration. After removal of the solvent, the crude product is purified by chromatography (silica gel, hexane) to yield 0.28 g (22%) of TBC as a yellow solid.

References

1. M. Nakagawa, *The Chemistry of the Carbon–Carbon Triple Bond*, Part 2, S. Patai, (ed.) Wiley: Chichester, 1978, pp. 635–712.
2. H. Irngartinger, L. Seiserowitz, and G. M. J. Schmidt. *Chem. Ber.*, **103**, 1119 (1979)
3. (a) J. D. Ferrara, C. A. Tessier, and W. J. Youngs, *J. Am. Chem. Soc.* **107**, 6719 (1985). (b) J. D. Ferrara, A. A. Tanaka, C. Fierro, C. A. Tessier, and W. J. Youngs, *Organometallics*, **8**, 2089 (1989). (c) J. D. Ferrara, C. A. Tessier, and W. J. Youngs, *Organometallics*, **6**, 676 (1987).
4. J. D. Ferrara, A. Djebli, C. A. Tessier, and W. J. Youngs, *J. Am. Chem. Soc.*, **110**, 647 (1988).
5. (a) A. Djebli, J. D. Ferrara, C. A. Tessier, and W. J. Youngs *J. Chem. Soc., Chem. Commun.*, 548 (1988). (b) J. D. Ferrara, C. A. Tessier, and W. J. Youngs, *Inorg. Chem.*, **27**(13), 2201 (1988).
6. (a) R. D. Stephens and C. E. Castro, *J. Org. Chem.*, **28**, 3313 (1963). (b) I. D. Campbell, G. Eglinton, W. Henderson, and R. A. Raphael, *J. Chem. Soc., Chem. Commun.*, 87 (1966). (c) H. H. Staab and K. Neunhoeffer, *Synthesis*, 424 (1979).
7. (a) H. A. Staab and F. Graf, *Tetrahedron Lett.*, **7**, 751 (1966). (b) H. A. Staab and F. Graf, *Chem. Ber.*, **103**, 1107 (1970). (c) H. A. Staab and R. Bader, *Chem. Ber.*, **103**, 1157 (1970).
8. C. Huynh and G. Linstrumelle, *Tetrahedron*, **44**(20), 6337 (1988).
9. D. Solooki, J. D. Bradshaw, C. A. Tessier, W. J. Youngs, R. F. See, M. R. Churchill, and D. Ferrara, *J. Organomet. Chem.*, **470**, 231 (1994).
10. H. Hommes, H. D. Verkruijsse, and L. Brandsma, *Tetrahedron Lett.*, 2495 (1981).
11. D. Solooki, V. O. Kennedy, C. A. Tessier, and W. J. Youngs, *Synlett*, 427 (1990).
12. W. J. Youngs, J. D. Kinder, J. D. Bradshaw, and C. A. Tessier, *Organometallics*, **12**, 2406 (1993).
13. D. F. Shriver, *The Manipulation of Air-Sensitive Compounds*, Krieger, Malabar, FL, 1982.
14. W. G. Kofron and L. M. Baclawski, *J. Org. Chem.*, **41**, 1879 (1976).
15. W. C. Still, M. Kahn, and A. Mitra, *J. Org. Chem.*, **43**, 2923 (1978).
16. C. E. Castro, E. J. Gaughan, and D. C. Owsley, *J. Org. Chem.*, **31**, 4061 (1966).

20. (CHLOROMETHYLENE)BIS[TRIMETHYLSILANE] [BIS(TRIMETHYLSILYL)CHLOROMETHANE]

SUBMITTED BY RICHARD A. KEMP* and ALAN H. COWLEY[†]
CHECKED BY HARMUT SCHULZ[‡] and MALCOLM H. CHISHOLM[‡]

$$CH_2Cl_2 + 2(CH_3)_3SiCl \xrightarrow{2n\text{-BuLi}} \text{``}[(CH_3)_3Si]_2CCl_2\text{''} + 2LiCl + 2C_4H_{10}$$

$$\text{``}[(CH_3)_3Si]_2CCl_2\text{''} + n\text{-BuLi} \rightarrow \text{``}[(CH_3)_3Si]_2C(Cl)Li\text{''} + n\text{-BuCl}$$

$$\text{``}[(CH_3)_3Si]_2C(Cl)Li\text{''} + CH_3CH_2OH \rightarrow [(CH_3)_3Si]_2CHCl + LiOCH_2CH_3$$

* Union Carbide Corporation, 3333 Highway 6 South, Houston TX 77082.
† Department of Chemistry and Biochemistry, University of Texas, Austin, TX 78712.
‡ Department of Chemistry, Indian University, Bloomington, IN 47405.

Use of the sterically demanding (chloromethylene)bis[trimethylsilane] ligand has allowed the isolation of many kinetically stabilized main group and transition metal complexes.[1] The most common method for the introduction of this ligand is by treatment of $Li[CH[Si(CH_3)_3]_2]$ with an active halide. The precursor of this lithium reagent is the chloride, $[(CH_3)_3Si]_2CHCl$. Previous literature preparations of $[(CH_3)_3Si]_2CHCl$ have several disadvantages. One common route requires the prior synthesis and isolation of $[(CH_3)_3Si]_2CCl_2$[2] from CH_2Cl_2, a low-yield (50–65%) process, followed by conversion to $[(CH_3)_3Si]_2CHCl$ by reaction with n-BuLi and a protic quench. The other commonly employed route involves the deprotonation of $[(CH_3)_3Si]CH_2Cl$ with s-BuLi followed by treatment with $(CH_3)_3SiCl$.[3] It was desirable to develop an inexpensive, reproducible, high-yield synthesis of $[(CH_3)_3Si]_2CHCl$ which could easily be scaled up to large quantities. This present synthesis is a modification of the initial literature route which now allows $[(CH_3)_3Si]_2CHCl$ to be prepared in a simple "one-pot" reaction and does not require isolation of any intermediates, thus significantly raising the yield of the desired compound.[4] The (chloromethylene)bis(trimethylsilane) can be converted into $[(CH_3)_3Si]_2CHLi$ by treating the chloride with excess powdered lithium metal (20-μm size) in refluxing diethyl ether at 55°C for 24 h. The resulting LiCl and lithium can be then separated from the organolithium product by filtration through a medium-porosity frit covered with Celite[R] 545 filter agent.

Procedure

Caution. *Butyllithium reacts violently with water and should be handled under an inert atmosphere. Until the ethanol quench, all manipulations are performed under anhydrous, oxygen-free conditions. To minimize chances for an explosion, diethyl ether should always be tested for peroxides prior to distillation from $LiAlH_4$. Additionally, under no circumstances should the distillation still be allowed to dry.[5]*

The preparation of $[(CH_3)_3Si]_2CHCl$ utilizes standard inert atmosphere techniques. All solvents must be rigorously dried and degassed prior to use. Tetrahydrofuran is freshly distilled from CaH_2, diethyl ether is distilled from $Li[AlH_4]$ (see caution above), and CH_2Cl_2, hexane, and pentane are each distilled from P_4O_{10} immediately prior to use. During the cooling procedures, a strong nitrogen purge must be present to avoid sucking the oil from the mineral oil bubbler back into the reaction flask. Into a nitrogen-filled, 2-L, three-necked flask equipped with a magnetic stirbar and a mineral oil bubbler are placed 250 mL of THF, 90 mL of diethyl ether, 40 mL of pentane, 78.6 g (0.725 mol, 92.2 mL) of chlorotrimethylsilane, and 29.8 g (0.35 mol, 22.4 mL) of CH_2Cl_2. The contents are cooled to -110°C by means of an ethanol/liquid nitrogen slush bath.[6] A precooled solution (− 78°C) of n-BuLi (453 mL, 1.6 M in hexane, 0.75 mol) is then added dropwise to the cooled solution under positive nitrogen pressure via a 16-gauge stainless-steel cannula. This addition should occur over a 40–60 min period. After this addition, the solution is allowed to warm to room temperature, during which time the precipitation of LiCl is clearly evident. The reaction mixture is then recooled to − 110°C, and the second stage of the synthesis is accomplished by adding a precooled solution (− 78°C) of n-BuLi (219 mL, 1.6 M in hexane, 0.35 mol) dropwise via a 16-gauge stainless-steel cannula. The light tan color characteristic of the $\{[(CH_3)_3Si]_2CCl\}^-$ anion becomes evident as the addition occurs. After stirring at − 100°C for 45 min, the reaction mixture is treated with 100 mL of 95% ethanol and is allowed to begin warming to room temperature. When the temperature reaches − 40°C, 100 mL of a 6.0-M hydrochloric acid solution is added to quench the lithium ethoxide which is present. After warming to room temperature, the aqueous layer is separated from the organic layer and extracted twice with 75 mL portions of hexane. The organic solutions are combined and the solvents are removed by means of a rotary evaporator, leaving $[(CH_3)_3Si]_2CHCl$ as a pale yellow liquid. Fractional vacuum distillation through a 10-cm Vigreaux column affords 52.8 g (0.272 mol) of pure $[(CH_3)_3Si]_2CHCl$, bp 68–80°C/25 torr. The overall yield of $[(CH_3)_3Si]_2CHCl$ based on CH_2Cl_2 consumption is 78%.

Properties

(Chloromethylene)bis[trimethylsilane] is a colorless, water-stable liquid. ^1H NMR (CD_2Cl_2): δ 0.15, s (18H); δ 2.45, s (1H). ^{13}C NMR (CD_2Cl_2): δ 2.49, s (CH_3); δ 35.16, s (CH).

Anal. Calcd. for $C_7H_{19}ClSi_2$: C, 43.1; H, 9.8. Found: C, 41.9; H, 9.3.

References

1. For example, see (a) E. Niecke, M. Leuer, and M. Nieger, *Chem. Ber.*, **122**, 453 (1989); (b) T. Fjeldberg, A. Haaland, B. E. R. Schilling, M. F. Lappert, and A. J. Thorne, *J. Chem. Soc., Dalton Trans.*, 1551 (1986); and (c) A. H. Cowley, R. A. Kemp, and J. C. Wilburn, *J. Am. Chem. Soc.*, **104**, 331 (1982).
2. R. Appel, *Inorg. Synth.*, **24**, 117 (1986).
3. C. Burford, F. Cooke, E. Ehlinger, and P. Magnus, *J. Am. Chem. Soc.*, **99**, 4536 (1977); T. J. Barton and S. K. Hoekman, *J. Am. Chem. Soc.*, **102**, 1584 (1980).
4. A. H. Cowley and R. A. Kemp, *Synth. React. Inorg. Met.-Org. Chem.*, **11**, 591 (1981).
5. For a cautionary note, see *Inorg. Synth.*, **12**, 317 (1970).
6. D. F. Shriver and M. A. Drezdzon, *The Manipulation of Air-Sensitive Compounds*, 2nd ed., Wiley, New York, 1986, p. 109.

21. S,S-CHIRAPHOS[(S,S)-(−)-(1,2-DIMETHYL-1,2-ETHANEDIYL)BIS(DIPHENYLPHOSPHINE)]

SUBMITTED BY STEVEN H. BERGENS,* JOHN WHELAN,*
and B. BOSNICH*
CHECKED BY ROBERT J. DONOVAN† and IWAO OJI MA†

The bidentate chiral bis(phosphine), S,S-CHIRAPHOS [$Ph_2PCH(CH_3)$-$CH(CH_3)PPh_2$] is the simplest chiral bis(phosphine) possessing a twofold rotation axis which has been used in asymmetric catalysis. It was first employed as its rhodium(I) complexes for asymmetric catalytic hydrogenation of amino acid precursors giving consistently high ee's.[1] Subsequent to this, it has found numerous applications in asymmetric catalysis.[2] The original preparation of S,S-CHIRAPHOS involved lithium diphenylphosphide displacements of the tosyl groups of (R,R)-1,2-dimethyl-1,2-ethanediylbis(toluenesulfonate). The yield was low in this step (∼ 20%) and the workup was cumbersome.[1] Replacing the ditosylate by the dimesylate produced, under similar conditions, only a slight increase in the yield.[3] The present procedure involving the (R,R)-1,2-dimethyl-1,2-ethanediyl bis(methanesulfonate) and potassium diphenylphosphide is technically refined and provides pure S,S-CHIRAPHOS in high yield.

Starting Materials

All manipulations should be carried out in a well-ventilated fume hood; protective gloves and goggles should be worn. All glassware is oven dried at 90°C for 10 h, assembled hot, and cooled under a stream of dry argon or dry nitrogen. Dichloromethane is distilled from calcium hydride under argon before use. Triethylamine is similarly dried over potassium and distilled before use. Methanesulfonyl chloride (98%, Aldrich) is distilled (bp 86–88°C/80 mm) and stored under argon. Tetrahydrofuran is dried and distilled under argon before use and kept oxygen free. Diethyl ether is distilled from lithium aluminum hydride under argon and kept oxygen free. Diphenylphosphine is used as obtained (Aldrich) and is stored and handled under argon. Potassium hydride, a 35% dispersion in mineral oil, is used as obtained (Aldrich). (R,R)-(−)-2,3-Butanediol (98%, Aldrich) is used without further purification. Although none of the procedures is hazardous, due care should be taken when handling diphenylphosphine and its anion; both

* Department of Chemistry, The University of Chicago, 5735 Ellis Avenue, Chicago, IL 60637.
† Department of Chemistry, The State University of New York at Stony Brook, Stony Brook, NY 11794.

materials should be handled under argon as described. Diphenylphosphine can be quenched by dilute hydrogen peroxide in acetone solution. Trifluoroacetic acid (99 + %, Aldrich) and NaCN (95 + %, Aldrich) are used as obtained. (■ **Caution.** *NaCN is highly toxic. Care should be taken to avoid direct contact of the chemical or its solution with the skin and impervious gloves should be worn to handle the reagent.*) Pentane and methanol, "Baker Analyzed" (J. T. Baker), are used as obtained. The degassed solvents; distilled water, 95% ethanol (Midwest Grain Products), and absolute ethanol (Midwest Grain Products) are prepared by vigorously bubbling argon (or nitrogen) through the solvent (used as obtained) for ~ 0.5 h and are stored under argon (or nitrogen). All solvents and solutions that come into contact with uncomplexed S,S-CHIRAPHOS must be degassed and manipulated under an inert atmosphere.

A. (R,R)-2,3-DIMETHYL-1,2-ETHANEDIYL BIS(METHANESULFONATE)

$$CH_3CH(OH)CH(OH)CH_3 + 2CH_3SO_2Cl + 2(C_2H_5)_3N \xrightarrow{CH_2Cl_2}$$

$$CH_3CH(OSO_2CH_3)CH(OSO_2CH_3)CH_3 + 2(C_2H_5)_3NHCl$$

Procedure

The apparatus for the reaction consists of a 250-mL, three-necked, round-bottomed flask containing a Teflon-coated magnetic stirring bar. One side neck is capped with a rubber septum and the other contains a ground glass tube carrying inlet and outlet tubes, one of which is connected by plastic hose to the inlet argon; the other has a hose connected to a gas bubbler.* The argon flow is maintained throughout the reaction. The center neck of the flask is fitted with a 50-mL pressure-equalizing dropping funnel capped with a rubber septum. The flask is charged with (R, R)-(−)-2,3-butanediol (5.0 g, 55.5 mmol) and dry dichloromethane (90 mL) via syringe. The flask and its contents are placed in a − 35°C Dry-ice/2-propanol bath and the stirring is begun. Dry triethylamine (23.2 mL, 166.5 mmol) is added via syringe to the cold solution. The dropping funnel is charged, by syringe, with a solution of methanesulfonyl chloride (9.5 mL, 122.7 mmol) in dry dichloromethane (10 mL). When the temperature of the stirred liquid in the flask has reached − 35°C, the solution of methanesulfonyl chloride is added to it dropwise over a 10-min period. There is immediate formation of a white precipitate

---

* The checkers recommend the use of a gas-tight mechanical stirrer instead of a Teflon stirring bar.

of triethylamine hydrochloride. It is essential to keep the mixture stirring vigorously as the solid builds up. The stirred mixture is allowed to warm to room temperature over 2 h and is stirred at this temperature for a further 1 h.

The reaction flask is placed in an ice bath and the stirred reaction mixture is quenched by the addition of crushed ice (50 g) followed by ice-cold water (50 mL). The cooling bath is removed and stirring is continued until the ice melted. The mixture is then transferred to a 500-mL separatory funnel; the flask is washed with 75 mL of dichloromethane which is added to the separatory funnel. The mixture is then shaken, the organic layer is separated, and the aqueous layer is extracted with two 25-mL portions of dichloromethane. The combined organic material is then sequentially washed with two 100-mL ice-cold portions of 10% aqueous hydrochloric acid, 100 mL of saturated brine, 100 mL of cold saturated aqueous sodium hydrogen carbonate, and 100 mL of saturated brine. The organic layer is dried over anhydrous magnesium sulfate and filtered through a 140-mL glass filter with 25- to 50-μ porosity and the magnesium sulfate is washed with dichloromethane. The combined filtrates are evaporated to dryness at 40°C on a rotary evaporator to yield a tacky solid of the *product*.* This mixture is slurried with methanol (15 mL) and then is placed in a freezer (− 25°C) for 15 h. The mixture is filtered through a 60-mL glass filter with 25- to 50-μ porosity and the white crystalline solid is washed with three 10-mL portions of methanol and then with three 10-mL portions of diethyl ether. It is dried in a vacuum desiccator yielding 11.93 g (87.3%)* of the desired product as a white crystalline solid (mp 124–125°C). A further 0.28 g of slightly yellow product is obtained by evaporation of the filtrates to give an oil which deposits crystals upon the addition of about 2 mL of methanol and cooling as before.

The white crystalline solid product is pure enough for the next step, but an analytical sample is recrystallized as follows. A sample (0.45 g) is dissolved in dry dichloromethane (5 mL) and the solution is filtered through a 2-mL glass filter with 25- to 50-μ porosity. The solution is then diluted with methanol (3 mL) and the dichloromethane is carefully distilled off on a steam bath. Crystallization begins almost immediately on cooling. After 2 h at room temperature, the mixture is placed in the freezer (− 25°C) for 15 h. The mixture is filtered through a 2-mL glass filter with 25- to 50-μ porosity and the white crystalline solid is washed with three 5-mL portions of methanol

---

* The checkers report that the product is a yellow solid whose mp is 118–120°C.
* The checkers obtained a 72–77% yield of white crystalline product (mp 122–124°C). The authors obtained yields of 83–87% on reactions run on 5 and 6 g of diol. On larger-scale reactions, (10 and 16 g of diol), using external stirring, the yields obtained varied between 75 and 77%.

and three 5-mL portions of diethyl ether. It is dried in a vacuum desiccator yielding 0.40 g of analytically pure (*R*,*R*)-2,3-dimethyl-1,2-ethanediylbis-(methanesulfonate) which melts at 125–126°C (lit.[3] 115–117°C). The compound displays a small but positive specific rotation of $[\alpha]_D^{25} + 2.3 \pm 0.2°$ (C = 1.4, CHCl$_3$). The specific rotation is also small and positive when measured at the mercury emission lines.

Anal. Calcd. for C$_6$H$_{14}$O$_6$S$_2$: C, 29.3; H, 5.7. Found: C, 29.5; H, 5.8.

B. Ni(*S*,*S*-CHIRAPHOS)(NCS)$_2$]

$$CH_3CH(OSO_2CH_3)CH(OSO_2CH_3)CH_3 + 2KPPh_2 \xrightarrow{\text{THF}}$$
$$CH_3CH(PPh_2)CH(PPh_2)CH_3 + 2KOSO_2CH_3$$

$$CH_3CH(PPh_2)CH(PPh_2)CH_3 + Ni(NCS)_2 \xrightarrow{\text{C}_2\text{H}_5\text{OH}}$$
$$[Ni(S,S\text{-chiraphos})(NCS)_2]$$

Comment

The simplest and the least technically demanding method of isolating *S*,*S*-CHIRAPHOS from the reaction mixture is by the formation of its very insoluble bis(thiocyanato-*N*)nickel(II) complex. The nickel selectively separates the *S*,*S*-CHIRAPHOS from other phosphine species in the reaction mixture.

Procedure

The apparatus consists of a 500-mL, three-necked, round-bottomed flask. One side neck is capped with a rubber septum and the other contains a reflux condenser with a gas inlet connected to an argon flow line and an exit-gas bubbler. The center neck is fitted with a gas-sealed mechanical stirrer with a Teflon paddle. The flask is charged with potassium hydride (10.6 g; 35% mineral oil dispersion; 92.5 mmol). Dry tetrahydrofuran is added by cannula and the mixture is mechanically stirred as diphenylphosphine (15.5 mL, 89.0 mmol) is added by syringe slowly over a period of 15 min. After stirring for a further 0.5 h, the clear red solution is immersed in a − 5°C Dry-ice/2-propanol bath. When the red solution has cooled, finely powdered (*R*,*R*)-2,3-dimethyl-1,2-ethanediyl bis(methanesulfonate) (10.0 g, 40.6 mmol) is added in one portion and the stirring is continued at this temperature for 1 h to give a viscous orange-red slurry. The stirred mixture is then quenched by the addition of ammonium chloride (27 g) dissolved in degassed water (80 mL) via cannula. The cooling bath is removed and the stirring is stopped. A clear colorless organic upper layer forms.

A one-necked, 1000-mL, round-bottomed flask (not under argon) is equipped with a Teflon-coated magnetic stirring bar and is charged with nickel acetate tetrahydrate (11.12 g, 44.7 mmol) and sodium thiocyanate (14.48 g, 178.6 mmol). To the solids is added warm (50°C) 95% ethanol (250 mL). The mixture is stirred until all of the solids have dissolved to give a green solution.

The organic layer obtained from the phosphide reaction still under argon is rapidly transferred by cannula into the stirred-nickel(II) solution. A yellow-brown precipitate forms immediately. The aqueous layer from the phosphide reaction, still in the reaction flask under argon, is extracted* with three successive 50-mL portions of diethyl ether added via syringe, and the diethyl ether extracts are transferred by cannula into the nickel(II) mixture. The stirring of the mixture is continued for about 5 min[†] and then it is filtered through a 170-mL glass filter with 25- to 50-μ porosity and washed sequentially with two 60-mL portions of 95% ethanol and three 60-mL portions of pentane. The air-stable yellow-brown precipitate is then sucked dry on the frit.

The dried product on the glass frit is slurried with dichloromethane (100 mL) with a glass rod. Trifluoroacetic acid (20 mL) is then added to the stirred slurry, whereupon the solid dissolves to give a deep brown-red solution. After a few minutes of stirring the solution is filtered into a one-necked, 1000-mL, round-bottomed flask containing a Teflon-coated stirring bar. The last traces of the solid remaining on the frit are solubilized by the addition of two 50-mL portions of dichloromethane containing 1 mL of trifluoroacetic acid. The combined filtrates are stirred and methanol (5 mL) is added, whereupon yellow-brown crystals form. More methanol (∼ 300 mL) is added in portions over 0.5 h. The mixture is filtered through a 170-mL glass filter with 25- to 50-μ porosity and is successively washed with two 60-mL portions of 95% ethanol and three 60-mL portions of pentane. The solid is sucked dry on the frit* to give 18.8 g (77.0%)[†] of the fine crystalline

* The diethyl ether extractions are carried out in the reaction vessel in the following manner. Diethyl ether (50 mL) is added by syringe to the aqueous reaction mixture and the two-phase mixture is vigorously stirred, still under argon, for a few minutes. The layers are allowed to separate and the organic layer is quickly cannulated into the nickel solution. The procedure is repeated twice.

[†] The total time between the formation of the nickel complex and the filtration should not exceed 0.5 h.

* The authors passed copious quantities of hot solvents through the glass filters to try to clean them, but had to resort to aqua regia in the end.

[†] The checkers obtained a 47–55% yield of the nickel complex. The authors obtained yields of the nickel complex between 67 and 77%, except for a reaction run at room temperature when the yield dropped to 52%.

yellow-brown [Ni(*S,S*-chiraphos)(NCS)$_2$] complex. This complex is indefinitely stable in air.

Anal. Calcd. for C$_{30}$H$_{28}$N$_2$P$_2$S$_2$Ni: C, 59.9; H, 4.7; N, 4.7. Found: C, 59.9; H, 4.9; N, 4.7.

C. *S,S*-CHIRAPHOS

[Ni(*S,S*-chiraphos)(NCS)$_2$] + 4NaCN →

$$S,S\text{-CHIRAPHOS} + Na_2[Ni(CN)_4] + 2NaNCS$$

Procedure

The apparatus for this reaction consists of a 500-mL, three-necked, round-bottomed flask equipped with a Teflon-coated magnetic stirring bar. One side neck is capped with a rubber septum and the other is equipped with a reflux condenser with a gas inlet connected to an argon flow line and an exit-gas bubbler. The center neck is fitted with a graduated 100-mL pressure-equalizing dropping funnel capped with a rubber septum. The solid nickel complex [Ni(*S,S*-chiraphos)(NCS)$_2$] (18.8 g, 31.3 mmol) is added to the flask and is then suspended in 95% degassed ethanol (190 mL). To the dropping funnel is added via syringe a solution of sodium cyanide (9.39 g, 191.5 mmol) dissolved in degassed water (45 mL). The reaction mixture is immersed in a heated oil bath and vigorous stirring is started. When the mixture is refluxing, 25 mL of the aqueous sodium cyanide solution is added dropwise over 5 min. The resulting blood-red solution is refluxed for 5 min and then the remaining aqueous sodium cyanide is added dropwise over 10 min. During the final cyanide addition, the color fades to light yellow and crystals of *S,S*-CHIRAPHOS begin to form. Degassed water (45 mL) is then added to the boiling yellow-brown solution over 10 min, after which the reaction vessel is removed from the oil bath and is allowed to cool for 3 h. The mixture is filtered through a 60-mL glass filter with 25- to 50-μ porosity and the off-white solid is washed with four 50-mL portions of degassed water. The solid is dried in a vacuum desiccator giving 12.8 g (95.9%) of *S,S*-CHIRAPHOS as an off-white crystalline solid (mp 108–109°C), [α]$_D^{25}$ − 186° (C = 1.5, acetone).* This material is sufficiently pure[†] for more purposes, but it may be recrystallized under argon as follows.

The solid (12.8 g) is placed in a 250-mL, two-necked, round-bottomed flask containing a Teflon-coated magnetic stirring bar. One neck is capped with a rubber septum and the other is equipped with a reflux condenser with a gas inlet adapter connected to an argon flow line and an exit-gas bubbler.

Degassed absolute ethanol (66 mL) is added via syringe and the reaction flask is immersed in a 120°C oil bath and the stirring is started. After a few minutes of reflux, a light amber solution is produced. The stirring is stopped, the oil bath is removed, and the solution is left to cool for 2 h as crystals form. It is then capped with a rubber septum and is placed in a freezer (− 25°C) for 18 h.[‡] The mixture is filtered through a 60-mL glass filter with 25- to 50-μ porosity and the solid is washed quickly with two 15-mL portions of cold (− 25°C) degassed absolute ethanol, and is dried in a vacuum dessiccator, resulting in 12.24 g (91.8%) of S,S-CHIRAPHOS as white sheets. The S,S-CHIRAPHOS melts at 109–111°C (lit.[1] 108–109°C,[3] 101–103°C) and has a specific rotation of $[\alpha]_D^{25}$ − 190° (C = 1.5, acetone).*

Anal. Calcd. for $C_{28}H_{28}P_2$: C, 78.9; H, 6.6; P, 14.5. Found: C, 79.0; H, 6.7; P, 14.8.

Properties

The ligand S,S-CHIRAPHOS can be handled for short periods as a solid in air, but should be stored under argon in a refrigerator. Under these conditions it can be kept for long periods (years) without noticeable change. It is soluble in acetone, tetrahydrofuran, dichloromethane, chloroform, diethyl ether, toluene, and hot ethanol and is insoluble in hexanes, methanol, and cold ethanol. In solution it slowly oxidizes in the presence of oxygen. Thus the ligand can be handled as a solid without any special precautions and in solution it requires a minimum of care. ^{1}H NMR (500 MHz, CDCl₃)

* The checkers report that they obtained S,S-CHIRAPHOS, Run 1, 99% yield, mp 101–103°C, $[\alpha]_D^{25}$ − 181° (C = 1.5, acetone); Run 2, 93% yield, mp 107–108°C, $[\alpha]_D^{25}$ − 198° (C = 1.5, acetone).

† The S,S-CHIRAPHOS contains traces of a nickel complex (a concentrated solution appears yellow-brown) which is removed by a single recrystallization, which also increases the magnitude of the optical rotation by 3–4°.

‡ The checkers suggested that cooling to − 25°C was unnecessary. The authors have also cooled the solution to 0°C with similar results.

* The authors have obtained material with specific rotations ($[\alpha]_D$) in acetone between − 188° and − 191° (one crystallization), unchanged by further recrystallizations [Strem Chemicals, Inc. gives $[\alpha]_D$ − 190° (C = 1.5, acetone) and Aldrich gives $[\alpha]_D^{22}$ − 191° (C = 1.5, CHCl₃) and for R,R-CHIRAPHOS $[\alpha]_D^{22}$ + 195° (C = 1.5 CHCl₃)]. This material has been used to prepare a chiral hydrogenation catalyst which has been used to make chiral compounds with essentially 100% ee's,[1] attesting to the optical purity of the S,S-CHIRAPHOS. The checkers report that an attempted recrystallization of S,S-CHIRAPHOS [mp 107–108°C, $[\alpha]_D^{25}$ − 198° (C = 1.5, acetone)] from degassed (freeze and thaw, 6 times) "100% ethanol" produced an 86% yield of material with mp 107.5–108.5°C and $[\alpha]_D^{25}$ − 145° (C = 1.5, acetone). The authors are unable to account for this substantial decrease in optical purity.

δ 7.37–7.18 (m, 20H, ArH), 2.46 [q, J = 7.0 Hz, 2H, –PCH(CH$_3$)–], 1.17 (dd, J$_{P-H}$ = 14.1 Hz, J$_{H-H}$ = 7.0 Hz, 6H, –CH$_3$) relative to internal TMS.

References

1. M. D. Fryzuk and B. B. Bosnich, *J. Am. Chem. Soc.*, **99**, 6262 (1977).
2. I. Ojima, N. Clos, and C. Bastos, *Tetrahedron*, **45**, 6901 (1989) and references cited therein.
3. N. C. Payne and D. W. Stephan, *J. Organometallic Chem.*, **221**, 203 (1981).

22. β-KETOPHOSPHINES: LIGANDS OF CATALYTIC RELEVANCE

SUBMITTED BY DOMINIQUE MATT*, MICHAEL HUHN*, and PIERRE BRAUNSTEIN†
CHECKED BY REMY E. SAMUEL and ROBERT H. NEILSON‡

A recent review has highlighted the extensive and interesting chemistry of phosphine–ketone ligands of the type R$_2$PCH$_2$C(O)R'.[1] These bidentate ligands are suitable for the formation of dissymetric *P,O* chelate complexes.[2] In some of them the chelating ligand may display hemilabile behavior: the specific property of these ligands lies in the ability of the keto group to dissociate easily and reversibly from the metal center, the phosphine thus remaining bound only through the phosphorus atom.[3] Of particular interest[4] are complexes containing the corresponding enolato-ligands [R$_2$PCH=C(O)R']$^-$ which usually act as chelating three-electron *P,O* bidentates. These react selectively at the carbon atom in the α position to the phosphino group with electrophilic species such as isocyanates or alkynes to yield new dissymmetric chelating systems.[5] The nickel(II) enolato complex [Ni(Ph){Ph$_2$PCH=C(O)Ph}PPh$_3$] is a catalyst precursor in the so-called SHOP process for the low-pressure oligomerization of ethylene. With this complex α-olefins, of which 99% are linear, are produced with a selectivity of 98%.[6] Another phosphino-enolato complex, namely [Ni(Ph)-{Ph$_2$PCH=C(O)Ph}(Ph$_3$P=CH$_2$)], was shown to convert catalytically

* Ecole Européenne des Hautes Études des Industries Chimiques de Strasbourg (EHICS), 1 Rue Blaise Pascal, F-67008 Strasbourg Cedex, France.
† Laboratoire de Chimie de Coordination, Université Louis Pasteur, 4 rue Blaise Pascal, F-67070 Strasbourg Cedex, France.
‡ Department of Chemistry, Texas Christian University, Fort Worth, TX 76129.

acetylene to polyacetylene.[7] The following syntheses, based on two original papers,[8, 9] detail the preparations of two β-ketophosphines, $Ph_2PCH_2C(O)Ph$ and $[\{Ph_2PCH_2C(O)(C_5H_4)\}Fe(C_5H_5)]$, which are simple to carry out and make use of commercially available starting materials. An easy preparation of an enolate complex, namely *cis*-[Pd{$Ph_2PCH = C(O)Ph\}_2]$, is also given.

General Procedures

■ **Caution.**   *Because of the toxicity of chlorodiphenylphosphine, the reactions must be carried out in an efficient fume hood.*
All reactions and manipulations described here must be carried out under an atmosphere of dry nitrogen using Schlenk tube techniques. Tetrahydrofuran and diethyl ether used for the reaction with *n*-BuLi are dried over sodium benzophenone. Hexane and toluene are dried over sodium metal and distilled under nitrogen. Dichloromethane is dried and distilled over P_2O_5, pentane over a sodium-potassium alloy. Diisopropylamine is dried over activated molecular sieves (4 Å) and distilled under nitrogen. Acetophenone should be dried over molecular sieves and degassed. The following detailed procedure is appropriate for the preparation of both phosphines. Precise information concerning reaction times, solvent, and quantity of reagents, is given for each of the two syntheses. Both procedures A and B consist of three steps: (1) preparation of lithium diisopropylamide (LDA); (2) preparation of the ketone-derived Li-enolate; and (3) reaction of the Li-enolate with chlorodiphenylphosphine. Each phosphine synthesis requires two 500-mL Schlenk flasks, a 100-mL Schlenk tube, and a glass frit filter connectable to the Schlenk flasks.

A 500-mL Schlenk flask containing a magnetic stirring bar is connected to a standard nitrogen gas-vacuum line and filled with nitrogen. Using a piston pipette, diisopropylamine in about 5% excess (with respect to the theoretical quantity of butyllithium) is added to the flask, followed by tetrahydrofuran or diethyl ether. The solution is cooled to $-78°C$ by means of a dry ice–acetone cooling bath. A 1.6-M *n*-BuLi–hexane solution is then added dropwise via a syringe to the solution containing the amine. Butane gas is produced in the course of this reaction. After 0.5 h, a precooled (ca. $-50°C$) THF solution of the appropriate ketone is added over a period of 3 min to the LDA solution. This is best performed via a Teflon canula joining the two flasks equipped with serum caps. After 1 h of stirring, using the same transfer technique as that described above, a THF (or Et_2O) solution of freshly distilled dichlorophosphine at $-78°C$ is added slowly to the enolate solution. The temperature is then raised and the mixture is stirred for several hours. The solvent is removed in vacuo. Addition of toluene with a pipette and filtration of the resulting mixture through a glass frit (porosity 4) allows the separation of LiCl. At this stage, a practically pure phosphine is usually obtained which may, if necessary, be purified further (see below). The preparation of the enolato-palladium complex requires two 100-mL Schlenk tubes and a Schlenk filter apparatus.

A. 2-(DIPHENYLPHOSPHINO)ACETOPHENONE

$$H_3CC(O)Ph + LiN(i\text{-}C_3H_7)_2 \rightarrow CH_2=C(OLi)Ph + HN(i\text{-}C_3H_7)_2$$

$$Ph_2PCl + CH_2=C(OLi)Ph \rightarrow Ph_2PCH_2C(O)Ph$$

Procedure

Following the general procedure described above, a 1.55-M hexane solution of *n*-BuLi (32.3 mL, 50.0 mmol, Janssen) is added slowly to a solution of diisopropylamine (5.26 g, 52.0 mmol) in THF (30 mL) at $-78°C$. After the mixture has been stirred for 0.5 h, a solution of dry acetophenone (6.00 g, 50.0 mmol) in THF (20 mL) is added slowly. After 1 h, while maintaining the temperature at $-78°C$, a solution of Ph_2PCl (11.03 g, 50.0 mmol, Aldrich) in THF (50 mL) is added. After stirring for 12 h at room temperature, the solvent is removed in vacuo. The residue is treated with toluene (200 mL), and the resulting suspension is filtered. The filtrate is then evaporated to dryness yielding a white (sometimes yellowish) residue which is washed with cold ($-20°C$) methanol (20 mL). Dissolution in THF (25 mL) and precipitation with hexane (120 mL) yields an analytically pure white product (14.45 g, 47.5 mmol, 95%).

Anal. Calcd. for $C_{20}H_{17}OP$: C, 78.93; H, 5.64. Found: C, 78.80; H, 5.70.

Properties

The phosphine is a white powder (mp 71°C) which should be stored under N_2. It is soluble in CH_2Cl_2, THF, and Et_2O. It can be manipulated in air for a short period of time without noticeable oxidation of the compound. The ^1H NMR spectrum (200 MHz) shows no $^2J(PH)$ coupling, whereas for complexes containing this phosphine a $^2J(PH)$ coupling constant of about 10 Hz is usually observed. The IR (KBr) spectrum shows a strong $\nu(C=O)$ band at 1670 cm^{-1}. ^1H NMR (CDCl$_3$): δ 3.80 [s, 2H, CH$_2$, $^2J(PH) = 0$ Hz], 7.2–8.1 (15 H, aromatic H). $^{31}P\{^1H\}$ NMR (CDCl$_3$): δ $-17.1(s)$. $^{13}C\{^1H\}$ NMR (CDCl$_3$): δ 40.54 (d, CH$_2$, $^1J(PC) = 21.5$ Hz), 128.37–137.57 (aromatic C), 196.94 [d, C=O, $^2J(PC) = 8.1$ Hz].

B. [(DIPHENYLPHOSPHINO)ACETYL]FERROCENE

$$[\{H_3CC(O)(C_5H_4)\}Fe(C_5H_5)] + LiN(i\text{-}C_3H_7)_2 \rightarrow$$
$$[\{CH_2=C(OLi)(C_5H_4)\}Fe(C_5H_5)] + HN(i\text{-}C_3H_7)_2$$
$$Ph_2PCl + [\{CH_2=C(OLi)(C_5H_4)\}Fe(C_5H_5)] \rightarrow$$
$$[\{Ph_2PCH_2C(O)(C_5H_4)\}Fe(C_5H_5)]$$

Procedure

Following the general procedure described above, a 1.55-M hexane solution of *n*-BuLi (5.2 mL, 8.00 mmol, Janssen) is added slowly to a solution of diisopropylamine (0.850 g, 8.40 mmol) in Et_2O (50 mL) at $-78°C$. After the mixture has been stirred for 0.5 h, a solution of acetylferrocene (1.825 g, 8.00 mmol, Aldrich) in Et_2O (30 mL) is added. The mixture is stirred for 1 h at $-78°C$ and a precooled ($-78°C$) solution of Ph_2PCl (1.765 g, 8.0 mmol, Aldrich) in Et_2O (20 mL) is added. The temperature is then raised and a brick-red precipitate gradually appears. After stirring the mixture for 3 h at room temperature, the solvent is removed in vacuo. The residue is then treated with toluene (ca. 60 mL). Filtration through a glass frit (porosity 4) and cooling of the filtrate to $-20°C$ yields analytically pure crystals of the phosphine (3.035–3.200 g, 92–97%).

Anal. Calcd. for $C_{24}H_{21}FeOP$: C, 69.92, H, 5.13. Found: C, 69.85; H, 5.20.

Properties

When obtained from a toluene solution, the product forms brick-red crystals (space group $P\bar{1}$) of hexagonal prismatic shape (mp 142–143°C).[10] As revealed by a crystal structure determination, the two cyclopentadienyl rings are eclipsed. The compound can be handled in air, but is best stored under nitrogen for long periods. In dichloromethane solution, this phosphine decomposes after a week. The cyclic voltammogram shows a reversible oxidation step at 0.750 mV vs. SCE (CH_2Cl_2, $[NBu_4]ClO_4$). As in the case of the previous phosphine, the $^2J(PH)$ value is zero. The IR (KBr) spectrum shows a strong $v(C=O)$ band at 1642 cm^{-1}. ^1H NMR (acetone-d_6): δ 3.72 [s, 2H, PCH_2, $^2J(PH) = 0$ Hz], 4.20 (s, 5H, C_5H_5), 4.54 and 4.85 (two triplets corresponding to an AA'BB' spin system, 4H, C_5H_4), 7.37–7.57 (10 H, aromatic H). $^{31}P\{^1H\}$ NMR ($CDCl_3$): δ $-19.1.$(s).

C. *CIS*-BIS[2-(DIPHENYLPHOSPHINO)-ACETOPHENONATO-*P,O*]PALLADIUM(II)

$$1/3[Pd(O_2CCH_3)_2]_3 + 2Ph_2PCH_2C(O)Ph \rightarrow$$

$$cis\text{-}[Pd\{Ph_2PCH=C(O)Ph\}_2] + 2CH_3CO_2H$$

A 100-mL Schlenk tube is purged with nitrogen and filled with $Ph_2PCH_2C(O)Ph$ (0.330 g, 1.08 mmol) and CH_2Cl_2 (30 mL). To the resulting solution solid palladium acetate (0.122 g, 0.54 mmol Pd, Strem) is added. A color change to yellow occurs instantly and a smell of acetic acid can be detected. After stirring for 10 min, the volume of the solution is reduced to approximately 5 mL. Addition of pentane (80 mL) to the solution precipitates the product as yellow microcrystals. These are filtered off and washed twice with pentane (2 × 10 mL). Slow diffusion of pentane (30 mL) into a CH_2Cl_2 solution (15 mL) gives, after cooling at − 20°C, analytically pure crystals of the complex (0.325 g, 85%).

Anal. Calcd. for $C_{40}H_{32}O_2P_2Pd$: C, 67.38; H, 4.52. Found: C, 67.30; H, 4.45.

Properties

This air-stable, yellow crystalline compound (mp 215°C dec) is soluble in CH_2Cl_2, THF, toluene, MeOH, and Et_2O. The IR spectrum (KBr pellet) shows typical enolate absorption bands at 1513 and 1484 cm^{-1}. ^1H NMR (C_6D_6): δ 4.75 [d, 2H, PCH, ^2J(PH) = 2 Hz], 6.75–8.39 (30 H, aromatic H). $^{31}P\{^1H\}$ NMR ($CDCl_3$): δ 38.0 (s). This complex reacts at room temperature with various electrophilic species.[5] Such reactions involve both enolate arms.

References

1. A. Bader and E. Lindner, *Coord. Chem. Rev.*, **108**, 27 (1991).
2. See, for example, H. D. Empsall, S. Johnson, and B. L. Shaw, *J. Chem. Soc. Dalton Trans.*, 302 (1980); P. Braunstein, D. Nobel, D. Matt, F. Balegroune, S.-E. Bouaoud, and D. Grandjean, *J. Chem. Soc. Dalton Trans.*, 353 (1988); L. Douce and D. Matt, *C. R. Séances Hebd. Acad. Sci. Paris*, **310 II**, 721 (1990).
3. This terminology was introduced by Jeffrey and Rauchfuss: J. C. Jeffrey, and T. B. Rauchfuss, *Inorg. Chem.*, **18**, 2658 (1979). Further examples of hemilabile ligands are given in following publications: D. M. Roundhill, R. A. Bechtold, and S. G. N. Roundhill, *Inorg. Chem.*, **19**, 284 (1980); M. C. Bonnet, I. Tkatchenko, R. Faure, and H. Loiseleur, *Nouv. J. Chim.*, **7**, 601 (1983); P. Braunstein, D. Matt, and Y. Dusausoy, *Inorg. Chem.*, **22**, 2043 (1983).
4. W. Keim, A. Behr, B. Gruber, B. Hoffmann, F. H. Kowaldt, U. Kürschner, B. Limbäcker, and F. P. Sistig, *Organometallics*, **5**, 2356 (1986); U. Klabunde, T. H. Tulip, D. C. Rose, and S. D. Ittel, *J. Organometal. Chem.*, **334**, 141 (1987); J. Dupont, T. M. Gomes Carneiro, M. Luke, and D. Matt, *Quimica Nova*, **11**, 215 (1988).

5. F. Balegroune, P. Braunstein, T. M. Gomes Carneiro, D. Grandjean, and D. Matt, *J. Chem. Soc. Chem. Commun.*, 582 (1989); S. E. Bouaoud, P. Braunstein, D. Grandjean, D. Matt, and D. Nobel, *Inorg. Chem.*, **27**, 2279 (1988).
6. W. Keim, *New J. Chem.*, **11**, 531 (1987).
7. K. A. Ostoja Starzewski and J. Witte, *Angew. Chem. Int. Ed. Engl.*, **6**, 839 (1988).
8. S. E. Bouaoud, P. Braunstein, D. Grandjean, D. Matt, and D. Nobel, *Inorg. Chem.*, **25**, 3765 (1986).
9. P. Braunstein, T. M. Gomes Carneiro, D. Matt, F. Balegroune, and D. Grandjean, *J. Organomet. Chem.*, **367**, 117 (1989).
10. The compound may also crystallize in another space group (space group *P*2$_1$/c, orange crystals), for example, when the crystals are grown from dichloromethane–hexane solutions: F. Balegroune, Thèse de doctorat, 9 Nov. 1990, Université de Rennes, Rennes (France).

23. *N,N*-DIISOBUTYL-2-(OCTYLPHENYLPHOSPHINYL)-ACETAMIDE (CMPO)*

SUBMITTED BY RALPH C. GATRONE,[†] LOUIS KAPLAN,[†]
and E. PHILIP HORWITZ[†]
CHECKED BY ROBERT M. HANDS[‡] and ARLAN D. NORMAN[‡]

(A) $ClCH_2COCl + 2(i\text{-}Bu)_2NH \rightarrow ClCH_2C(O)N(iBu)_2 + (i\text{-}Bu)_2NH_2Cl$

(B) $PhPO(H)OH + (EtO)_3P \rightarrow PhPO(H)OEt + (EtO)_2P(O)H$

(C) $C_8H_{17}Br + Mg \rightarrow C_8H_{17}MgBr$

$2C_8H_{17}MgBr + PhPO(H)OEt \rightarrow Ph(C_8H_{17})P(O)MgBr$

$$+ C_8H_{18} + MgBr(OEt)$$

$Ph(C_8H_{17})P(O)MgBr + ClCH_2C(O)N(iBu)_2 \rightarrow$

$$Ph(C_8H_{17})P(O)CH_2C(O)N(iBu)_2 + MgBrCl$$

Because of the interest aroused by the properties of the title extractant, more commonly referred to as CMPO, for use in chromatographic columns[1] and the TRUEX process,[2] a more detailed description than previously published[3] of the preparation, purification, and physical properties of *N,N*-diisobutyl-2-

* Work performed under the auspices of the Office of Basic Energy Sciences, Division of Chemical Sciences, U.S. Department of Energy under contract number W-31-109-ENG-38. Accordingly the U.S. Government retains a nonexclusive royalty-free license to publish or reproduce the published form of this contribution or allow others to do so, for U.S. Government purposes.
† Chemistry Division, Argonne National Laboratory, Argonne, IL 60439.
‡ Department of Chemistry and Biochemistry, University of Colorado at Boulder, Boulder, CO 80300.

(octylphenylphosphinyl)acetamide is reported. The procedures detailed for the title compound can be applied to the preparation of the *N,N*-dialkyl-[2-alkyl(phenyl)phosphinyl]acetamides for which preliminary syntheses have been reported in the earlier citations.

A. 2-CHLORO-*N,N*-DIISOBUTYLACETAMIDE[4]

Procedure

A 3-L, three-necked, round-bottomed flask equipped with a nitrogen inlet, a pressure-equalizing dropping funnel, a thermometer, and an overhead stirrer is charged with 226 g (2 mol) of chloroacetyl chloride (■ **Caution.** *Chloroacetyl chloride is toxic, corrosive, a lachrymator and is moisture sensitive. All manipulations should be performed in a good fume hood wearing gloves.*) in 500 mL of anhydrous diethyl ether. After cooling to − 20°C (acetone/dry ice), a solution of 516 g (4 mol) of diisobutylamine in 500 mL of anhydrous diethyl ether is added dropwise over 1.5 h at a rate such that the reaction temperature is maintained at 0°C. After addition is complete, the acetone/dry ice bath is removed and replaced with an ice bath. Stirring is continued for an additional 1.5 h, while the temperature is allowed to rise to 10°C. One liter of ice water is added to the reaction mixture and the organic phase is washed successively with 400-mL portions of cold H_2O, H_2O, 0.1 M HCl, H_2O, 1 M K_2CO_3, and saturated NaCl. After drying over anhydrous sodium sulfate, the solvent was concentrated at reduced pressure. The residue was distilled (bath temperature of 80–100°C/0.05 torr) to provide 375–382 g of product (91–93%).

B. ETHYL PHENYLPHOSPHINATE

Procedure

A 2-L, round-bottomed flask equipped with a thermometer and a nitrogen source is charged with 284 g (2 mol) of phenylphosphinic acid (Aldrich) (■ **Caution.** *Phenylphosphinic acid and its solutions are very corrosive. The powder is irritating to mucous membranes and should be avoided. The toxicological properties of the acid and its esters is unknown and it should be handled with gloves in a good fume hood.*) and 365 g (2.2 mol) of triethyl phosphite (■ **Caution.** *Triethyl phosphite is irritating to the eyes, mucous membranes, and upper respiratory tract. A good fume hood and gloves must be used while handling.*) After stirring overnight at 50°C, diethyl phosphite and triethyl phosphite are removed at reduced pressure (bath temperature 80°C/0.05

torr). (■ **Caution.** *The phosphorus-containing waste must be discarded in an acceptable manner.*) The resultant residue is distilled (bath temperature 90–100°C/0.05 torr) to yield 308 g of product (91%).

C. *N,N*-DIISOBUTYL-2-(OCTYLPHENYLPHOSPHINYL)-ACETAMIDE (CMPO)

Procedure

A 3-L, three-necked, round-bottomed flask equipped with a reflux condenser, a pressure-equalizing dropping funnel, an overhead stirrer, and a nitrogen source is charged with 32.1 g (1.32 mol) of magnesium turnings. A solution of 5 mL of 1-bromooctane in 100 mL of anhydrous diethyl ether (or anhydrous tetrahydrofuran) is added and the flask is warmed gently with a heat gun until the reaction is initiated. (The use of iodine crystals is not recommended to initiate the reaction because a lower yield with greater purification difficulties is observed.) An additional 500 mL of diethyl ether (or THF) containing 244 g (1.26 mol) of 1-bromooctane is added dropwise to maintain a steady reflux. After the addition is complete, the resultant dark solution is refluxed for 40 min using a heating mantle. After cooling to room temperature, a solution of 102 g (0.6 mol) of ethyl phenylphosphinate in 100 mL of ether is added dropwise. After addition is complete, 400 mL of benzene (■ **Caution.** *Benzene is a suspected carcinogen. A good fume hood and gloves are mandatory. As the purpose of the benzene is to raise the boiling point and to increase the solubility of the phosphorus products, toluene may be substituted to reduce exposure to carcinogenic chemicals.*) are introduced and the resultant solution is refluxed (< 80°C) for 4 h. After cooling to room temperature, a solution of 123 g (0.6 mol) of 2-chloro-*N,N*-diisobutylacetamide in 100 mL of benzene (or toluene) is added dropwise over 30 min. The resultant reaction mixture is refluxed for 4 h.

After cooling to room temperature, 100 mL of H_2O is slowly introduced, followed by 420 mL of 2 M HCl. After separating the phases, the organic phase is washed with 250 mL of 1 M HCl followed by water until the resultant aqueous phase is neutral. The organic phase is concentrated at reduced pressure. The residue is stirred at room temperature for several hours with 200 mL of 2 M NaOH to hydrolyze any organophosphorus acids and to remove several unidentified acidic by-products. After dilution with 200 mL of water and 600 mL of heptane, the organic phase is washed with 200 mL of 1 M NaOH, followed by water, until the aqueous phase is neutral. The organic phase is concentrated at reduced pressure (0.01 torr) to provide 169 g of a viscous yellow oil, which is not of satisfactory purity for use in a chromatographic or extraction process.[5]

Purification

Crystallization. The crude CMPO (138 g) is dissolved in 500 mL of heptane and is cooled to $-15°C$ in a freezer. Several seed crystals are added after a few hours. After several days the crystals are collected by vacuum filtration on a Buchner funnel and washed with 200 mL of cold heptane. Residual solvent is removed at reduced pressure in a desiccator overnight to provide a 119 g of white, somewhat deliquescent, solid (mp 44–45°C). This is the preferred method of purifying the product.

Distillation. The crude CMPO (169 g) is distilled through a short, wide, connecting tube into a dry-ice cooled receiver at reduced pressure (bath temperature of 220°C/0.001 torr). A significant degree of decomposition is observed during the distillation. The impurities from the distillate are removed by washing a heptane solution with 5% Na_2CO_3.

Mercury Salt.[6] The impure CMPO (60 g) is dissolved in a solution of 150 mL of benzene (■ **Caution.** *Benzene is a suspected carcinogen. A good fume hood and gloves are mandatory.*) and 150 mL of heptane. The resultant solution is contacted with 225 mL of an aqueous solution containing 75 g mercuric nitrate dihydrate (■ **Caution.** *Mercury compounds are toxic. A good fume hood and gloves are mandatory.*) and 15 mL of concentrated HNO_3 for 7 h at 50°C. After cooling to -5 to 0°C overnight, the liquid phases are discarded by filtration (or centrifugation/decantation). The solid salt is washed with a mixture of 200 mL HNO_3, 150 mL of benzene, and 50 mL of heptane. The liquid phases are discarded (■ **Caution.** *Traces of mercury salt might be present in the solutions and all waste must be discarded in a manner appropriate for heavy metal contamination.*) The solid salt is washed with a warm solution of benzene and heptane while stirring with a spatula. After cooling to -5 to 0°C, the mixture is centrifuged and the liquid phase is discarded.

The resultant solid mercury salt is contacted with a mixture of 23 g potassium cyanide (■ **Caution.** *Cyanide compounds are extremely poisonous.*) and 48 g potassium carbonate in 350 mL of water and 200 mL of heptane. After stirring for several hours, the phases were separated. The heptane phase is stirred vigorously with 12.5 g KCN and 24 g K_2CO_3 in 175 mL water. After separating the phases, the organic phase is washed successively with 250 mL portions of 0.5 M K_2CO_3, H_2O, 0.5 M HNO_3, 0.5 M NaOH, and H_2O. The solvent is concentrated at reduced pressure to provide 52 g of the title compound.

Properties

NMR: The decoupled phosphorus-31 signal is recorded in $CDCl_3$ at 37.67 ppm downfield from external 85% phosphoric acid.[7] The proton spectrum is recorded in CDCl$_3$ with resonances at 7.8 (ortho H), 7.49 (meta and para H), 3.27–3.00 (P–C$\underline{H}_2$–CO and (H$_2$C–N–C$\underline{H}_2$), 2.30 (heptyl-C$\underline{H}_2$–P), 1.8 (C$\underline{H}$), 1.6–1.1 (methylene envelope), and 0.85–0.79 ppm (C$\underline{H}_3$) downfield from internal trimethylsilane. Proton-decoupled carbon-13 resonances in CDCl$_3$ were observed at 165.9 (C=O), 131.9 (aromatic C1, J_{C-P} = 94.3 Hz), 131.5 (para), 130.5 (meta, $J_{C-C-C-P}$ = 12.1 Hz), 128.2 (ortho, $J_{C-C-C-P}$ = 11.3 Hz), 56.6 (syn-amide CH$_2$), 54.0 (anti-amide CH$_2$), and 37.6 ppm (P–CH$_2$–CO, J_{C-P} = 60.4 Hz). IR: Absorbances were recorded on a thin film at 3047 (phenyl), 2950, 2920, 2860 (CH), 1630 (C=O), 1462, 1435, 1200 cm^{-1}. UV: In hexane, λ_{max} 271, 264 (ε_{max} 660), 258, 251 nm. MS: m/e 407 (M$^+$, 11%), 353, 308, 295, 279, 252, 154 (100%) at 70 eV. GC: Analyses are performed on a Hewlett-Packard 5890 gas chromatograph equipped with a flame ionization detector, a split-splitless injector, and a Hewlett-Packard 3393 computing integrator. A fused-capillary column, DB-5 (J&W Scientific) 15 m × 0.25 mm i.d. (0.25 μm film thickness) is employed. Helium is the carrier gas at 2 mL/min. The temperature profile used is 100°C for 1 min, increased to 260°C, at 10°C/min, and held for 10 min. The detector and injector temperatures are set at 300°C. Injections of 1 μL are made using a split ratio of 35:1. Under these conditions, CMPO has a retention time (t_R) of 17.54 ± 0.02 min.

All samples (100 mg) are derivatized with a diethyl ether solution of diazomethane. (■ **Caution.** *Diazomethane is an explosive gas. It is prepared by the base decomposition of a commercially available reagent (Diazald^{TM}) as an ether solution in fire-polished glassware. A good fume hood and gloves are mandatory.*) The solvent and excess diazomethane are evaporated in a warm water bath. The derivatized sample is prepared for analysis by adding 1 mL of CH$_2$Cl$_2$, quantitatively transferring the sample to a 10-mL volumetric flask, and diluting to 10 mL with CH$_2$Cl$_2$. LC: High-performance liquid chromatography is performed on a Waters Associates liquid chromatograph equipped with a multisolvent delivery pump, a differential refractive index detector (sensitivity-64), a programmable multiwavelength UV-VIS detector (262 nm). A reversed-phase μBondapak (Waters) stainless-steel (30 cm × 3.9 mm i.d.o. column maintained at 32°C is used. The mobile phase (70:29.5:0.5 acetonitrile: water: triethylamine) is continuously sparged with helium. A flow rate of 1 mL/min is used. These conditions provide a retention time (t_R) of 9.20 min for CMPO.

References

1. E. P. Horwitz, M. L. Dietz, D. M. Nelson, J. J. LaRosa, and W.D. Fairman, *Anal. Chem. Acta*, **238**, 263 (1990).
2. D. G. Kalina, E. P. Horwitz, L. Kaplan, and A. C. Muscatello, *Sep. Sci. Technol.*, **17**, 859 (1981), E. P. Horwitz, D. G. Kalina, L. Kaplan, G. W. Mason, and H. Diamond, *Sep. Sci. Technol.*, **18**, 1261 (1982).

3. R. C. Gatrone, L. Kaplan, and E. P. Horwitz, *Solvent Extr. Ion Exch.*, **5**, 1075 (1987).
4. M. Neeman, *J. Chem. Soc.*, 2525 (1955).
5. E. P. Horwitz, R. Chiarizia, and R. C. Gatrone, *Solvent Extr. Ion Exch.*, **6**, 93 (1988).
6. N. C. Schroeder, L. D. Mclsaac, and J. F. Krupa, *ENICO*-1026, Idaho Falls, ID, 1980.
7. R. C. Gatrone and P. R. Rickert, *Solvent Extr. Ion Exch.*, **5**, 1117 (1987).

24. ARSENIC(III) CHLORIDE

SUBMITTED BY SUSHIL K. PANDEY,* ALEXANDER STEINER,*
and HERBERT W. ROESKY*
CHECKED BY SMURUTHI KAMEPALLI[†] and ALAN H. COWLEY[†]

Arsenic(III) chloride is an important starting material for synthetic inorganic and element-organic chemistry. Usually it is prepared by passing a chlorine gas stream over arsenic metal[1] or by the reaction of arsenic(III) oxide, disulfurdichloride, and chlorine gas.[2] Industrially, arsenic(III) chloride is obtained by the reaction of arsenic(III) oxide and hydrochloric acid.[3] These methods are time consuming and complicated on a laboratory scale.

   In view of the foregoing, the reaction of As_2O_3 with $SOCl_2$ (sulfinyl chloride) has been investigated and found to be a suitable method for the preparation of pure arsenic(III) chloride. The yield is quantitative and the preparation can be accomplished in a short time on laboratory scale. Although this reaction has been described previously, the product was not obtained pure and the yield was not quantitative.

Procedure

$$As_2O_3 + 3SOCl_2 \rightarrow 2AsCl_3 + 3SO_2 \uparrow$$

$$197.8 \quad 3 \times 118.9 \quad 2 \times 181.3 \quad 3 \times 64.0$$

■ **Caution.** *Because of the poisonous nature of arsenic compounds and sulfur dioxide, all manipulations must be carried out in an efficient fume hood.* 100 g (0.505 mol) of As_2O_3 are placed in a 500-mL, two-necked, round-bottomed flask, or a one-necked flask fitted with a Claisen head with a dropping funnel, a water-cooled reflux condenser, and a magnetic stirrer

* Institut für Anorganische Chemie der Universität Göttingen, Tammannstr. 4, 37077 Göttingen, Germany.
† Department of Chemistry and Biochemistry, The University of Texas at Austin, Austin, TX 78763.

(Fig. 1). A current of dry nitrogen should be maintained during all manipulations. The reflux condenser is connected with a drying tube containing calcium chloride and three washing flasks; two are filled with a solution of calcium hydroxide and one is left empty for safety purposes, (caused by back suction during the final stages of reaction, see figure). The SOCl$_2$ (200 mL, 2.75 mol) (in excess) is added portion-wise through the dropping funnel with stirring. After the addition of 50 mL of SOCl$_2$ the reaction should have started, before any additional SOCl$_2$ is added (an excess of SOCl$_2$ is necessary because it is also used here as a solvent). If the reaction does not start, the reaction mixture must be heated to 50–60°C, whereupon the reaction starts within 5–10 min, as indicated by the evolution of SO$_2$ gas. The reaction is completed when the solution becomes clear. To complete the reaction, it is stirred further for 1 h. The solution is fractionally distilled over a 20-cm Vigreux column. The fraction collected between 127 and 133°C is pure arsenic(III) chloride. The fraction below 120°C is mostly excess of SOCl$_2$, which can be used again for further preparations of AsCl$_3$. The reaction yielded 155.6 g (85 %) of arsenic(III) chloride.

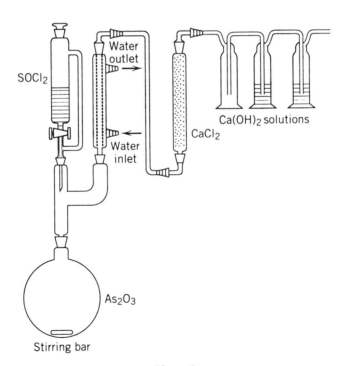

Figure 1

Anal. Found (calc.): As, 41.31 (41.33); Cl, 58.34 (58.67).

Disposal. The reaction residue of $AsCl_3$ is treated with hot water under a well-ventilated fume hood to give solid As_2O_3, which is recovered by filtration.

$$2AsCl_3 + 3H_2O \rightarrow As_2O_3 + 6HCl$$

Properties. Molecular weight 181.28, colorless viscous liquid, volatile, gives white fumes when exposed to air, extremely poisonous, it solidifies between 20 and $-13°C$ as colorless shiny crystals (bp 131.4°C). It is soluble in water and HCl. The density is 2.16 g/mL.

References

1. G. P. Baxter, W. E. Shaefer, M. J. Dorcas, and E. W. Scripture, *J. Am. Chem. Soc.*, **57**, 851 (1935).
2. J. R. Paddington, *J. Chem. Soc. (London)*, 2577 (1929).
3. W. Biltz and A. Sapper, *Z. Anorg. Allgem. Chem.*, **203**, 277 (1932).
4. H. B. North and A. M. Hageman, *J. Am. Chem. Soc.*, **35**, 352 (1913).
5. G. Brauer, *Handbuch der Präparativen Anorganischen Chemie*, Enke Stuttgart, 572 (1981).

25. TRIS(TRIMETHYLSILYL)ARSINE AND LITHIUM BIS(TRIMETHYLSILYL)ARSENIDE

SUBMITTED BY RICHARD L. WELLS,* MARK F. SELF,*
JAMES D. JOHANSEN,* JANEEN A. LASKE,* STEVEN R. AUBUCHON,*
and LEONIDAS J. JONES, III*
CHECKED BY A. H. COWLEY[†] and S. KAMEPALLI[†]

Tris(trimethylsilyl)arsine and lithium bis(trimethylsilyl)arsenide are valuable reagents for dehalosilylation and salt elimination reactions, respectively. Each compound reacts with a wide variety of metal halides to form metal–arsenic bonds.[1, 2] The syntheses reported herein are a modification of the published procedures of Becker et al.,[3] designed to minimize the use of Schlenk techniques and to allow researchers to prepare two very useful compounds with a minimum of danger. However, since these *compounds do*

* Department of Chemistry, Duke University, Durham, NC 27708.
[†] Department of Chemistry and Biochemistry, The University of Texas at Austin, Austin, TX 78712.
[‡] DME = 1,2-dimethoxyethane.

burn spontaneously in air and are very toxic, great care must be taken in their synthesis and subsequent manipulations. Adequate knowledge of both Schlenk and vacuum techniques is required.[4]

The following procedure can readily be adapted for the preparation of $(Me_3Si)_3P$ and ether-free $LiP(SiMe_3)_2$.[5, 6]

A. TRIS(TRIMETHYLSILYL)ARSINE

$$3\,Na/K + As \xrightarrow{DME^{\ddagger}} \text{``}(Na/K)_3As\text{''} \xrightarrow[-3\,(Na/K)Cl]{3\,Me_3SiCl} (Me_3Si)_3As$$

■ **Caution.** *Sodium–potassium alloy reacts violently with water and may ignite upon exposure to air. Tris(trimethylsilyl)arsine is a toxic, highly pyrophoric liquid which may form highly toxic arsine gas (AsH₃) upon reaction with air or water. This procedure and the following cleanup must be performed in a well-ventilated fume hood. A Class D metal fire extinguisher should be on hand throughout the synthesis since an excess of sodium–potassium alloy is employed. The flammable solvent, DME, must be dried and deoxygenated by distillation from sodium benzophenone under nitrogen. Although this synthesis utilizes a minimum of Schlenk techniques, one must still be familiar with handling air-sensitive/pyrophoric compounds.*

The reaction vessel consists of a 1-L flask with 200-mL and 250-mL round-bottomed flask blanks attached via Teflon® valves* and a gas inlet (see Fig. 1). The neck of this apparatus consists of an inner 34.5 RODAVISS® joint† with a drip tip (the RODAVISS® joints are threaded and allow the components of the apparatus to be securely attached). A fritted filter is constructed with a 34.5 outer RODAVISS® joint inlet and a 24.4 inner RODAVISS® joint outlet with a drip tip. The receiving flask is a 2-L, round-bottomed flask which had been formed into a "mushroom" shape by heating the upper half and allowing it to partially collapse (see Fig. 2). A Teflon® valve is attached at the upper curve of the "mushroom." Each time this reaction is performed, it is imperative that the o-rings on both the Teflon® valves and the RODAVISS® joints be replaced because they deteriorate, and also that the glassware be annealed by a qualified glassblower to ensure integrity of the components.

---

* Kontes, 1022 Spruce Street, Vineland, NJ 08360.
† Chemglass, Inc., 3861 North Mill Road, Vineland, NJ 08360.

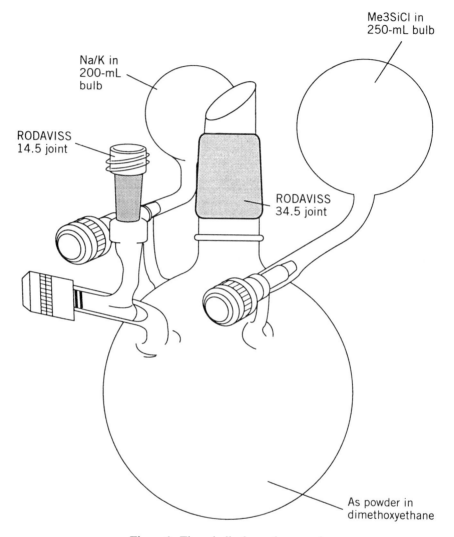

Me3SiCl in
250-mL bulb

Na/K in
200-mL
bulb

RODAVISS
14.5 joint

RODAVISS
34.5 joint

As powder in
dimethoxyethane

Figure 1. Three-bulbed reaction vessel.

Procedure

All glassware and reagents must be scrupulously dried and deoxygenated prior to use. All ground-glass connections should be lubricated with an inert and highly heat-stable grease such as Apiezon T-grease.[‡] A continuous and vigorous flow of argon must be maintained when the stirring rod is removed,

[‡] Apiezon Products, M&I Materials Ltd, P.O. Box 136, Manchester M60 1AN, England.

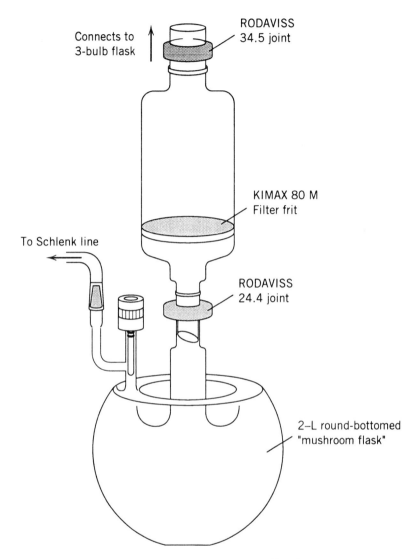

Figure 2. "Mushroom" flask with attached filtration apparatus.

the filtration glassware is assembled to the reaction flask, and the filtration glassware is removed from the receiving flask.

In an inert-atmosphere glove box under an argon atmosphere, the 200-mL upper bulb is charged with 20 g of sodium–potassium alloy (44%: 56%)[§]

§ Strem Chemicals, Inc., 7 Mulliken Way, Newburyport, MA 01950.

using a disposable Pasteur pipette. The liquid alloy should be added quickly and in small amounts to the bulb to ensure that it does not back up into the valve stem. To the 250-mL bulb is added 77 g of chlorotrimethylsilane.[¶] After closing the valves to the upper bulbs, finely powdered arsenic (99.9995% pure)[@] (14 g) is placed in the 1-L flask and covered with 800 mL DME. A TRUBORE®[#] stirrer equipped with a glass blade (a glass blade must be employed as Teflon will decompose in the reaction mixture) is fitted to the flask and lubricated with Stir-Lube®.[#] To minimize vibration and alignment problems, the strirrer shaft is connected to the motor unit via a FLEXIBLE SHAFT®.[#]

■ **Caution.** *The reaction must be monitored continually while it is stirred, as the stir rod must be lubricated periodically to prevent it from freezing up.* The apparatus is placed in an oil bath in a fume hood and a condenser, flushed with argon, is attached to the gas inlet (see Fig. 3). Under a slight flow of argon, the solvent is stirred vigorously and heated to 76°C, at which point the sodium–potassium alloy is added over a 1.5-h period. The addition is controlled by carefully heating the gas above the alloy with a "heat gun" and allowing the increased pressure to force down the alloy. The solution becomes black before addition of the alloy is complete. The solution is heated at reflux and stirred for 24 h.

The solution is allowed to cool to room temperature, then the flask is placed in an ice water bath. The stir rate is increased and chlorotrimethyl-silane[¶] is added over a 1.5-h period while adding ice to the water bath to maintain a temperature of 22°C. Upon nearing complete addition of the chlorotrimethylsilane, the solution becomes extremely viscous and the color changes from black to pale gray. The mixture is heated at reflux and stirred vigorously for 24 h.

Upon cooling the reaction mixture to room temperature, the gas inlet on the vessel is closed and the condenser removed. Under a steady flow of argon, the TRUBORE® stir rod is removed and an inner RODAVISS® 34.5 cap is placed over the mouth of the flask. After the assembled frit and receiving flask have been evacuated and back-filled with argon five times, under a vigorous flow of argon, both the stopper in the RODAVISS® 34.5 joint of the frit and the cap on the matching joint of the reaction flask are removed and the frit assembled to the flask. The entire filtration apparatus is then inverted. After the first filtration is complete, the entire assembly is evacuated and the reaction flask is partially placed in a liquid nitrogen bath. The mushroom-shaped receiving flask allows the filtration apparatus to be laid horizontally

[¶] Aldrich Chemical Company, Inc., 1001 West Saint Paul Avenue, Milwaukee, WI 53233.
[@] Atomergic Chemmetals Corp., 222 Sherwood Avenue, Farmingdale, NY 11735.
[#] Ace Glass, Inc., P.O. Box 688, 1430 Northwest Boulevard, Vineland, NJ 08360.

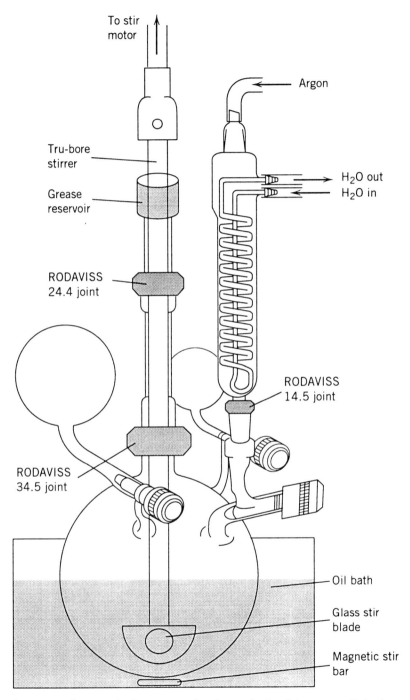

To stir
motor

Tru-bore
stirrer

Grease
reservoir

RODAVISS
24.4 joint

RODAVISS
34.5 joint

Argon

H$_2$O out
H$_2$O in

RODAVISS
14.5 joint

Oil bath

Glass stir
blade

Magnetic stir
bar

Figure 3. Fully assembled apparatus for the synthesis of the organosilylarsine.

without allowing the pale yellow filtrate to flow back into the neck of the frit. The DME which condenses in the reaction flask is thawed and used to wash this flask and the salt cake on the frit. This procedure is repeated three additional times with approximately 150-mL portions of DME until the salt cake is clean. Under a blanket of argon, the fritted filter is removed from the receiving flask and the vessel is capped. The solvent is removed at room temperature in vacuo (2.5×10^{-3} torr) and trapped at $-196°C$ to leave a pale yellow liquid, impure tris(trimethylsilyl)arsine. Under a constant flow of argon, a Teflon® stirring bar is placed in the vessel and a 20-cm Vigereux fractional distillation column is equipped with a water-cooled condenser and thermometer, and an attached 200-mL one-bulbed receiving flask equipped with a Teflon® joint sidearm is connected to the reaction vessel via the RODAVISS® 34.5 joint. The sidearm of the receiving flask is attached to the vacuum line. The entire assembly is evacuated and the contents of the reaction vessel are carried through three freeze (liquid nitrogen)–pump–thaw cycles to insure that there are no dissolved gases (*Note: The presence of dissolved gases causes the tris(trimethylsilyl)arsine to "bump" seriously during the distillation.*) The reaction vessel is placed in an oil bath atop a stirring hot plate and with adequate stirring, the tris(trimethylsilyl)arsine is purified by fractional distillation (bp $65–71°C/2.5 \times 10^{-3}$ torr). Yield: 34 g (62%).*

■ **Caution.** *The residues on the glassware and the frit may contain sodium–potassium alloy and/or tris(trimethylsilyl)arsine and should not be exposed to air until disposal. tert-Butyl alcohol and isopropyl alcohol can be used to destroy these pyrophoric compounds; however, this should be performed in a fume hood, as arsine gas may be generated. The solid on the frit can be mixed with class D metal fire extinguisher and subsequently treated with either of the aforementioned alcohols.*

Properties

Tris(trimethylsilyl)arsine is a colorless liquid which melts slightly below room temperature, and is stable at room temperature under an inert atmosphere or in a degassed, flame-sealed ampoule. The ^1H NMR spectrum in C_7D_8 (reference δ 2.09) consists of a single resonance at δ 0.30 and the ^{13}C NMR spectrum shows a single peak at δ 4.31. Solubility: very soluble in benzene, toluene, pentane, THF, and diethyl ether.

---

* The checkers obtained 59%.

B. LITHIUM BIS(TRIMETHYLSILYL)ARSENIDE

$$(Me_3Si)_3As \xrightarrow[- Me_4Si]{MeLi/THF} LiAs(SiMe_3)_2$$

■ **Caution.** *Lithium bis(trimethylsilyl)arsenide is very sensitive to moisture and can react explosively upon exposure to air. Methyllithium is corrosive and reacts violently with water. Both compounds should be manipulated in a dry argon or nitrogen atmosphere. Lithium bis(trimethylsilyl)arsenide should not be heated above 80°C as it begins to decompose.*

The reaction flask consists of a 100-mL round-bottomed flask blank attached to a 250-mL round-bottomed flask through a Teflon® valve. The 250-mL flask has a 24/40 ground glass joint and a gas inlet equipped with a Teflon® valve.

Procedure

The upper 100-mL bulb is charged with 4.85 mL of 1.4 M methyllithium in diethyl ether. This is best accomplished in an inert-atmosphere glove box, but can also be done in a fume hood using Schlenk transfer techniques. Approximately 10 mL of THF, freshly distilled from sodium benzophenone, is added to this solution to ensure a quantitative transfer. After closing the valve to the upper bulb, 2 g of tris(trimethylsilyl)arsine in 40 mL of THF is added to the main bulb and a stir bar is added. A greased 24/40 male stopper is used to seal the vessel and it is then attached to a vacuum line. The solution is cooled in an ice-water bath to 0°C with stirring. The methyllithium solution is added dropwise over a 15-min period, producing a bright yellow solution. After stirring for 2 h, the solution is allowed to warm to room temperature and stirred for an additional 2 h. The volatiles are then removed in vacuo, leaving a yellow, solid residue. The flask is placed in an oil bath and heated to 70°C under dynamic vacuum (2.5×10^{-3} torr). The solid slowly becomes bright white and after 48 h, THF-free lithium bis(trimethylsilyl)arsenide is recovered in nearly quantitative yield. Yield: 1.45 g (94%).

Properties

Lithium bis(trimethylsilyl)arsenide is a white powder which is stable at room temperature under an inert atmosphere. It is important for stoichiometric reasons to ascertain whether there is any residual THF. This can be readily determined by ^{1}H NMR spectroscopy. The ^{1}H NMR spectrum in C_6D_6 (reference δ 7.15) consists of a single resonance at δ 0.62. Solubility: very soluble in THF and diethyl ether, and minimally soluble in nonethereal solvents.

References

1. R. L. Wells, *Coord. Chem. Rev.*, **112**, 273 (1992).
2. A. H. Cowley and R. A. Jones, *Angew. Chem. Int. Ed. Engl.*, **28**, 1208 (1989).
3. G. Becker, G. Gutekunst, and H. J. Wessely, *Z. Anorg. Allg. Chem.*, **462**, 113 (1980).
4. D. F. Shriver, *The Manipulation of Air-Sensitive Compounds*, McGraw-Hill, New York, 1969.
5. G. Becker and W. Holderich, *Chem. Ber.*, **108**, 2484 (1975).
6. G. Becker, H. Schmidt, G. Uhl, and W. Uhl, *Inorg. Syn.*, **27**, 243 (1990).

26. STERICALLY HINDERED ARENE CHALCOGENOLS

SUBMITTED BY MANFRED BOCHMANN*‡ and KEVIN J. WEBB*
CHECKED BY M. A. MAILK†, J. R. WALSH,† and P. O'BRIEN†

Arene chalcogenols ArEH (E = S, Se, Te) are excellent starting materials for introducing chalcogenolato ligands into a wide variety of metal complexes, particularly if high steric hindrance is required to produce complexes of restricted coordination number and low degrees of association. However, whereas thiols such as $2,4,6-t-Bu_3C_6H_2SH$ are well known,[1] the oxidation-sensitive selenols have been less readily available, while arene tellurols are thermally unstable and appear not to have been isolated[2] until recently.[3] The synthetic procedures described here for 2,4,6-trisubstituted arene chalcogenols are equally applicable for the synthesis of thiols, selenols, and tellurols in high yields and are easily scaled up if required. The thiol $2,4,6-t-Bu_3C_6H_2SH$ was first prepared from the Grignard reagent.[1] However, difficulties are sometimes encountered in the preparation of $t-Bu_3C_6H_2MgBr$; these are eliminated by using the aryllithium route described here. Although arene selenols and tellurols can, in principle, be obtained by treating aryllithium solutions with elemental chalcogen followed by acidification, as described here for $t-Bu_3C_6H_2SH$, this procedure is not satisfactory for selenols and tellurols. The preferred route to these compounds, therefore, uses diaryl dichalcogenides as starting materials which undergo clean reductive cleavage to ArE^- under very mild conditions.

General Procedure

■ **Caution.** *Chalcogenophenols have an unpleasant odor and are toxic.* H_2S, H_2Se, *or* H_2Te *may be liberated on treatment with acid or exposure*

* School of Chemical Sciences, University of East Anglia, Norwich, NR4 7TJ, UK.
† Department of Chemistry, Queen Mary and Westfield College, University of London, London E1 4NS, UK.
‡ New address: School of Chemistry, University of Leeds, Leeds LS2 9JT, UK.

to the open air. Compounds should therefore be handled under inert gas in a well-ventilated hood. Contaminated glassware should be treated with sodium hypochlorite (bleach) solution for several hours and thoroughly rinsed with water and acetone before removal from the fume hood.

Unless otherwise stated, all operations are carried out in flame-dried glassware under an inert gas atmosphere using standard Schlenk techniques.[4] Solvents were dried over sodium (light petroleum, bp 40–60°C) or sodium-benzophenone (diethyl ether, THF). Lithium triethylhydroborate (1 −) and hydrogen tetrafluoroborate (1 −) diethyl ether complex were obtained from Aldrich and used as supplied.

A. 2,4,6-TRI-*tert*-BUTYLBENZENETHIOL*

$$Ar''Li + 1/8\ S_8 \rightarrow Ar''SLi$$

$$Ar''SLi + HCl \rightarrow Ar''SH$$

$$Ar'' = 2,4,6\text{-}t\text{-}Bu_3C_6H_2$$

Procedure

A 250-mL, three-necked reaction flask equipped with a 100-mL pressure-equalized dropping funnel and a magnetic stirring bar and connected via a stopcock adaptor to the inert gas supply of the vacuum line is charged with bromo-2,4,6-tri-*tert*-butylbenzene[5, 6] (8.0 g, 22.9 mmol). Light petroleum (50 mL) and THF (12 mL) are added, and the solution is cooled to − 78°C. Butyllithium in hexane (2.5 M solution, 10 mL, 25 mmol) is added dropwise with rapid stirring to give a white precipitate of Ar''Li. Stirring is continued for 2.5 h at − 78°C before the mixture is allowed to warm to room temperature. The solvent is then removed in vacuo and the white residue washed with light petroleum (2 × 25 mL), dried in vacuo for 2 h, and subsequently dissolved in THF (100 mL). Yields of the lithium reagent are typically 80–90%. The solution is cooled to − 78°C and sulfur (0.79 g, 24.6 mmol) is added in small portions to give an orange-brown solution which is warmed to room temperature and heated to reflux for 1 h. Small amounts of Li[AlH₄] are added to the hot solution until the mixture is almost colorless. After cooling to room temperature, dilute hydrochloric acid (0.1 M, 90 mL) is added dropwise.

■ **Caution.** *Care should be taken on adding the first drops of the acid because any remaining Li[AlH₄] reacts vigorously with liberation of hydrogen.* The solvent is removed in vacuo, leaving a light brown solid residue which is extracted with diethyl ether (3 × 100 mL) and filtered. The combined filtrates

---

* See cautionary note under *General Procedure*.

are evaporated in vacuo to give Ar″SH as an air-stable white solid (5.8 g, 85%). The compound is recrystallized from hot ethanol (150 mL) to give white crystals (4.9 g, 72% based on Ar″Br). More may be recovered from the mother liquor.

Anal. Calcd. for $C_{18}H_{30}S$: C, 77.63; H, 10.86; S, 11.51. Found: C, 77.73; H, 10.88; S, 11.37; mp 175–176°C (lit.[1]: 180–181°C). ^{1}H NMR (CCl$_4$): δ 1.30 (s, 9H), 1.58 (s, 18H), 3.35 (s, 1H), 7.28 (s, 2H).

B. 2,4,6-TRI-*tert*-BUTYLBENZENESELENOL*

$$Ar''_2Se_2 + 2LiBHEt_3 \rightarrow 2Ar''SeLi + 2BEt_3 + H_2$$

$$Ar''SeLi + HBF_4 \cdot Et_2O \rightarrow Ar''SeH + LiBF_4$$

$$Ar'' = 2,4,6\text{-}t\text{-}Bu_3C_6H_2$$

A 100-mL, three-necked flask equipped with a magnetic stirring bar and connected to the inert gas supply of the vacuum line via a stopcock adaptor is charged with $Ar''_2Se_2{}^6$ (0.685 g, 1.06 mmol). A rubber septum is attached to one neck of the flask, THF (40 mL) is injected, and the solution is kept, as far as possible, in the dark. A solution of Li[BHEt$_3$] in THF (Super-Hydride®, 1 M, 2.1 mL, 2.1 mmol) is added via syringe. The color turns from orange to pale yellow. After stirring at room temperature for 30 min, the mixture is cooled to $-78°C$ and treated dropwise by syringe with hydrogen tetra-fluoroborate $(1-)$ diethyl ether complex (1.0 g, 6 mmol). Stirring is continued while the mixture is allowed to warm to room temperature. The solvent is removed in vacuo and the residue is taken up in light petroleum (30 mL) and filtered through a stainless-steel cannula equipped with a small filter paper into a second 100-mL flask under inert gas. The solution is concentrated to about 10 mL and cooled to $-20°C$ overnight to give Ar″SeH as white crystals (0.55 g, 1.69 mmol, 80%). The crude petroleum extract of this compound may be used without isolation for further reactions with metal amides (see Sections 5 and 6).

Anal. Calcd. for $C_{18}H_{30}Se$: C, 66.44; H, 9.29. Found: C, 66.86; H, 9.34; mp 172–173°C.

Properties

In contrast to Ar″SH, Ar″SeH is air-sensitive and must be handled and stored under inert gas. Air oxidation or treatment with oxidizing agents such

---

* See cautionary note under *General Procedure*.

as aqueous alkali $K_3[Fe(CN)_6]$ gives the diselenide $Ar_2''Se_2$. ^1H NMR of t-$Bu_3C_6H_2SeH$ ($CDCl_3$): δ 1.40 (s, 9H), 1.66 (s, 18H), 1.92 (s, 1H), 7.35 (s, 2H).

C. 2,4,6-TRI-*tert*-BUTYLBENZENETELLUROL*

The same apparatus and procedure as in Section 25.B is used. To a stirred solution of $Ar_2''Te_2$[6] (0.43 g, 0.58 mmol) in THF (30 mL) at room temperature is added via syringe a solution of $Li[BHEt_3]$ in THF (1 M, 1.2 mL, 1.2 mmol). The color changes from deep red to green. After stirring for another 15 min the solvent is removed in vacuo and the solid residue suspended in light petroleum (60 mL), stirred, and cooled to $-78°C$. Hydrogen tetrafluoroborate(1 $-$) diethyl ether complex (85% solution, 1 g, 5.2 mmol) is added slowly via syringe. Stirring is continued at $-78°C$ for 10 min before the mixture is allowed to warm to $-45°C$. When the green solid has dissolved, the solution is filtered cold via a stainless-steel filter cannula into a second 100-mL flask cooled to $-78°C$. The solution is concentrated at $-40°C$ to approximately 10 mL. Colorless crystals of $Ar''TeH$ are formed. More solid precipitates on keeping the solution at $-78°C$ for several hours. After filtration and drying in vacuo at $-40°C$, the product is obtained as a colorless to pale-pink crystalline solid. 2,4,6-Trimethylbenzenetellurol and 2,4,6-triisopropylbenzenetellurol are made by analogous procedures from $(Me_3C_6H_2Te)_2$ and $(i$-$Pr_3C_6H_2Te)_2$, respectively.[3] The i-$Pr_3C_6H_2TeH$ is highly soluble and forms a solid only after complete removal of any residual solvent.

Properties

Arene tellurols are thermally sensitive and decompose slowly above $-30°C$ and rapidly above $0°C$. Direct light should be excluded. Although the compounds can be stored at $-78°C$, solutions in light petroleum are best prepared without isolation and used immediately for subsequent reactions with metal complexes. While protonation of $Ar''TeLi$ with $H[BF_4] \cdot Et_2O$ proceeds smoothly in light petroleum to give $Ar''TeH$, extensive decomposition is observed if THF is used as solvent for this reaction. The purity of the products is checked by ^1H NMR in $CDCl_3$ below $-20°C$; characteristic are the high-field signals for Te-H.[3] t-$Bu_3C_6H_2TeH$: δ -1.25 (s, 1H), 1.24 (s, 9H), 1.58 (s, 18H), 7.40 (s, 2H).

---

* See cautionary note under *General Procedure*.

References

1. W. Rundel, *Chem. Ber.*, **101**, 2956 (1968).
2. H. Gysling in S. Patai, and Z. Rappoport, *The Chemistry of Organic Selenium and Tellurium Compounds*, Vol. 1, Wiley, Chichester 1986, p. 650; J. E. Drake and R. T. Hemmings, *Inorg. Chem.*, **19**, 1879 (1980).
3. M. Bochmann, A. P. Coleman, K. J. Webb, M. B. Hursthouse, and M. Mazid, *Angew. Chem.*, **103**, 975 (1991); *Angew. Chem. Int. Ed. Engl.*, **30**, 973 (1991).
4. D. F. Shriver and M. A. Drezdzon, *The Manipulation of Air-Sensitive Compounds*, 2nd ed., Wiley, New York, 1986.
5. D. E. Pearson, M. G. Frazer, V. S. Frazer, and L. C. Washburn, *Synthesis*, 621 (1976).
6. For preparative details see preceding section.

27. TRIS(TRIMETHYLSILYL)SILYL LITHIUM TRIS(TETRAHYDROFURAN), LITHIUM TRIS(TRIMETHYLSILYL)SILYLTELLUROLATE BIS(TETRAHYDROFURAN), AND TRIS(TRIMETHYLSILYL)SILYLTELLUROL*

CHECKED BY MANFRED BOCHMANN‡ and
GABRIEL C. BWEMBYA‡
SUBMITTED BY PHILIP J. BONASIA† and JOHN ARNOLD†

At present there are few examples of isolable, well-characterized sources of tellurolate anions (RTe⁻).[1] Although insertion of elemental tellurium into reactive metal–carbon bonds has been known for many years, the resulting solutions contain a mixture of compounds in addition to the RTe⁻ species of interest.[2] Alkali metal phenyltellurolate salts, prepared via metal reduction of diphenyl ditelluride in liquid ammonia, were first isolated by Klar and co-workers.[3] More recently Lange and Du Mont reported the synthesis of the bulky aryl tellurolate $(THF)_3Li[Te(2,4,6-t-Bu_3C_6H_2)]$,[4] and Sladky described the in situ formation of a bulky alkyl tellurolate via reaction of tellurium with $LiC(SiMe_3)_3$.[5] Acidification of aryltellurolate anions affords thermally sensitive tellurols (RTeH) that are stable only below room temperature.[6]

Here we describe the high-yield preparations of sterically hindered silyltellurolate derivatives resulting from insertion of tellurium into the lithium

---

* The systematic names for these compounds are lithium 1,1,1,3,3,3-hexamethyl-2-(trimethylsilyl)-2-trisilanetellurolate-bis(tetrahydrofuran) and 1,1,1,3,3,3-hexamethyl-2-(trimethylsilyl)-2-trisilanetellurol.

† Department of Chemistry, University of California, Berkeley, CA 94720.

‡ School of Chemical Sciences, University of East Anglia, Norwich, NR4 7TJ, UK.

silicon bond in $(THF)_3Li[Si(SiMe_3)_3]$.[7] [This material is prepared from $Si(SiMe_3)_4$ which may be purchased from Aldrich Chemical Company or prepared according to Gilman's procedure: H. Gilman and C. L. Smith *J. Organomet. Chem.*, **8**, 245 (1967)]. In contrast to most known alkyl- and aryltellurolate derivatives, the lithium silyltellurolate and the corresponding silyltellurol derivatives are stable, crystalline materials that are simple to prepare and purify on large scales. Both compounds are useful reagents for the synthesis of a wide range of stable metal–tellurolate derivatives via metathesis and tellurolysis reactions.[8]

A. TRIS(TRIMETHYLSILYL)SILYL LITHIUM TRIS(TETRAHYDROFURAN)

$$Si(SiMe_3)_4 + MeLi \rightarrow (THF)_3LiSi(SiMe_3)_3 + SiMe_4$$

■ **Caution.** *Solid $(THF)_3Li[Si(SiMe_3)_3]$ is mildly pyrophoric. Care should be taken in disposing of any materials coated with this solid. The following compounds are air-sensitive and all operations must be conducted under an inert atmosphere of dry nitrogen or argon.*[9]

*Procedure**

The following is a modification of the procedure described by Gutekunst and Brook.[7] A 1-L, round-bottomed Schlenk flask, equipped with a 250-mL pressure-equalizing addition funnel and magnetic stirring bar, is charged with 78.3 g (0.239 mol) of $Si(SiMe_3)_4$. Dry THF (400 mL) is added in two portions via the addition funnel and stirring is initiated. A solution of halide-free MeLi (141 mL of a 1.7-M solution in diethyl ether, 0.24 mol) is added over 3 h, and the mixture is stirred for 12 h. The volatile components are removed under reduced pressure and the sticky solid is left to dry under vacuum for 12 h to yield 110 g (97%) of beige product. This material may be crystallized from hexane, but the crude product is of sufficient purity for use in the following step.

Properties

Tris(trimethylsilyl)silyl lithiumtris(tetrahydrofuran) is a colorless, air-sensitive solid that is soluble in all hdyrocarbon solvents, diethyl ether, and THF. ^{1}H NMR (300 MHz, C_6D_6): δ 3.46 (m, 12H), 1.34 (m, 12H), 0.63 (s, 27H).

---

* The checkers repeated these procedures on 1/4 scale.

B. LITHIUM TRIS(TRIMETHYLSILYL)SILYLTELLUROLATE BIS(TETRAHYDROFURAN)

$$(THF)_3Li[Si(SiMe_3)_3] + Te \rightarrow (THF)_2Li[TeSi(SiMe_3)_3]$$

■ **Caution.** *Tellurium powder is highly toxic and extreme care should be exercised when handling this solid. All procedures are best carried out in properly ventilated fume hoods, and adequate safety clothing should be worn at all times. (THF)$_3$Li[Si(SiMe$_3$)$_3$] and (THF)$_2$Li[TeSi(SiMe$_3$)$_3$] are mildly pyrophoric solids. Care should be taken in disposing of any materials coated with these solids. The tellurium compounds are relatively nonvolatile and can be manipulated on a vacuum line outside of a fume hood without noticeable problems; nevertheless, we recommend that all handling be carried out either in a glovebox or well-ventilated hood whenever possible. The following compounds are air-sensitive and all operations must be conducted under an inert atmosphere of nitrogen or argon.*

Procedure

The compound $(THF)_3LiSi(SiMe_3)_3$ (16 g, 34 mmol) and tellurium powder (200 mesh, Strem, 4.3 g, 34 mmol) are combined in a one-necked, 500-mL, round-bottomed Schlenk flask with a magnetic stir bar. Tetrahydrofuran (200 mL) is added via cannula and stirring is initiated immediately. As the tellurium rapidly dissolves, a dark orange-colored mixture is formed which is left to stir for 24 h. The solvent is removed under reduced pressure at 20°C and the pale-yellow solid is dried in vacuo for at least 6 h. The residue is extracted with two portions of hexane (300 mL then 100 mL) and the dark orange solution is filtered (using either a filter-stick[9] or fine-porosity Schlenk frit) into a large Schlenk flask. The filtrate is concentrated under reduced pressure at 20°C until it is saturated (approximately 150–200 mL), then it is cooled to − 40°C overnight to afford large, pale yellow crystals (14.7 g, 82%).*

Properties

Lithium tris(trimethylsilyl)silyltellurolate bis(tetrahydrofuran) is a pale yellow, air-sensitive solid that is soluble in hydrocarbons, diethyl ether, and THF. The finely powdered solid oxidizes (sometimes pyrophorically) in air to give the dark green ditelluride $(SiMe_3)_3SiTeTeSi(SiMe_3)_3$.[8] The lithium tellurolate is dimeric in the solid state as shown by X-ray crystallography[10]

---

* The checkers obtained a 79% yield after crystallizing at − 16°C.

(mp 169–171°C). ^{1}H NMR (300 MHz, C_6D_6): δ 3.75 (m, 8H), 1.43 (m, 8H), 0.45 (s, 27H). ^{125}Te{^{1}H} NMR (500 MHz, 0.3 M in C_6D_6): δ − 1622 ($\Delta v_{1/2}$ = 80 Hz) (referenced to external Me_2Te at δ 0 ppm). IR (CsI, Nujol): 1236(m), 1048(m), 917(w), 893(w), 834(s), 737(w), 686(m), 623(m) cm^{-1}.

Anal. Calcd. for $C_{17}H_{43}Si_4TeLiO_2$: C, 38.8; H, 8.36. Found: C, 38.5; H, 8.36.

C. TRIS(TRIMETHYLSILYL)SILYLTELLUROL

$$(THF)_2Li\,[TeSi(SiMe_3)_3 + F_3CSO_2OH \rightarrow HTeSi(SiMe_3)_3 + F_3CSO_2OLi$$

Procedure

Method A. A 100-mL, round-bottomed Schlenk flask equipped with a magnetic stir bar is charged with $(THF)_2LiTeSi(SiMe_3)_3$ (1.00 g, 1.90 mmol). Hexane (40 mL) is added, and stirring is initiated. After the starting material is completely dissolved, trifluoromethanesulfonic acid (0.17 mL, 1.9 mmol) is added via syringe.

■ **Caution.** *Trifluoromethanesulfonic acid (triflic acid) is one of the strongest acids known and is exceedingly corrosive. Extreme care should be used and protective clothing must be worn when handling this compound.*

The solution gradually darkens as the reaction proceeds; stirring should be continued for at least 2 h as triflic acid is only sparingly soluble in hexane. The solvent is removed under reduced pressure and the grey solid residue is transferred to a sublimator. The colorless waxy product sublimes slowly between 40 and 70°C/10^{-3} torr onto a cold finger cooled to − 78°C with dry ice/acetone. Yield: 0.648 g (91%). The reaction has been scaled up by a factor of 12 without diminution in yield.*

Method B. $(THF)_3Li[Si(SiMe_3)_3]$ (5.0 g, 11 mmol) and tellurium powder (1.3 g, 11 mmol) are combined in a 100-mL, round-bottomed Schlenk flask equipped with a magnetic stir bar. The flask is cooled to 0°C, THF (50 mL) is added via cannula, and stirring is initiated. The ice bath is removed and the orange mixture is stirred at room temperature for 1 h. Trifluoromethanesulfonic acid (0.94 mL, 11 mmol) is added dropwise and the resulting black mixture is stirred for 2 h. Using the workup and purification procedure described in Method A yields 2.2 g (56%) of colorless, waxy product.

---

* The checkers obtained a 72% yield and suggest carrying out the sublimation under a static vacuum to reduce loss.

Properties

Tris(trimethylsilyl)silyltellurol is a white, wax-like, air-sensitive solid that is extremely soluble in hydrocarbons, ethers, and chlorinated solvents. In air the material oxidizes rapidly to the dark green ditelluride $(SiMe_3)_3SiTeTeSi$-$(SiMe_3)_3$. Under nitrogen at ambient temperatures and under normal room lighting, the solid discolors over several days, while remaining spectroscopically and analytically pure. To prevent discoloration, the compound may be stored in a freezer (mp 128–130°C). ^1H NMR (300 MHz, C_6D_6) δ 0.23 (s, 27H), -8.82 (s, 1H, J_{HTe} = 74 Hz). ^{125}Te NMR (500 MHz, 0.5 M in C_6D_6): δ -955 (d, J_{TeH} = 74 Hz, $\Delta v_{1/2}$ = 27 Hz) (relative to external Me_2Te at δ 0 ppm). IR (CsI, Nujol): 2017(m), 1397(m), 1311(w), 1256(m), 1245(s), 837(s), 745(m), 691(s), 623(s) cm^{-1}. MS (EI, 70 eV) m/z 378 ($M^{\oplus}$), 304, 289, 73 (base peak).

Anal. Calcd. for $C_9H_{28}Si_4Te$: C, 28.7; H, 7.50. Found: C, 29.0; H, 7.52.

References

1. H. J. Gysling, *The Chemistry of Organic Selenium and Tellurium Compounds*, S. Patai and Z. Rappoport, (eds.), Wiley, New York, 1986, p. 679.
2. K. J. Irgolic, *The Organic Chemistry of Tellurium*, Gordon and Breach, New York, 1974.
3. J. Liesk, P. Schulz, and G. Klar, *Z. Anorg. Allg. Chem.*, **435**, 98 (1977).
4. L. Lange and W. -W. Du Mont, *J. Organomet. Chem.*, **286**, C1 (1985).
5. F. Sladky, B. Bildstein, C. Rieker, A. Gieren, H. Betz, and T. Hubner, *J. Chem. Soc., Chem. Commun.*, 1800 (1985).
6. M. Bochmann, A. P. Coleman, K. J. Webb, M. B. Hursthouse, and M. Mazid, *Angew. Chem., Int. Ed. Engl.*, **30**, 973 (1991).
7. G. Gutekunst and A. G. Brook, *J. Organomet. Chem.*, **225**, 1 (1982).
8. J. Arnold, *Prog. Inorg. Chem.*, **43**, 353 (1995).
9. J. P. McNally, V. S. Leong, and N. J. Cooper, *Experimental Organometallic Chemistry*, A. L. Wayda and M. Y. Darensbourg (eds.), A.C.S., Washington, D.C., 1985, p. 6.
10. P. J. Bonasia, D. E. Gindelberger, B. O. Dabbousi, and J. Arnold, *J. Am. Chem. Soc.*, **114**, 5209 (1992). P. J. Bonasia and J. Arnold, *Inorg. Chem.*, **31**, 2508 (1992).

28. METAL COMPLEXES OF THE LACUNARY HETEROPOLYTUNGSTATES
$[B\text{-}\alpha\text{-}PW_9O_{34}]^{9-}$ AND $[\alpha\text{-}P_2W_{15}O_{56}]^{12-}$

SUBMITTED BY WILLIAM J. RANDALL,* MICHAEL W. DROEGE,[†]
NORITAKA MIZUNO,[‡] KENJI NOMIYA,[§] TIMOTHY J. R. WEAKLEY,[‖]
and RICHARD G. FINKE[¶]
CHECKED BY NANCY ISERN,** JOSÉ SALTA,** and JON ZUBIETA**

Heteropolytungstates are complexes of the general formula $[XM_pO_q]^{n-}$, such as $[PW_{12}O_{40}]^{3-}$ or $[P_2W_{18}O_{62}]^{6-}$. Removal of one or more MO_y group gives rise to *lacunary heteropolytungstates* [1] such as $Na_9[A\text{-}PW_9O_{34}]$ or $Na_{12}[\alpha\text{-}P_2W_{15}O_{56}]\cdot24H_2O$. Such lacunary anions are key *synthons* for the preparation of more complicated, metal-substituted heteropolyanions. For this reason, lacunary heteropolyanion syntheses have been described in an earlier volume of this series,[2] including the $[PW_9O_{34}]^{9-}$ thermal isomer designated[3] as $[\Delta\text{-}PW_9O_{34}]^{9-}$ and the preparation[4] of $[\alpha\text{-}P_2W_{15}O_{56}]^{12-}$, precursors of central importance to the preparations described herein. (For a discussion of the α and β nomenclature,[5] see reference 16.)

In the present syntheses $[\Delta\text{-}PW_9O_{34}]^{9-}$ is used to prepare $(K, Na)_{10}$-$[M_4(H_2O)_2(B\text{-}\alpha\text{-}PW_9O_{34})_2]\cdot nH_2O$ where M is Co^{2+}, Cu^{2+}, and Zn^{2+}. Similarly $[\alpha\text{-}P_2W_{15}O_{56}]^{12-}$ is used to prepare $Na_{16}[M_4(H_2O)_2$-$(\alpha\text{-}P_2W_{15}O_{56})_2]$, where M is Co^{2+}, Cu^{2+}, and Zn^{2+}. These complexes are members of a new subclass of heteropolyoxoanions.[6, 7] Important details that are improvements in the scale or preparations of the lacunary precursors are also reported herein. Infrared spectroscopy is used throughout as a convenient and a definitive spectroscopic method of identification of the resultant heteropolyanions. Therefore, the IR spectrum of the phosphate and metal oxide region ($1200–500\ cm^{-1}$) of each intermediate and product is included. The products have been characterized by elemental analysis, molecular weight measurements, IR, ^{31}P and ^{183}W NMR[2] and, where possible, by X-ray crystallography.[7, 8]

It is important to pay close attention to the exact details of the preparation that follow. In general, it is well known by workers in the area that conditions such as the following are often crucial: pH, concentration, counter-ion type

* Lewis and Clark College, Portland, OR 97219.
[†] Lawrence Livermore Laboratories, Livermore, CA 94550.
[‡] Catalysis Research Center, Hokkaido University, Sapporo, 060, Japan.
[§] Kanagawa University, Hiratsuka, Kanagawa, 259-12, Japan.
[‖] University of Oregon, Eugene, OR 97403.
[¶] Department of Chemistry, Colorado State University, Fort Collins, CO 80523.
** Department of Chemistry, Syracuse University, Syracuse, NY 13244.

and concentration, temperature, timing, method, and rate of cooling to effect precipitation. This important point is echoed by the editor of a recent volume of *Inorganic Syntheses* who noted that heteropolyanion and polyoxoanion syntheses represent, " ... the synthesis of a class of compounds that are notably difficult to prepare in pure form. The successful preparation of many of these compounds requires close attention to the details of the procedure, and in several cases critical details taken for granted by the submitters were brought out by the checkers."[9] Especially crucial to the present syntheses is the solid-state, thermal rearrangement step of the $[A\text{-}PW_9O_{34}]^{9-}$ isomer to mostly the $[B\text{-}PW_9O_{34}]^{9-}$ form,[3] the mixture of which is labeled as $[\Delta\text{-}PW_9O_{34}]^{9-}$ ("thermalized $[PW_9O_{34}]^{9-}$").

The degree of hydration of the products from these preparations and the water content given by analytical procedures depends upon the heat treatment (method and history) of the product. A sample subjected to TGA (thermal gravimetric analysis) looses water almost continually from room temperature until it becomes the completely anhydrous heteropolytungstate salt at about 400°C. On the other hand, these crystals lose some lattice water rapidly upon removal from the mother liquor and exposure to air even at room temperature.

Complexes of the Nonatungstophosphate Ion, $[B\text{-}\alpha\text{-}PW_9O_{34}]^{9-}$

The *Inorganic Syntheses* preparation[10] of $Na_8H[A\text{-}PW_9O_{34}]$ was followed *until the crucial drying step* (see elsewhere for the original discovery of the importance of the drying step[6a,c]). It was then modified as follows: After obtaining the $Na_8H[A\text{-}PW_9O_{34}]$ as a white, wet precipitate, it is heated in an oven at 162°C from 2 days to a week, or until the odor of acetic acid from the hot substance is no longer apparent. This gives maximum conversion of the A form to the B form,[6a] as monitored by IR spectroscopy. *This conversion is essential for the successful completion of the following syntheses (A through C).* The IR spectrum should be exactly as represented in Fig. 1 to assure successful preparations.*

---

* The preparation of the KBr pellet is standardized to use about 0.20 g of warm, dry, IR-spectroscopic grade KBr and about 2 mg of sample which will yield a pellet approximately 0.5 mm thick and 13 mm in diameter. The typical percent transmittance values of the 1200- to 500-cm^{-1} region of the spectrum range from a 70–85% base line to 5–30% for the peaks. This set of characteristics adequately allows the resolution of those peaks that can be resolved. The spectra given herein were taken on a Nicolet DX FTIR System. The resolution was 2.0 cm^{-1}, the gain was set to 1, the interferometer velocity was 20, and 30 scans were taken.

† The IUPAC name for these complexes is illustrated by the name for the zinc complex ion: Diaquatetratriconta-μ-oxotetra-μ_3-oxo-octadecaoxobis$[\mu_{12}$-phosphato$(3-)$-O: O: O: O': O': O': O'': O'': O''': O''': O'''](tetrazincoctadecatungsten)ate$(10-)$.

* A sample name for the zinc analogue is given as a footnote under preparation A.

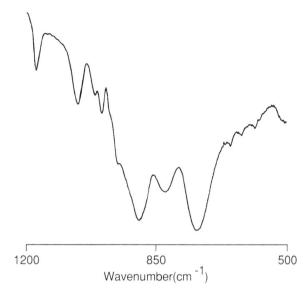

1200   850   500
Wavenumber(cm^{-1})

Figure 1. The IR spectrum of $Na_8H[\Delta\text{-}PW_9O_{34}]\cdot 19H_2O$ (KBr pellet).

A. PREPARATION OF $K_{10}[Zn_4(H_2O)_2(B\text{-}\alpha\text{-}PW_9O_{34})_2]\cdot 20H_2O^\dagger$

$$4ZnCl_2 + 2Na_8H[\Delta\text{-}PW_9O_{34}]\cdot 19H_2O + 10KCl \rightarrow$$

$$K_{10}[Zn_4(H_2O)_2(B\text{-}\alpha\text{-}PW_9O_{34})_2]\cdot 20H_2O + 16NaCl + 16H_2O + 2HCl$$

Procedure

Anhydrous zinc chloride (0.938 g of, 7.21 mmol) is dissolved in 30 mL of distilled water in a 100-mL beaker and solid $Na_8H[\Delta\text{-}PW_9O_{34}]\cdot 19H_2O$ (10.00 g, 3.60 mmol) is added *all at one time* as a dry, *finely crushed* powder. If the method of addition of the $Na_8H[\Delta\text{-}PW_9O_{34}]\cdot 19H_2O$ is not followed, different yields and different side products may result from the degradation of $[\Delta\text{-}PW_9O_{34}]^{9-}$ ion in solution. This mixture is stirred and heated to almost boiling (ca. 90°C). The hot, faintly turbid solution is gravity filtered through Whatman #1 paper. An excess of solid KCl (10.0 g, 134 mmol) is added to the clear, warm filtrate, which immediately gives a white granular precipitate. This suspension is warmed to 90°C for 5–10 min, cooled to 5°C in an ice bath (solid forms almost immediately after placing the beaker in the cooling medium), and the solid is collected on a medium, sintered-glass, filter funnel. The precipitate is washed with distilled, room-temperature water (3 × 5 mL), air-dried with aspiration, transferred to a 100-mL beaker, and dissolved in

50 mL of distilled water by heating on a hot plate to approximately 90°C. The hot solution, which is slightly turbid at this point, is gravity filtered through Whatman #5 paper. This mixture is cooled at 5°C for 12–15 h during which time a colorless, crystalline product of the title compound forms (6.31 g of air-dried material, 63% yield). These crystals are further purified by repeating the recrystallization procedure, followed by cooling slowly to room temperature for 24 h (or longer). Well-formed clear or white crystals[6, 7] are formed (4.98 g after air drying for 24 h in a 50°C oven, 49% overall yield).

Anal. Calcd. for $K_{10}[Zn_4(H_2O)_2(B-\alpha-PW_9O_{34})_2] \cdot 20H_2O$: K, 7.10; P, 1.12; W, 60.08; Zn, 4.75. Found: K, 6.77; P, 1.13; W, 60.01; Zn, 4.94; TGA indicates 20 molecules of volatile H_2O.

Properties

This pure compound forms colorless, clear, monoclinic crystals, or a white, crushed powder, which is quite insoluble in water at room temperature. Metathetical exchange of the potassium by lithium with lithium perchlorate in boiling water, followed by cooling to 2–5°C and then gravity filtration through Whatman #5 paper to separate the $KClO_4$ and the excess complex from the filtrate, gives a sufficient anion concentration in a $H_2O:D_2O$ solution (1:1) to observe the NMR spectra of the ^{31}P and ^{183}W nuclei. The ^{31}P NMR spectrum shows only one peak at -4.5 ppm (upfield from 85% H_3PO_4 using the substitution method). The ^{183}W NMR spectrum shows five peaks (relative integrated areas) at: $-90.3(1)$, $-105.4(2)$, $-116.5(2)$, $-129.6(2)$, $-134.9(2)$ ppm (upfield relative to a saturated sodium tungstate solution by the substitution method). The IR spectrum shown in Fig. 2 has peaks at: 1034(s), 973(sh), 966(sh), 948, 941, 883, 825(sh), 763(br), 713(br), 588(w), and 510 cm^{-1}. The IR spectrum is displayed in Fig. 2. It is worth noting that the crude product shows very weak shoulders at 1077 and 813 cm^{-1} which are not present in the recrystallized, pure product. The crystals used to make the KBr pellet for this spectrum were as clear and homogeneous to ordinary and plane-polarized light as could be found in the filtered, air-dried, final product.

B. PREPARATION OF $K_{10}[Co_4(H_2O)_2(B-\alpha-PW_9O_{34})_2] \cdot 20H_2O$*

$4Co(NO_3)_2 \cdot 6H_2O + 2Na_8H[\Delta-PW_9O_{34}] \cdot 19H_2O + 10KCl \rightarrow$

$K_{10}[Co_4(H_2O)_2(B-\alpha-PW_9O_{34})_2] \cdot$

$\qquad\qquad 20H_2O + 8NaCl + 40H_2O + 2HCl + 8NaNO_3$

* A sample name for the zinc analogue is given as a footnote under preparation A

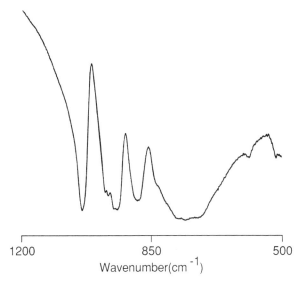

Figure 2. The IR spectrum of $K_{10}[Zn_4(H_2O)_2(B-\alpha-PW_9O_{34})_2] \cdot 20H_2O$ (KBr pellet).

Procedure

A solution of cobalt(II)nitrate hexahydrate (2.17 g, 7.46 mmol) dissolved in 30 mL of distilled water is prepared in a 100-mL beaker. (The nitrate salt is recommended, although the acetate and chloride salts have been successfully employed.[7]) Solid $Na_8H[\Delta-PW_9O_{34}] \cdot 19H_2O$ (10.00 g, 3.60 mmol) is added to this solution *all at one time* as a dry, *finely crushed* powder. If this method of addition of the $Na_8H[\Delta-PW_9O_{34}] \cdot 19H_2O$ is not followed, different yields and different side products may result from the degradation of the $[\Delta-PW_9O_{34}]^{9-}$ ion in solution. This mixture is stirred and heated to 80°C until a clear, dark burgundy solution results (5–10 min). An excess of solid KCl (10.0 g, 134 mmol) is added to the clear solution, which immediately gives a deep purple, granular precipitate. This suspension is cooled in an ice bath at 5°C (for 2–3 h) or a refrigerator (overnight) and solid product is then collected on a medium-frit, sintered-glass, filter funnel. The precipitate is washed with ethanol (3 × 10 mL) and with diethyl ether (3 × 10 mL), air dried with aspiration, and finally air dried at 50°C overnight (10.24 g of dried material, which contains the desired product, a side product, and some KCl). These crystals are purified by recrystallization from 80 mL of distilled water heated to 80°C, filtered hot though Whatman #5 paper by gravity, and then cooled slowly to room temperature. It is necessary to wash away a pink side product with several rinses of distilled water at room temperature (10–15 portions of 10 mL each). The desired product is not very soluble and is dense enough to remain in the beaker while the side product is decanted away. Well-formed, dark purple crystals result. The overall yield is 6.1 g (65%).

Anal. Calcd. for $K_{10}[Co_4(H_2O)_2(B-\alpha-PW_9O_{34})_2] \cdot 20H_2O$: K, 7.13; P, 1.12; W 60.08; Co, 4.29. Found: K, 7.02; P, 1.15; W, 60.13; Co, 4.05; TGA indicates 20 molecules of volatile H_2O.

Properties

The $K_{10}[Co_4(H_2O)_2(B-\alpha-PW_9O_{34})_2] \cdot 20H_2O$ crystals are dark purple to almost black in color. The UV/visible spectrum show λ_{max} values at 323, 339, 498(sh), 513(sh), and 571 nm with ε values of 4540, 4360, 104(est.), 124(est.), and 181 $L \cdot mol^{-1} \cdot cm^{-1}$, respectively. The IR spectrum shown in Fig. 3 has peaks at: 1064(w, sh), 1034(s), 958(sh), 939(s), 884(s), 778(br), 734(br), 588(w), and 509(m). These crystals are sparingly soluble in water at room temperature, but dissolve readily in water at 90°C. The crystal space group[7] is $P2_1/n$, where $a = 15.794$, $b = 21.360$, and $c = 12.312$ Å; $\beta = 91.96°$. The calculated density is 4.38 $g \cdot cm^{-3}$.

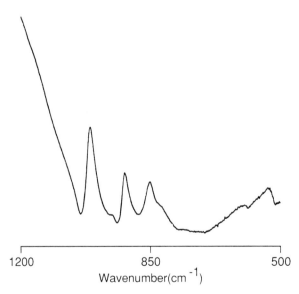

Figure 3. The IR spectrum of $K_{10}[Co_4(H_2O)_2(B-\alpha-PW_9O_{34})_2] \cdot 20H_2O$ (KBr pellet).

C. PREPARATION OF $K_7Na_3[Cu_4(H_2O)_2(B-\alpha-PW_9O_{34})_2]\cdot 20H_2O$*

$$4CuCl_2\cdot 2H_2O + 2Na_8H[\Delta-PW_9O_{34}]\cdot 19H_2O + 7KCl \rightarrow$$

$$K_7Na_3[Cu_4(H_2O)_2(B-\alpha-PW_9O_{34})_2]\cdot 20H_2O + 13NaCl + 24H_2O + 2HCl$$

This preparation is the most demanding of this set and updates a previous synthesis.[6a] The seemingly least significant variation often leads to a different proportions of products. In fact, the final desired product is obtained pure only after a recrystallization from an initial mixture of products.

Procedure

Copper(II)chloride (1.55 g, 9.09 mmol) is dissolved in 30 mL of distilled water in a 100-mL beaker. (It is essential to use the chloride salt in this preparation. The nitrate salt will usually lead to several other compounds.[11]) Solid $Na_8H[\Delta-PW_9O_{34}]\cdot 19H_2O$ (12.5 g, 4.50 mmol) is added *all at one time* as a dry, *finely crushed* powder. The mixture is stirred vigorously with a magnetic stir bar on a magnetic hot plate/stirrer to yield a clear yellow-green solution. A dark green precipitate forms within 10 min. This material is separated from the green solution by cooling to 4°C, and after 20 min of total elapsed time, collection on a medium-frit, sintered-glass, filter funnel. The yield of this product after it is dried overnight at 50°C is about 0.50 g. The filtrate is returned to a clean 100-mL beaker and allowed to warm to room temperature. Solid KCl (1.05 g, 14.1 mmol, 6.25 equiv.) is added in one step and stirred for 10–15 min. A dark green precipitate forms immediately. The suspension is cooled in an ice bath to 5°C while stirring on a magnetic stirrer. The stirring is stopped, the magnetic stirring bar is removed, and the solid is collected using a medium-frit, sintered-glass, filter funnel and washed successively first with ethanol (3×10 mL) and then with diethyl ether (3×10 mL). During each washing the precipitate is thoroughly mixed with the wash solution before aspiration is reapplied. The resulting precipitate is green and, after drying overnight at 50°C, weighs 9.22 g (74% yield). If the filtrate is allowed to sit, covered, at room temperature for several days, an additional side product precipitates.[6a] All 9.22 g of the crude product mixture can be recrystallized from 30 mL of distilled water at 55–60°C to yield 1.89 g (15%) of a pure, crystalline product suitable for X-ray structural analysis.[8a] To obtain pure product, these crystals must be collected soon (6–9 h) after they begin to form. If the filtrate from these crystals is allowed to evaporate slowly for several weeks, most of the remaining mass of reagents will eventually precipitate as a mixture of at least two additional materials, one yellow-green (presently uncharacterized) and the other is the light-blue, crystalline

$K_{5.5}Na_{1.5}[Cu_2(H_2O)_2(PW_{10}O_{39})] \cdot 13H_2O$ that is a mixture of two isomers in which the copper centers are adjacent.[12]

Anal. Calcd. for $K_7Na_3[Cu_4(H_2O)_2(B-\alpha-PW_9O_{34})_2] \cdot 20H_2O$: K, 5.02; Na, 1.26; P, 1.14; W, 60.69; Cu, 4.66; water, 6.61. Found: K, 5.13; Na, 1.26; P, 1.15; W, 61.00; Cu, 4.69; water, 6.23; TGA indicates 20 molecules of volatile H_2O.

Properties

The complex, $K_7Na_3[Cu_4(H_2O)_2(B-\alpha-PW_9O_{34})_2] \cdot 20H_2O$, is green ($\lambda_{max} =$ 829 nm, $\varepsilon = 81$). It has a uniquely characteristic IR peak at 1010 cm^{-1} which the other reaction products do not. The IR spectrum shown in Fig. 4 has peaks at: 1045(s,vs), 1010(m,vs), 969(sh), 944, 889(br), 769(s,br), 718(br), 512(m) cm^{-1}. The product forms cubic, yellow-green, transparent, well-formed crystals upon recrystallization from water at 55–60°C. When removed from the mother liquor, these crystals loose water and crystallinity. The crystal space group[8a] is P$\bar{1}$ with $a = 12.369$, $b = 16.957$, $c = 11.736$; $\alpha = 108.64°$, $\beta = 98.47°$, and $\gamma = 82.42°$. The calculated density is 3.98 g·cm^{-3}.

Complexes of the Pentadecatungstodiphosphate Ion, $[\alpha-P_2W_{15}O_{56}]^{12-}$

A conversion of sodium tungstate to the $[\alpha-P_2W_{15}O_{56}]^{16-}$ ion appeared in an earlier volume of this series, and we wish to note that the original synthesis

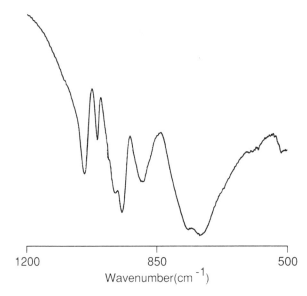

Figure 4. The IR spectrum of $K_7Na_3[Cu_4(H_2O)_2(B-\alpha-PW_9O_{34})_2] \cdot 20H_2O$ (KBr pellet).

of $[\alpha\text{-}P_2W_{15}O_{56}]^{16-}$ resulted from the efforts of others.[13, 14a] Our scaled-up and independently refined version of this synthesis is given[4, 6a] here and provides sufficient starting material to accomplish the several syntheses (E–G) that follow. The synthesis provided below for $Na_{12}[\alpha\text{-}P_2W_{15}O_{56}]\cdot18H_2O$ was also (and should be) modified, depending on the exact yields for each intermediate (70–90%) in a given repeat of this preparation, by changing proportionately the reagent quantities required in each subsequent step. One exception is that the solution is always reduced in volume to about 1.8–2.0 L before the addition of the potassium chloride in the formation of the pure $K_6[\alpha\text{-}P_2W_{18}O_{62}]\cdot14H_2O$.

D. PREPARATION OF PURE DODECASODIUM PENTADECATUNGSTODIPHOSPHATE OCTADECAHYDRATE, $Na_{12}[\alpha\text{-}P_2W_{15}O_{56}]\cdot18H_2O$

The preparation of the dodecasodium α-pentadecatungstodiphosphate octadecahydrate is accomplished from sodium tungstate dihydrate in three steps. The conversion of the isomeric mixture of $K_6[\alpha,\beta\text{-}P_2W_{18}O_{62}]\cdot xH_2O$ to the pure $K_6[\alpha\text{-}P_2W_{18}O_{62}]\cdot14H_2O$ isomer is accomplished by base degradation of $[\alpha,\beta\text{-}P_2W_{18}O_{62}]^{6-}$ to mixed isomers of $[\alpha,\beta\text{-}P_2W_{17}O_{61}]^{10-}$ as a suspension of potassium salts at a pH of 9.0, followed by acidification, which yields the isomerically pure[14] $[\alpha\text{-}P_2W_{18}O_{62}]^{6-}$ as the sole product at a pH of 4. Starting with the isomerically pure $[\alpha\text{-}P_2W_{18}O_{62}]^{6-}$ and base degradation to form the $[\alpha\text{-}P_2W_{15}O_{56}]^{12-}$, syntheses E–G yield the desired $[M_4(H_2O)_2(\alpha\text{-}P_2W_{15}O_{56})_2]^{16-}$ complexes, where M is Co^{2+}, Cu^{2+}, and Zn^{2+}.

1. Preparation of $K_6[\alpha,\beta\text{-}P_2W_{18}O_{62}]\cdot xH_2O$ (x is 14 for the α, and 19 for the β isomer[4])

$$18Na_2WO_4\cdot2H_2O + 32H_3PO_4 + 6KCl \rightarrow$$

$$K_6[\alpha,\beta\text{-}P_2W_{18}O_{62}]\cdot xH_2O + (54-x)H_2O + 30NaH_2PO_4 + 6NaCl$$

Procedure

The reaction vessel for the first part of this preparation is a 2- or 3-L flask set in a heating mantle, which is regulated by a potentiostat and placed on top of a magnetic stirrer, fitted with a reflux condenser, and a 500-mL Teflon-stoppered, pressure-equalizing, dropping funnel. Solid $Na_2WO_4\cdot2H_2O$ (300 g), distilled water (1050 mL) and a large magnetic stirring bar are placed in a 2- or 3-L flask. The sodium tungstate is dissolved with stirring and heating and the solution is brought to a boil. While the mixture is coming to a boil, 450 mL of concentrated H_3PO_4 solution (85%, density of 1.7 g $\cdot$ cm^{-3}) is placed in the dropping funnel, and the funnel is placed in one neck of the flask. After the solution in the flask boils, the heating is reduced and the phosphoric acid drip is started at the rate of 2–3 drops/sec.

■ **Caution.** *This is a concentrated acid being introduced into a hot, strongly basic solution. Do not rush the addition or a violent steam explosion is possible.*

The slow addition of the acid is obviously necessary for safety reasons but also appears to avoid the formation of an unwanted side product, which, if present, will be seen upon the addition of the KCl in the next precipitation step. The acid is allowed to drip into the reaction flask at this rate until all of it is added, usually about 1–2 h. After almost 50 mL of the acid is added, the solution turns from colorless to a light yellow-green, and remains this color throughout the reaction. When the addition of the acid is complete, the dropping funnel is removed, that neck is stoppered, and the reaction mixture is allowed to reflux for 15–24 h. Higher yields of this intermediate qualitatively correlate to longer reflux times. The hot solution is transferred to a 4-L beaker and allowed to cool to room temperature in an ice bath with stirring. Solid, granular KCl (300 g) is added slowly and continuously over 10–20 sec to this solution. If no unwanted side product is present, no precipitation will occur until almost 20% of the KCl has been added. If the unwanted side product is present, a white, granular product will precipitate immediately upon the first addition of the KCl. (This white side product was identified[4] as $K_{14}Na[P_5W_{30}O_{110}] \cdot xH_2O$, but we find that the IR spectrum is different from that published.[15] This unwanted product will be removed in the synthesis of the next intermediate.) After the addition of the KCl, the solution is stirred for approximately 1 h, cooled with additional stirring to approximately 5°C in an ice bath for 2–3 h, and the solid is collected on a 600-mL, medium-frit, sintered-glass, filter funnel with aspiration. The last of the solid product is washed from the 4-L beaker with small portions of the mother liquor using a wash bottle. This wet precipitate is dried on the funnel with aspiration until no more liquid is seen to come through the funnel (this takes from 2–3 h), then transferred to a 400-mL beaker and dissolved in boiling water.

■ **Caution.** *Use only plastic or ceramic spatulas when handling these wet products, inasmuch as metal (even stainless-steel) spatulas will cause reduction of the tungstates to a blue, reduced form.*

Boiling water (75–80 mL) is poured over the fresh precipitate and the mixture is brought to a boil. More boiling water is added in small portions with stirring until a clear, boiling solution is obtained using a *minimum* amount of boiling water (usually 100–150 mL). The beaker is covered with a watch glass, cooled to room temperature, and placed in a refrigerator at 5°C for at least overnight. The product is filtered, transferred to a tared, 10-cm, plastic, weighing pan and dried in an oven at 50°C overnight to yield 190–220 g (ca. 77–90% yield). This product is approximately 80% $K_6[\alpha\text{-}P_2W_{18}O_{62}] \cdot 14H_2O$ and 20% $K_6[\beta\text{-}P_2W_{18}O_{62}] \cdot 19H_2O$ by ^{31}P NMR.[16]

2. Preparation of $K_6[\alpha\text{-}P_2W_{18}O_{62}] \cdot 14H_2O$

$$K_6[\alpha,\beta\text{-}P_2W_{18}O_{62}] \cdot xH_2O \rightarrow$$

$$K_6[\alpha\text{-}P_2W_{18}O_{62}] \cdot 14H_2O + (x - 14)H_2O$$

Procedure

If 220 g of $K_6[\alpha,\beta\text{-}P_2W_{18}O_{62}] \cdot xH_2O$ from the previous synthesis is used for this conversion, it is placed in a 3- or 4-L beaker and dissolved in 825 mL of distilled water at room temperature; about 0.2–0.4 mL of liquid bromine is added with a Pasteur pipet (to assure complete oxidation of the tungstate to the VI oxidation state), upon which the solution turns yellow.

■ **Caution.** *Bromine is corrosive and a health hazard. This procedure should be completed in a good fume hood!*
A solution of 10% (w/v) $KHCO_3$ (1.10 L) is added slowly over 3–5 min. After approximately 750 mL of the $KHCO_3$ solution has been added, a white granular precipitate forms. After the solution has stirred at room temperature for about 1 h, 210 mL of 6 M HCl solution is slowly, but continuously added over 5–10 min to produce a *clear*, yellow solution. The solution must be clear, but do not add more than just enough HCl to accomplish this neutralization. The pH at this point should be around 3–4 (as tested by Hydrion paper). This solution is reduced in volume to approximately 1.8–2 L with gentle boiling and slow stirring on a magnetic stirrer/heater (ca. 3–4 h). Solid KCl (275 g) is added and the solution is stirred to dissolve all of the KCl. The solution should be clear and yellow. If the hot mixture contains a white, insoluble, granular precipitate at this stage, an undesired side product is present and the solution should be gravity-filtered through a 600-mL, medium-frit, sintered-glass, filter funnel, or through a large fine-paper filter. The clear, yellow solution is allowed to cool to room temperature and then placed in a refrigerator at 5°C for at least overnight. This cold suspension is filtered and the solid is collected in a 600-mL, medium-frit, sintered-glass filter funnel and thoroughly air dried with aspiration (2–3 h). The resultant pale-yellow precipitate is transferred to a 400-mL beaker and boiling, distilled water (50–60 mL) is poured over the solid. The beaker with the suspension is placed on a hot plate and returned to boiling, then small portions of boiling water are added until

a clear solution forms (usually a total of 75–125 mL of water is needed). This solution is covered with a watch glass, allowed to cool on the bench until it has reached room temperature, then placed in a refrigerator at 5°C for at least overnight. (A higher yield of the pure alpha isomer is obtained if the solution is left in the refrigerator longer.) This cold suspension is broken up by a heavy ceramic spatula, and the precipitate is collected in a 150-mL (or larger) medium-frit, sintered-glass, filter funnel, thoroughly dried with aspiration (2–3 h), transferred to a tared 10-cm, plastic, weighing pan, and dried in a 50°C drying oven overnight. The resulting light yellow, powdered $K_6[\alpha\text{-}P_2W_{18}O_{62}]\cdot14H_2O$ weighs 186 g (76% yield). The yield through this step ranges from 150 to 186 g (61–76% based upon the original 300 g of sodium tungstate). The characteristic IR spectrum is given in Fig. 5. This intermediate also has a single ^{31}P NMR peak at -13.11 (relative to 1% H_3PO_4 as 0.00 ppm by the substitution method).

3. Preparation of $Na_{12}[\alpha\text{-}P_2W_{15}O_{56}]\cdot18H_2O$

$$K_6[\alpha\text{-}P_2W_{18}O_{62}]\cdot14H_2O + 6NaClO_4 \rightarrow$$
$$Na_6[\alpha\text{-}P_2W_{18}O_{62}]_{(aq)} + 14H_2O + 6KClO_{4(s)}$$
$$Na_6[\alpha\text{-}P_2W_{18}O_{62}]_{(aq)} + 6Na_2CO_3 + 24H_2O \rightarrow$$
$$Na_{12}[\alpha\text{-}P_2W_{15}O_{56}]\cdot18H_2O + 3Na_2WO_4\cdot2H_2O + 6CO_{2(g)}$$

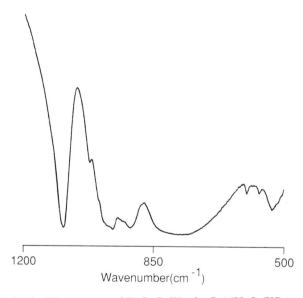

$$\begin{array}{ccc} 1200 & 850 & 500 \\ & \text{Wavenumber(cm}^{-1}) & \end{array}$$

Figure 5. The IR spectrum of $K_6[\alpha\text{-}P_2W_{18}O_{62}]\cdot16H_2O$ (KBr pellet).

Procedure

All of the product from the previous step is used to prepare the Na_{12} $[\alpha\text{-}P_2W_{15}O_{56}] \cdot 18H_2O$. The $K_6[\alpha\text{-}P_2W_{18}O_{62}] \cdot 14H_2O$ (186 g) is dissolved in 620 mL of distilled water at room temperature in a 1-L beaker. Solid, crystalline $NaClO_4$ (265 g) is added to this clear solution and the suspension is stirred for 20–30 min. The beaker is then placed in an ice bath and stirring is continued for about 2 h at which time the temperature of the suspension is 2–4°C. The $KClO_4$ metathesis product is removed from the solution by filtration through a medium-frit, sintered-glass, filter funnel and the filtrate is placed in a 4-L beaker. A solution of 1 M Na_2CO_3 is added (approximately 1700 mL) until the pH of the suspension is 9.0 ± 0.05 as monitored by a calibrated pH meter with a combination glass and calomel electrode. The pH is maintained at this value with vigorous stirring for 1 h. The total amount of sodium carbonate solution needed is approximately 1700 mL. The solid precipitate is collected in a 600-mL, medium-frit, sintered-glass, filter funnel and thoroughly dried with aspiration. The solid is first washed with a saturated NaCl solution at room temperature (3 × 50 mL), then with absolute ethanol (3 × 50 mL), and finally with anhydrous diethyl ether (3 × 50 mL). During these washing steps the solid is thoroughly mixed with the rinsing solution before aspiration is reapplied. The resulting powder is placed on a tared, 10-cm, plastic, weighing pan and dried overnight in a drying oven at 50°C to give 129 g of the desired $Na_{12}[\alpha\text{-}P_2W_{15}O_{56}] \cdot 18H_2O$ (78% conversion). The characteristic IR spectrum is found in Fig. 6.

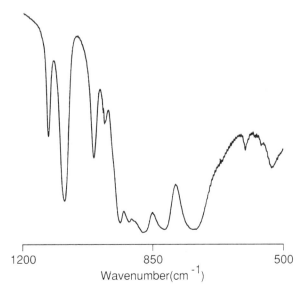

Figure 6. The IR spectrum of $Na_{12}[\alpha\text{-}P_2W_{15}O_{56}] \cdot 18H_2O$ (KBr pellet).

E. PREPARATION OF $Na_{16}[Zn_4(H_2O)_2(\alpha\text{-}P_2W_{15}O_{56})_2]$*

$$4ZnCl_2 + 2Na_{12}[\alpha\text{-}P_2W_{15}O_{56}] \cdot 18H_2O \rightarrow$$
$$Na_{16}[Zn_4(H_2O)_2(\alpha\text{-}P_2W_{15}O_{56})_2] + 8NaCl + 16H_2O$$

Procedure

A clear solution of zinc chloride (0.34 g, 2.49 mmol) and sodium chloride (2.92 g, dissolved in 50 mL of water, 1 M) is prepared in a 100-mL beaker at room temperature. The sodium salt of the complexing anion, $Na_{12}[\alpha\text{-}P_2W_{15}O_{56}] \cdot 18H_2O$ (5.00 g, 1.25 mmol), is added to this solution all at one time as a dry, powdered solid with vigorous stirring using a magnetic stirring bar on a heater/stirrer. The solution is heated with stirring to about 80°C to yield a turbid solution. This *hot* solution is rapidly, gravity filtered through Whatman #5 paper. (The filter funnel and paper are preheated by passing about 50–100 mL of boiling water through them just prior to this filtration.) The clear filtrate is collected in a clean 100-mL beaker and placed in the refrigerator at 5°C overnight. The product is retrieved on a medium-frit, filter funnel and washed first with absolute ethanol (3 × 10 mL) and then with anhydrous diethyl ether (3 × 10 mL). The product is then dried overnight on an open watch glass, or in a weighing pan, in a drying oven at 50°C overnight to yield a white granular product which weighs 3.61 g (71% yield).

Anal. Calcd. for $Na_{16}[Zn_4(H_2O)_2(\alpha\text{-}P_2W_{15}O_{56})_2]$: Na, 4.54; P, 1.53; W, 68.12; Zn, 3.23. Found: Na, 4.24; P, 1.49; W, 67.99; Zn, 3.02; water is < 0.01% after drying to constant weight at 200°C.

Properties

The title compound is crystalline, but loses its crystallinity upon drying (crystals grown from a 1-M NaCl solution are microcrystalline, diamond-shaped platelets which are very thin and extremely fragile). The complex is easily characterized by its IR spectrum, given in Fig. 7, which exhibits the following peaks: 1086(vs), 1048(vs), 1014(w), 948(sh), 936(br), 919(sh), 881, 831(sh), 781(br), 735(br), 600(w,s), and 527(m) cm^{-1}. A solution of this compound in a 1:1 mixture of $H_2O:D_2O$ gives a ^{31}P NMR spectrum of two equally intense peaks at -4.54 and -14.54 ppm (upfield relative to 85% H_3PO_4 at 0.00 ppm by the substitution method). The same solution gives

---

* The IUPAC name for this class of complexes is illustrated for the zinc species: Diaquado-hexaconta-μ-oxotetra-μ_3-oxo-triacontaoxotetrakis[μ_9-phosphato(3 −)-O: O: O: O': O': O'': O'': O''': O'''](tetrazinctriacontatungsten)ate(16 −).

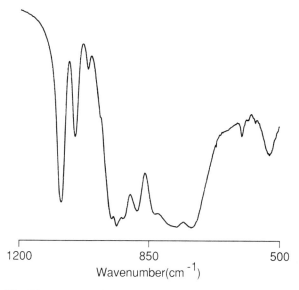

Figure 7. The IR spectrum of $Na_{16}[Zn_4(H_2O)_2(\alpha\text{-}P_2W_{15}O_{56})_2]$ (KBr pellet).

a ^{183}W NMR spectrum of eight peaks (relative integrated areas): $-149.5(1)$, $-158.0(2)$, $-161.2(2)$, $-178.0(2)$, $-183.2(2)$, $-236.1(2)$, $-241.6(2)$, and $-242.7(2)$ ppm (upfield referenced to sodium tungstate as 0.00 ppm in the same solvent mixture by the substitution method). These data demonstrate that the C_{2h} symmetry which was found for the anion by single-crystal X-ray crystallography[8b] is maintained in solution.

F. PREPARATION OF $Na_{16}[Co_4(H_2O)_2(\alpha\text{-}P_2W_{15}O_{56})_2]$*

$$4Co(NO_3)_2 \cdot 6H_2O + 2Na_{12}[\alpha\text{-}P_2W_{15}O_{56}] \cdot 18H_2O \rightarrow$$
$$Na_{16}[Co_4(H_2O)_2(\alpha\text{-}P_2W_{15}O_{56})_2] + 8NaNO_3 + 58H_2O$$

Procedure

A solution of cobalt(II)nitrate hexahydrate (1.83 g, 2.50 mmol) and sodium chloride (2.93 g dissolved in 50 mL of water, 1 M) is prepared in a 100-mL beaker. This mixture is stirred on a magnetic stirrer until homogeneous. The solid $Na_{12}[\alpha\text{-}P_2W_{15}O_{56}] \cdot 18H_2O$ (5.00 g, 1.25 mmol) is added all at one time with vigorous stirring. The solution turns to a fluorescent olive-green

---

* A sample name for the zinc analogue is given as a footnote under preparation E.

and eventually to a dark red-brown as more of the white solid dissolves. Finally, the solution is a deep-red to burgundy color with an olive-green reflectance. This solution is heated to 50°C, covered, and placed in a refrigerator at 5°C overnight. The product is filtered through a medium-frit, filter funnel, washed with absolute ethanol (3 × 10 mL) and with anhydrous diethyl ether (3 × 10 mL), and dried in an oven at 50°C to give an amorphous, brown, olive-green powder that weighs 3.52 g (70% yield).

Anal. Calcd. for $Na_{16}[Co_4(H_2O)_2(\alpha\text{-}P_2W_{15}O_{56})_2]$: Na, 4.56; P, 1.54; W, 68.34; Co, 2.92. Found: Na, 4.32; P, 1.46; W, 67.67; Co, 2.74.

Properties

This compound is characterized by its IR spectrum, shown in Fig. 8, which gives the following peaks: 1084(vs), 1048(m,vs), 1014(s,vs), 947(sh), 934(s), 917(sh), 878(m), 833(sh), 767(br), 736(br), 598(w,vs), 564(w,vs), and 520(m) cm^{-1}. It is sparingly soluble in water. The UV/visible spectrum in water shows a large peak at 578 nm with shoulders at 506 and 517 nm. The ε values are 157, 106(est.), and 109(est.) $L \cdot mol^{-1} \cdot cm^{-1}$, respectively.

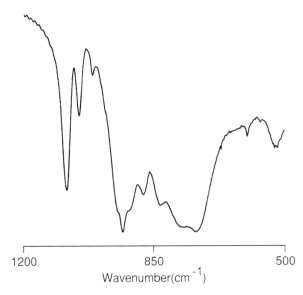

1200   850   500
Wavenumber(cm^{-1})

Figure 8. The IR spectrum of $Na_{16}[Co_4(H_2O)_2(\alpha\text{-}P_2W_{15}O_{56})_2]$ (KBr pellet).

G. PREPARATION OF $Na_{16}[Cu_4(H_2O)_2(\alpha\text{-}P_2W_{15}O_{56})_2]$ AND $Na_{14}Cu[Cu_4(H_2O)_2(\alpha\text{-}P_2W_{15}O_{56})_2]\cdot 53H_2O^*$

$$4CuCl_2\cdot 2H_2O + 2Na_{12}[\alpha\text{-}P_2W_{15}O_{56}]\cdot 18H_2O \rightarrow$$
$$Na_{16}[Cu_4(H_2O)_2(\alpha\text{-}P_2W_{15}O_{56})_2] + 8NaCl + 42H_2O$$

Procedure

A solution of copper(II) chloride dihydrate (0.43 g, 2.50 mmol) and sodium chloride (2.92 g dissolved in 50 mL of distilled water, 1 M) is prepared in a 100-mL beaker. Solid $Na_{12}[\alpha\text{-}P_2W_{15}O_{56}]\cdot 18H_2O$ (5.00 g, 1.25 mmol) is added all at one time and the solution is stirred and heated to boiling. The color changes from the blue of the hexaaquacopper(II) to a bright, lime green of the desired compound. The hot, turbid solution is gravity filtered through Whatman # 5 paper. What appears to be a substantial amount of lime green, insoluble material is retained on the filter paper, but when dried it is a small amount. This precipitate is insoluble in 0.5 M NaCl and remains uncharacterized. The clear, yellow-green solution yields a crystalline product within a few minutes. The suspension is allowed to cool to room temperature, covered, and placed in a refrigerator at 5°C overnight. The resulting lime green, crystalline product is filtered, washed with absolute ethanol (3×10 mL) and then with anhydrous diethyl ether (3×10 mL). The resulting lime green powder weighs 3.53 g (70% yield). The product need not be dried with ethanol and diethyl ether, but instead may be air-dried with aspiration (2–3 h). The product is the same in either case. The latter treatment yields a product that is quite crystalline in appearance, but which rapidly loses water and crystallinity when exposed to air.

Anal. (On a sample dried at 80°C under vacuum for ≤ 0.5 h). Calcd. for $Na_{16}[Cu_4(H_2O)_2(\alpha\text{-}P_2W_{15}O_{56})_2]$: Na, 4.55; P, 1.53; W, 68.18; Cu, 3.14. Found: Na, 4.36; P, 1.49; W, 67.92; Cu, 3.17.

Recrystallization of this product from 10 mL of a 0.5-M NaCl solution which is warmed to 40°C, cooled to room temperature, and allowed to evaporate to half its original volume gives a different compound, which has the general formula $Na_{14}Cu[Cu_4(H_2O)_2(\alpha\text{-}P_2W_{15}O_{56})_2]\cdot 53H_2O$ and is formed as pale, yellow-green, block-shaped crystals suitable for X-ray structure analysis.[8a] If this product is allowed to air dry, it loses water and crystallinity.

Anal. Calcd. for $Na_{14}Cu[Cu_4(H_2O)_2(\alpha\text{-}P_2W_{15}O_{56})_2]\cdot 53H_2O$: Na, 3.55; Cu, 3.51; P, 1.37; W, 60.9. Found: Na, 3.52; Cu, 3.51; P, 1.37; W, 61.0; weight loss following drying overnight in a 195°C oven indicates 53 (± 1) molecules of volatile H_2O per complex.

* A sample name for the zinc analogue is given as a footnote under preparation E.

Properties

The characteristic IR spectrum of $Na_{16}[Cu_4(H_2O)_2(\alpha\text{-}P_2W_{15}O_{56})_2]$ shown in Fig. 9 has peaks as follows: 1084(vs), 1055(m,vs), 1014(w,sh,vs), 950(sh),

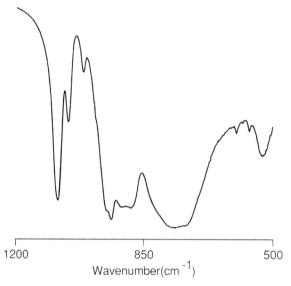

Figure 9. The IR spectrum of $Na_{16}[Cu_4(H_2O)_2(\alpha\text{-}P_2W_{15}O_{56})_2]$ (KBr pellet).

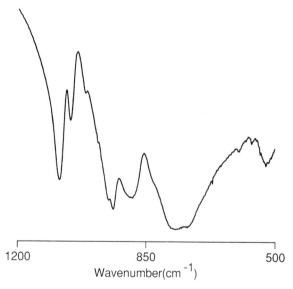

Figure 10. The IR spectrum of $Na_{14}Cu[Cu_4(H_2O)_2(\alpha\text{-}P_2W_{15}O_{56})_2]$ (KBr pellet).

939(s), 906(br), 888(br), 825(sh), 777(br), 739(br), 598(w,vs), 562(w,vs), and 523(m,br) cm^{-1}. The spectrum of this compound in water solution shows absorption peaks at 811 and 870 nm with ε values of 68 and 64 L $\cdot$ $mol^{-1} \cdot cm^{-1}$, respectively.

The IR spectrum of $Na_{14}Cu[Cu_4(H_2O)_2(\alpha-P_2W_{15}O_{56})_2]$ shown in Fig. 10 has peaks at: 1086(s,vs), 1055(s,m), 1014(w,s,sh), 950(w,s,sh), 937.5(s,vs), 889(w,sh), 769(s,br), 739(s,br), 598(w,s), 563(w,s), and 522(m,br) cm^{-1}. The UV/visible spectrum of this compound shows a maximum at 816 nm, with an ε value of 71 L $\cdot$ $mol^{-1} \cdot cm^{-1}$. The crystal space group[8a] for this product is $P\bar{1}$ with $a = 13.399$, $b = 25.017$, and $c = 13.339$ Å; $\alpha = 104.84°$, $\beta = 114.49°$, $\gamma = 82.61°$. The calculated density is 3.82 g $\cdot$ cm^{-3}.

References

1. M. T. Pope, *Heteropoly and Isopoly Oxometalates*, Springer-Verlag, New York, 1983, pp. 93–100.
2. W. G. Klemperer, *Inorg. Synth.*, **27**, 71 (1990).
3. P. J. Domaille, *Inorg. Synth.*, **27**, 100 (1990).
4. R. Contant, *Inorg. Synth.*, **27**, 108 (1990).
5. Y. Jeannin and M. Fournier, *Pure Appl. Chem.*, **59**, 1529 (1987).
6. (a) R. G. Finke, M. W. Droege, and P. J. Domaille, *Inorg. Chem.*, **26**, 3886 (1987) and references therein to the earlier work. (b) R. G. Finke, M. W. Droege, and P. J. Domaille, *Inorg. Chem.*, **22**, 1006 (1983). (c) R. G. Finke, M. Droege, J. R. Hutchison, and O. Gansow, *J. Am. Chem. Soc.*, **103**, 1587 (1981).
7. T. J. R. Weakley, H. T. Evans, Jr., J. S. Showell, G. F. Tourné, and C. M. Tourné, *J. Chem. Soc., Chem. Commun.*, 139 (1973); H. T. Evans, C. M. Tourné, G. F. Tourné, and T. J. R. Weakley, *J. Chem. Soc., Dalton Trans.*, 2699 (1986).
8. (a) T. J. R. Weakley and R. G. Finke, *Inorg. Chem.*, **29**, 1235 (1990). (b) R. G. Finke and T. J. R. Weakley, *J. Chem. Cryst.*, **24**, 123 (1994).
9. A. L. Ginsberg, *Inorg. Synth.*, **27**, viii (1990).
10. P. J. Domaille, *Inorg. Synth.*, **27**, 100 (1990); see the first four lines under *Properties*, p. 100.
11. W. H. Knoth, P. J. Domaille, and R. L. Harlow, *Inorg. Chem.*, **25**, 1577 (1986).
12. C. J. Gómez-Garcia, E. Coronado, P. Gómez-Romero, and N. Casañ-Pastor, *Inorg. Chem.*, **32**, 89 (1993).
13. R. Contant, *Inorg. Synth.*, **27**, 108 (1990). See also R. Contant and P. J. Ciabrini, *J. Chem. Res. Synop.*, 222 (1977); *J. Chem. Res.*, Miniprint, 2601 (1977).
14. (a) The preparation of pure $[\alpha-P_2W_{18}O_{62}]^{6-}$ was first described by H. Wu, *J. Biol. Chem.*, **43**, 189 (1920). (b) M. W. Droege, Ph.D. Dissertation, University of Oregon, 1984, pp. 78, 107.
15. Y. Jeannin and J. Martin-Frere, *Inorg. Synth.*, **27**, 115 (1990).
16. D. K. Lyon, W. K. Miller, T. Novet, P. J. Domaille, E. Evitt, D. C. Johnson, and R. G. Finke, *J. Am. Chem. Soc.*, **113**, 7209 (1991).

29. POLYOXOANION-SUPPORTED, ATOMICALLY DISPERSED IRIDIUM(I) AND RHODIUM(I): $Na_3[(C_4H_9)_4N]_5[Ir[\alpha\text{-}Nb_3P_2W_{15}O_{62}]\{\eta^4\text{-}C_8H_{12}\}]$ and $Na_3[(C_4H_9)_4N]_5[Rh[\alpha\text{-}Nb_3P_2W_{15}O_{62}]\{\eta^4\text{-}C_8H_{12}\}]$

SUBMITTED BY KENJI NOMIYA,* MATTHIAS POHL,[†]
NORITAKA MIZUNO,[‡] DAVID K. LYON,[†] and
RICHARD G. FINKE[§]
CHECKED BY MICHAEL H. DICKMAN,[∥] DIANA C. GLICKMAN,[∥]
ULI KORTZ,[∥] JOSEPH L. SAMONTE,[∥] FEIBO XIN,[∥] and
MICHAEL T. POPE[∥]

Polyoxoanions are soluble oxides that resemble discrete fragments of solid metal oxides. They are of considerable interest as robust, oxidation-inert, catalyst-support materials for both fundamental and practical reasons. The complexes $[Ti(\beta\text{-}V_3SiW_9O_{40})\{\eta^5\text{-}C_5H_5\}]^{4-}$, $[Rh(\beta\text{-}Nb_3SiW_9O_{40})\{\eta^5\text{-}C_5(CH_3)_5\}]^{5-}$, and $[Ir(\beta\text{-}Nb_3SiW_9O_{40})\{\eta^4\text{-}1,5\text{-}COD\}]^{6-}$ are prototypical examples of a transition metal supported on the surface oxygen atoms of a Keggin-type polyoxoanion.[1-3] Similarly, $[M(\alpha\text{-}Nb_3P_2W_{15}O_{62})\{\eta^4\text{-}1,5\text{-}COD\}]^{8-}$ (where 1,5-COD = 1,5-cyclooctadiene and M = Ir, Rh)[4-6] are both new compositions of matter and are presently the prototype examples of reactive, transition-metal organometallics supported on the Dawson-type, crystallographically characterized polyoxoanion $[\alpha\text{-}Nb_3P_2W_{15}O_{62}]^{9-}$.[7,8] The iridium(I)-polyoxoanion complex $[Ir(\alpha\text{-}Nb_3P_2W_{15}O_{62})\{\eta^4\text{-}1,5\text{-}COD\}]^{8-}$ serves as a potent precatalyst to an active, O_2-employing, and patented[9] oxidation catalyst[10] and, under hydrogen, to novel and patented[11] polyoxoanion and tetrabutylammonium-stabilized $Ir(0)_{\sim300}$ nanocluster hydrogenation catalysts.[12] The closely related rhodium(I) complex, $[Rh(\alpha\text{-}Nb_3P_2W_{15}O_{62})\{\eta^4\text{-}1,5\text{-}COD\}]^{8-}$, is also an effective oxidation[9,10] and hydrogenation[13] precatalyst. A review of the synthesis, characterization, catalysis, and mechanistic studies of $[Ir(\alpha\text{-}Nb_3P_2W_{15}O_{62})\{\eta^4\text{-}1,5\text{-}COD\}]^{8-}$ is available.[14]

In the present synthesis, the initial polyoxoanion product results from oligomerization, under acidic conditions (in unbuffered pH 4.6 water), to form one bridging Nb–O–Nb bond: $2[\alpha\text{-}Nb_3P_2W_{15}O_{62}]^{9-} + 2H^+ \rightarrow [Nb_6P_4W_{30}O_{123}]^{16-} + H_2O$. The Nb–O–Nb bridged product,

* Department of Materials Science, Faculty of Science, Kanagawa University, Hiratsuka, Kanagawa 259-12, Japan.
[†] Department of Chemistry, University of Oregon, Eugene, OR 97403.
[‡] Catalysis Research Center, Hokkaido University, Sapporo 060, Japan.
[§] Department of Chemistry, Colorado State University, Ft. Collins, CO 80523.
[∥] Department of Chemistry, Georgetown University, Washington, DC 20057.

$[(C_4H_9)_4N]_{12}H_4[Nb_6P_4W_{30}O_{123}]$, has been characterized by elemental analysis, solution ultracentrifugation molecular-weight measurement, IR, ^{31}P, and ^{183}W NMR spectroscopies.[7] Reversal of the oligomerization (cleavage of the Nb–O–Nb bond) is readily effected by the tetrabutylammonium hydroxide $[(C_4H_9)_4N]OH$.[7] The resultant Dawson-type, highly basic $[(C_4H_9)N]_9[\alpha-Nb_3P_2W_{15}O_{62}]$, which has three extra units of anionic surface-oxygen charge density (compared to the parent $[P_2W_{18}O_{62}]^{6-}$), has been characterized on the basis of solution molecular-weight measurement, IR,^{31}P, and ^{17}O NMR spectra[7] and, more recently, by ^{17}O NMR spectra.[5,6]

The synthesis and subsequent storage of the title complexes, $Na_3[(C_4H_9)_4-N]_5[Ir(\alpha-Nb_3P_2W_{15}O_{62})\{\eta^4-1,5-COD\}]$ and $Na_3[(C_4H_9N]_5[Rh(\alpha-Nb_3P_2W_{15}O_{62})\{\eta^4-1,5-COD\}]$, require strict oxygen-free conditions (a $\leq$ 1 ppm O_2 dry box). They are formed by the addition of $[Ir(1,5-COD)(CH_3CN)_2]^+$ or $[Rh(1,5-COD)(CH_3CN)_2]^+$, respectively, to a CH_3CN solution of $[Nb_3P_2W_{15}O_{62}]^{9-}$. The key to isolating them as pure solids is our recent method[2,4,6] employing mixed $[(C_4H_9)_4N^+/Na^+]$ salts and at least two reprecipitations from paper-filtered, homogeneous acetonitrile (CH_3CN) solution using ethyl acetate (EtOAc). They are characterized by full elemental analysis, solution molecular-weight measurement, and ^1H, ^{13}C, ^{31}P, ^{183}W, and ^{17}O NMR spectroscopies.[6] These NMR data show that, in the case of $[Ir(1,5-COD)]^+$, the designed $Nb_3O_9^{3-}$ support site on $[Nb_3P_2W_{15}O_{62}]^{9-}$ leads to one support-site regioisomer in which the $[Ir(1,5-COD)]^+$ is attached to the $Nb_3O_9^{3-}$ cap via three, directly detected[5] bridging Nb_2O-to-Ir bonds and in overall C_{3v} (average; pseudo)[5,6] symmetry.

A. PREPARATION OF $[(C_4H_9)_4N]_{12}H_4(\mu_2-O)[Nb_3P_2W_{15}O_{61}]_2]$

The preparation of the title compound* is based on the reaction of Nb(V) (in a peroxide form)[15] with the lacunary ion $[P_2W_{15}O_{56}]^{12-}$, which has been

---

* *Materials*: The following chemicals were obtained commercially from the indicated vendors and were reagent grade and used as received unless otherwise noted. From Aldrich: $Na_2[WO_4]\cdot 2H_2O$, Nb_2O_5, 30% H_2O_2, 37% HCl, $[(CH_3)_4N]Cl$, $RhCl_3\cdot xH_2O$ ($x = 2–3$), $Na[BF_4]$, $Ag[BF_4]$, 95% ethanol, 2-propanol, acetonitrile, ethyl acetate, diethyl ether (HPLC grade; glass-distilled under N_2 and filtered through 0.5-μm filters by the manufacturer), 1,5-cyclooctadiene, and 4,7,13,16,21,24-hexaoxa-1,10-diaza-bicyclo[8.8.8]hexacosane [Kryptofix® 2.2.2]. Purchased elsewhere and used as received: $[(NH_4)]_3[IrCl_6]$ (Johnson–Mathey); $[(C_4H_9)_4N]Br$ (Fluka); Ag_2O (Aesar); KCl and KOH (Malinckrodt); anhydrous K_2CO_3, $NaHSO_3$ (Baker). All solvents used in the preparation of air-sensitive compounds were distilled under N_2 (acetonitrile from CaH_2; ethyl acetate from anhydrous K_2CO_3) and stored in glassware that had been thoroughly cleansed, dried at 250°C, and allowed to cool under a dry N_2 flow. Deuterated NMR solvents (d_6-DMSO, CD_3CN; Cambridge Isotope Laboratories) used for air-sensitive NMR samples were degassed by purging with the dry box atmosphere or with argon (outside the drybox) for 0.5 h.

prepared by degradation of pure $K_6[\alpha\text{-}P_2W_{18}O_{62}]\cdot 14H_2O$ by aqueous Na_2CO_3. Excess H_2O_2 is added to the Nb(V) precursor[2] $[Nb_6O_{19}]^{8-}$ to inhibit the formation of Nb_2O_5 that would otherwise occur upon acidification of Nb(V) in aqueous solution. Workup consists of rapid addition of solid $NaHSO_3$ to destroy the peroxide (analogous to our earlier preparation of $[SiW_9Nb_3O_{40}]^{7-}$)[2] and then precipitation of crude $[(CH_3)_4N]_{12}H_4$-$[Nb_6P_4W_{30}O_{123}]$ using tetramethyl ammonium chloride. Purification is achieved by two reprecipitations from hot, unbuffered pH 4.6 water; conversion to the tetrabutylammonium salt is accomplished by precipitation of $[(C_4H_9)_4N]_{12}H_4[Nb_6P_4W_{30}O_{123}]$ with tetrabutylammonium bromide, and then thorough rinsing of the product with water followed by drying at 60°C for 12 h.

1. Preparation of $Na_{12}[\alpha\text{-}P_2W_{15}O_{56}]\cdot 18H_2O$

The preparation of the dodecasodium α-pentadecatungstodiphosphate octadecahydrate,[16] $Na_{12}[\alpha\text{-}P_2W_{15}O_{56}]\cdot 18H_2O$ (as its pure α-isomer[17]), is accomplished from sodium tungstate dihydrate via the four steps summarized below. Detailed procedures are provided elsewhere in this volume[16] (a smaller-scale preparation is also available[18]). The purity of the intermediates and of the $Na_{12}[\alpha\text{-}P_2W_{15}O_{56}]\cdot 18H_2O$ product was confirmed by IR or ^{31}P NMR spectroscopy in comparison to the spectral data available elsewhere.[16]

$$18Na_2WO_4\cdot 2H_2O + 32H_3PO_4 + 6KCl \rightarrow$$

$$K_6[\alpha,\beta\text{-}P_2W_{18}O_{62}]\cdot xH_2O + (39-x)\,H_2O + 30NaH_2PO_4 + 6NaCl$$

$$K_6[\alpha,\beta\text{-}P_2W_{18}O_{62}]\cdot xH_2O \rightarrow K_6[\alpha\text{-}P_2W_{18}O_{62}]\cdot 14H_2O + (x-14)H_2O$$

$$K_6[\alpha\text{-}P_2W_{18}O_{62}]\cdot 14H_2O + 6NaClO_4 \rightarrow$$

$$Na_6[\alpha\text{-}P_2W_{18}O_{62}]_{(aq)} + 14H_2O + 6KClO_{4(s)}$$

$$Na_6[\alpha\text{-}P_2W_{18}O_{62}]_{(aq)} + 6Na_2CO_3 + 24H_2O \rightarrow$$

$$Na_{12}[\alpha\text{-}P_2W_{15}O_{56}]\cdot 18H_2O + 3Na_2WO_4\cdot 2H_2O + 6CO_{2(g)}$$

2. Preparation of $K_7H[Nb_6O_{19}]\cdot 13H_2O$

$$3Nb_2O_5 + 7KOH + 10H_2O \rightarrow K_7H[Nb_6O_{19}]\cdot 13H_2O$$

The preparation and characterization of this complex generally follow previous reports.[19,20] The preparation employed here is similar to that described by Flynn and Stucky[20] and is essentially identical to our earlier preparation[2] except for different washing and reprecipitation steps. The present synthesis provides a simple means of obtaining the complex in good yield as a powder.

Procedure

The solids Nb_2O_5 (10 g, 37.6 mmol) and 85% KOH (35 g, 530 mmol) are placed in a nickel crucible. The crucible is heated cautiously in a hood, using a low flame (to prevent boilover and minimize splattering), until the contents occasionally boil. During heating, the reaction mixture changes color from white to gray through the formation of a transparent melt. The crucible is heated using a high flame for 10 min and then allowed to cool to room temperature (about 30 min). The crucible containing the solidified melt (now gray) is placed in a 250-mL beaker with 200 mL of distilled water. After the solidified melt has dissolved (this may be hastened by stirring with a glass rod), the crucible is removed and the warm solution is filtered through Celite® to remove any insoluble material. The solution is transferred to a beaker and, while stirring rapidly with a glass rod, 200 mL of 95% ethanol (EtOH) is added to the solution. The product separates first as an oil, which is readily transformed into a white powder with continued stirring (for about 15 min). After standing for 30 min, the product is collected on a medium-glass frit and washed twice with 25-mL portions of 95% EtOH. The reprecipitation from 200 mL of water by the addition of 200 mL of EtOH is repeated twice more. The white powder is spread in a 14-cm weighing dish and then dried at 60°C overnight. Yield: 9.3–13.4 g (6.8–9.8 mmol, 54–78%).

3. Preparation of $[(C_4H_9)_4N]_{12}H_4[Nb_6P_4W_{30}O_{123}]$

$$\{H[Nb_6O_{19}]\}^{7-} + 2[P_2W_{15}O_{56}]^{12-} + 19H^+ + 12[(CH_3)_4N]^+ \xrightarrow[\text{2) } NaHSO_3]{\text{1) } H_2O_2}$$

$$[(CH_3)_4N]_{12}H_4[Nb_6P_4W_{30}O_{123}] + 8H_2O$$

$$[(CH_3)_4N]_{12}H_4[Nb_6P_4W_{30}O_{123}] + 12[(C_4H_9)_4N]^+ \rightarrow$$

$$[(C_4H_9)_4N]_{12}H_4[Nb_6P_4W_{30}O_{123}] + 12[(CH_3)_4N]^+$$

$$\{H[Nb_6O_{19}]\}^{7-} + 2[P_2W_{15}O_{56}]^{12-} + 19H^+ + 12[(C_4H_9)_4N]^+ \rightarrow$$

$$[(C_4H_9)_4N]_{12}H_4[Nb_6P_4W_{30}O_{123}] + 8H_2O$$

Procedure[7b]

A pale-yellow solution is prepared by dissolving $K_7H[Nb_6O_{19}] \cdot 13H_2O$ (5.90 g, 4.30 mmol) in 650 mL of 0.5 M H_2O_2, which has been prepared by diluting 36.90 mL of 30% aqueous H_2O_2 to a volume of 650 mL with water. Next, 58 mL of 1 M HCl is added. Immediately thereafter, *finely powdered* $Na_{12}[P_2W_{15}O_{56}] \cdot 18H_2O$ (36.50 g, 8.46 mmol) is added *in a single step*. After complete dissolution of the $Na_{12}[P_2W_{15}O_{56}]$, $NaHSO_3$ (50 g) is added gradually over 10–15 min to destroy the peroxides, resulting in a clear, colorless solution.

■ **Caution.** *This reaction is quite exothermic and evolves SO_2; hence, it should be carried out with caution in a hood.*

The solution is stirred for 1–2 h until it has cooled to room temperature. A large excess (20 g) of solid tetramethylammonium chloride is added to the solution in a single step, causing a white precipitate to form. The white precipitate is collected on a Whatman No. 2 filter paper laid inside a Büchner funnel and dried in an oven at 60°C overnight. At this stage, the yield is typically about 40 g of crude material. The product is purified as follows by reprecipitation from a homogeneous solution of *hot* (> 80°C), unbuffered, pH 4.6 water. The compound is dissolved in pH 4.6 water, kept hot (> 80°C) in a steam bath, and then filtered once through a folded filter paper (Whatman No. 2). The volume of the clear filtrate is reduced to 300 mL, and the solution is cooled in an ice bath. The first crop of the product, $[(CH_3)_4N]_{12}H_4[Nb_6P_4W_{30}O_{123}] \cdot 16H_2O$, is collected on a Whatman No. 2 filter paper laid inside a Büchner funnel, washed several times with small portions of unbuffered (pH 4.6) water, and dried at room temperature under vacuum overnight. The combined filtrate and washings are reduced in volume to approximately 150 mL by rotary evaporation, and the solution is cooled in an ice bath. A second crop of product is collected, washed, and dried as above. Combined yield of the first and second crops: 28.0 g (3.00 mmol, 71%).

The $[(CH_3)_4N]_{12}H_4[Nb_6P_4W_{30}O_{123}] \cdot 16H_2O$ (15 g, 1.60 mmol) is dissolved in 500 mL of hot, unbuffered (pH 4.6) water; the clear solution is then allowed to cool to room temperature. Addition of solid tetrabutylammonium bromide (7 g, 21.7 mmol) *in a single step* results in the formation of a white precipitate. The precipitate is collected and washed on a coarse glass frit as follows* (this revised collection and washing procedure reduces the long

---

* An alternative procedure which requires less attention but more time is as follows. The suspension is placed in a refrigerator at 5°C overnight. The clear supernatant liquid (ca. 200 mL) is then removed carefully using a pipette so that the white precipitate at the bottom of the beaker is not disturbed by the pipetting/suction process. To this precipitate is added 400 mL of water; the resulting suspension is stirred vigorously for 30 min and then placed in a 5°C refrigerator overnight. The colloidal upper layer (ca. 400 mL) is again carefully removed using a pipette. After 200 mL of water is added and the suspension is vigorously stirred for 30 min, the white precipitate is collected on a Whatman No. 2 filter paper mounted on a Büchner funnel, washed with 400 mL of water (4 × 100 mL portions), and then thoroughly dried via prolonged aspiration. The product is then spread in a weighing dish and dried at 60°C overnight. Yield: 10.3 g (0.930 mmol, 58%).

filtration times inherent in our original procedure[7]: (1) The solution is placed on a coarse glass frit for 1 h without aspiration (weak aspiration is also acceptable); three layers appear (some water passed through the glass frit). (2) The top layer and then about one half of the second layer are carefully removed by a pipette that is connected to an aspirator. (3) Approximately 100–200 mL of water is added without stirring. (4) Steps 1–3 are repeated three or four times (i.e., with the precipitate on the same glass frit). (5) About 1050 mL (3×350 mL portions) of water are added, using a normal rinsing procedure of the precipitate employing suction via aspiration. This washing procedure usually requires 1–2 h. (If the amount of water used in the washing is smaller than that indicated, the yield is higher, but subsequent CH_3CN solutions of $[(C_4H_9)_4N]_{12}H_4[Nb_6P_4W_{30}O_{123}]$ are cloudy.) The sample is spread in a 14-cm weighing dish and dried at 60°C overnight. Yield: 10.0 g (0.903 mmol, 56%).

Anal. Calcd. for $C_{192}H_{436}N_{12}P_4W_{30}$-$Nb_6O_{123}$: C, 20.82; H, 3.97; N, 1.52; P, 1.12; Nb, 5.03; W, 49.79. Found: C, 20.50; H, 3.95; N, 1.50; P, 0.99; Nb, 5.16; W, 49.96. (Attempts to crystallize this product from acetone plus a few drops of neutral water led to a fine powder that, however, was *less pure* by elemental analysis, as documented elsewhere.[7])

Properties

The compound $[(C_4H_9)_4N]_{12}H_4[Nb_6P_4W_{30}O_{123}]$ is soluble in polar, aprotic solvents such as acetonitrile, dimethylsulfoxide, and *N,N*-dimethylformamide. Spectroscopic data (^{31}P and ^{183}W NMR spectroscopy and IR) are available[7]. The single Nb-O-Nb bridge in the dimeric $[Nb_6P_4W_{30}O_{123}]^{16-}$ is readily cleaved by $[(C_4H_9)_4N]OH$ to produce the monomeric $[Nb_3P_2W_{15}O_{62}]^{9-}$. The monomer–dimer equilibrium, $2[(C_4H_9)_4N]_9[Nb_3P_2W_{15}O_{62}] + 2H^+X^- \rightleftarrows [(C_4H_9)_4N]_{16}[Nb_6P_4W_{30}O_{123}] + 2[(C_4H_9)_4N]X$, lies to the right under acidic conditions and in nonaqueous solvent and is rather difficult to detect.[7] (Ultracentrifugation solution molecular-weight determinations under several conditions are available.[7]) Attempts to grow single crystals of this product suitable for a X-ray diffraction structural analysis have been unsuccessful.

B. PREPARATION OF $Na_3[(C_4H_9)_4N]_5[Ir(\alpha\text{-}Nb_3P_2W_{15}O_{62})\{\eta^4\text{-}C_8H_{12}\}]$

$$\{H_4[Nb_6P_4W_{30}O_{123}]^{12-} + 6OH^- \rightarrow 2[Nb_3P_2W_{15}O_{62}]^{9-} + 5H_2O$$

$$[Ir(C_8H_{12})Cl]_2 + 2Ag^+ + 4CH_3CN \rightarrow 2[Ir(C_8H_{12})(CH_3CN)_2]^+ + 2AgCl$$

$$2[Nb_3P_2W_{15}O_{62}]^{9-} + 2[Ir(C_8H_{12})(CH_3CN)_2]^+ + 10[(C_4H_9)_4N]^+ + 6Na^+$$

$$\rightarrow 2Na_3[(C_4H_9)_4N]_5[Ir[\alpha\text{-}Nb_3P_2W_{15}O_{62}]\{\eta^4\text{-}C_8H_{12}\}] + 4CH_3CN$$

$$\{H[Nb_6P_4W_{30}O_{123}]\}^{12-} + 6OH^- + [Ir(C_8H_{12})Cl]_2 + 2Ag^+ +$$

$$10^+[(C_4H_9)_4N]^+ + 6Na^+ \rightarrow 2Na_3[(C_4H_9)_4N]_5[Ir[\alpha\text{-}Nb_3P_2W_{15}O_{62}]$$

$$\{\eta^4\text{-}C_8H_{12}\}] + 5H_2O + 2AgCl$$

The deprotonated, organic solvent-soluble $[(C_4H_9)_4N]_9[\alpha\text{-}Nb_3P_2W_{15}O_{62}]$ is formed by cleavage of the $[(C_4H_9)_4N]_{12}H_4[Nb_6P_4W_{30}O_{123}]$ complex with six equivalents of tetrabutylammonium hydroxide, $[(C_4H_9)_4N]OH$.[7b] The resultant triniobium-substituted Dawson-type anion undergoes reaction with $[Ir(1,5\text{-}COD)(CH_3CN)_2]^+$ in CH_3CN in an oxygen-free atmosphere (i.e., in a ≤ 1 ppm O_2 dry box) to form a single isomer of the polyoxoanion-supported organometallic complex. The product is isolated in the best yield, and in analytically pure and solvate-free form, with the counterion composition $\{Na_3[(C_4H_9)_4N]_5\}^{8+}$. Purification of the desired product, and removal of the contaminating three equivalents of $[(C_4H_9)_4N]^+[BF_4]^-$ [not shown, but present, in the (simplified for clarity) third equation above], is accomplished by reprecipitating the product at least twice from acetonitrile using ethyl acetate (the contaminating $[(C_4H_9)_4N]^+[BF_4]^-$ is removed since it is very soluble in ethyl acetate whereas the title complex is not).

Procedure[7b]

A solution containing $[\alpha\text{-}Nb_3P_2W_{15}O_{62}]^{9-}$ is prepared by dissolving $[(C_4H_9)_4N]_{12}H_4[Nb_6P_4W_{30}O_{123}]$ (10.0 g, 0.903 mmol) in 30 mL of CH_3CN in a 100-mL, round-bottomed flask that includes a stirring bar, and adding dropwise an aqueous solution of tetrabutylammonium hydroxide, $[(C_4H_9)_4N]OH^*$ (0.415 M in water, 13.06 mL, 5.419 mmol). The resulting

---

* The purity of purchased aqueous $[(C_4H_9)_4N]OH$ varies in our experience; in particular, concentrated aqueous solutions of $[(C_4H_9)_4N]OH$ (40%, Aldrich) gradually deposit solid at the bottom of the bottle, especially when stored below 30°C [as seems advisable to prevent possible Hofmann elimination reactions producing $[(C_4H_9)_3N + 1\text{-butene} + H_2O]$. In some instances, simple warming of the bottle briefly in a steam bath redissolves the solid; however, it is better (as detailed below; see also p. 1695 and footnote 11 elsewhere[7]) to prepare fresh aqueous

clear solution is stirred for 30 min at room temperature. The solvent is then removed by rotary evaporation; drying is accomplished under vaccum at room temperature overnight. The residual colorless material, $[(C_4H_9)_4N]_9[\alpha\text{-}Nb_3P_2W_{15}O_{62}]$ (1.806 mmol), is taken into the dry box and used for the next preparation.

All further manipulations from here on must be done in an oxygen-free environment (a $\leq$ 1 ppm O_2 dry box as monitored by a Vacuum Atmospheres O_2 analyzer). Necessary items* for the drybox experiments should be assembled before the next step of the synthesis.

In a drybox, all of the solid $[(C_4H_9)_4N]_9 [Nb_3P_2W_{15}O_{62}]$ prepared above (11.32 g, 1.8 mmol) is placed in a 100-mL, round-bottomed flask and dissolved with stirring in 30 mL of distilled and degassed CH_3CN; a clear solution generally results. (If this solution is cloudy or contains a white precipitate, the yield is lower.) This solution is filtered through a folded filter paper (Whatman No. 2). The clear filtrate is collected in a 200-mL, round-bottomed flask that includes a stirring bar. The filter paper is washed with a small amount of CH_3CN and the washings are combined with the filtrate collected above. (By this step, a total of ca. 40 mL of CH_3CN have been used.) The combined filtrate is well stirred for the next step.

Approximately 17 mL of CH_3CN is then placed in a graduated cylinder. About 3–4 mL of the 17 mL of CH_3CN is placed into a test tube along with

---

* Solvents (all degassed by distillation under nitrogen): EtOAc (2 × 500 mL portions, distilled over K_2CO_3 and under nitrogen), diethyl ether (Et$_2$O; 500 mL; distilled under nitrogen by the manufacturer), and CH_3CN (500 mL, distilled over CaH$_2$ and under nitrogen). Reagents: [Ir(1,5-COD)Cl]$_2$ (0.6064 g, 0.903 mmol), Ag[BF$_4$] (0.3518 g, 1.807 mmol), and Na[BF$_4$] (0.5952 g, 5.421 mmol).

[(C$_4$H$_9$)$_4$N]OH directly from [(C$_4$H$_9$)$_4$N]X (X = Cl, Br or I) and Ag$_2$O,[21] to store the resultant solution in a refrigerator (followed by gentle warming before use if necessary to redissolve any solid), and to check the solution periodically by titration using methyl red and phenophthalein for both amine and total base content.

The aqueous [(C$_4$H$_9$)$_4$N]OH solution used herein is prepared as follows:[21] 29.4 g of Ag$_2$O is added to the aqueous solution of [(C$_4$H$_9$)$_4$N]Br (75 g in 250 mL of water, previously purged with N$_2$ for 1 h to remove any dissolved CO$_2$). This mixture is stirred at room temperature under N$_2$ for 24 h. Then ca. 2 g of charcoal is added to this solution which is then filtered through a Celite® analytical filter aid. This solution is diluted with water (which again was purged with N$_2$ to remove any dissolved CO$_2$) and then titrated with 1 M HCl solution to a phenolphthalein end point (the resulting titer should be ca. 0.415 M).

$[Ir(1,5-COD)Cl]_2{}^†$ (0.6064 g, 0.903 mmol); a yellow suspension results. Into this well-stirred suspension, a solution of $Ag[BF_4]$ (0.3518 g, 1.807 mmol; dissolved in a few additional milliliters of CH_3CN) is added using a pipette. The $[Ir(1,5-COD)Cl]_2$ dissolves completely as it reacts with the Ag^+, a white precipitate of AgCl deposits, and the supernatant solution is bright yellow. The test tube that contained the $Ag[BF_4]$ is now rinsed three times with a few mL of CH_3CN (to accomplish a quantitative transfer of the Ag^+; these rinses use up the remaining ca. 13 mL of CH_3CN in the graduated cylinder), and the washings are added to the well-stirred suspension which now contains all the $[Ir(1,5-COD)(CH_3CN)_2]BF_4$ and two equivalents of AgCl.

The white precipitate of AgCl is removed by filtration with a folded filter paper, and the clear filtrate is introduced directly into the clear, well-stirred solution of $[(C_4H_9)_4N]_9[\alpha\text{-}Nb_3P_2W_{15}O_{62}]$. Again, to accomplish a quantitative transfer, the precipitate on the filter paper is rinsed several times with a total of about 18 mL of CH_3CN, and the washings are added to the well-stirred solution described above which is now orange-red. The stirring is continued for an additional 15 min. (The total volume of the solution in the 200-mL, round-bottomed flask is ca. 75 mL at this point.) To this solution, three equivalents of solid $Na[BF_4]$ (0.5952 g, 5.421 mmol) is added quantitatively (i.e., the $Na[BF_4]$ left on the wall of the test tube is rinsed into the orange-red solution using a few milliliters of CH_3CN). The stirring is continued at room temperature until all of the $Na[BF_4]$ dissolves (ca. 20 min), as confirmed by the disappearance of white crystals of $Na[BF_4]$ on the bottom of the flask. ($Na[BF_4]$ itself is sparingly soluble in CH_3CN; however, it is soluble in the CH_3CN solution containing the $[(C_4H_9)_4N]^+$ salt of the polyoxoanion.)

---

† This complex is prepared by a slight modification of the directions given in the literature[22-24] (specifically, it is a midly air-sensitive complex and thus is collected, washed, dried, and stored under nitrogen, vide infra). Into a 50-mL, round-bottomed flask, equipped with a glass-blown stopcock side-arm, are placed water (20 mL), isopropanol (7 mL), and 1,5-cyclooctadiene (3.6 mL, treated for 1–2 h over alumina prior to use). This mixture is purged with nitrogen gas for 1 h to remove oxygen, then $(NH_4)_3[IrCl_6]$ (4 g, 8.72 mmol) is added. This mixture is refluxed under nitrogen for 8–19 h (somewhat higher yields have been recorded after the longer reflux times). The suspension is cooled to room temperature, and the reddish precipitate is collected by filtration under a stream of nitrogen on a medium glass frit. The precipitate is washed with 30 mL (3×10 mL) of cold (ca. 0°C) water and then with 50 mL (2×25 mL) of cold (ca. 0°C) ethanol (EtOH), both of which had previously been well purged with nitrogen. The reddish precipitate is then dried for at least 1 h under a nitrogen stream on the glass frit, followed by further drying for several hours under vacuum. The final product is stored under a nitrogen atmosphere (e.g., in a dry box). Yield: 1.7–2.0 g (2.59–2.98 mmol, 60–68%) of bright vermillion microcrystals. ^{1}H NMR (CD$_3$CN, 21°C): δ 1.54–1.57, 2.16–2.21, and 3.95 ppm; (CDCl$_3$, 21°C): δ 1.51–1.54, 2.16–2.24–2.47, and 4.23 ppm; ^{13}C NMR (CDCl$_3$, 21°C): δ 31.8 and 62.2 ppm.

The resulting reddish solution is filtered with a folded filter paper and the filtrate is collected in a 200-mL, round-bottomed flask. Next, the filtrate is rotary-evaporated thoroughly and then dried under vacuum for 5 h (all while still within the drybox). The residual solid, which contains air-sensitive $Na_3[(C_4H_9)_4N]_5[Ir(\alpha\text{-}Nb_3P_2W_{15}O_{62})\{\eta^4\text{-}C_8H_{12}\}]$ and four equivalents of $[(C_4H_9)_4N]BF_4$, is a darker yellow-orange.

Isolation and purification of the desired iridium–polyoxoanion product are performed using EtOAc reprecipitation as follows (again, all while in the drybox). The darkened orange solid in the 200-mL, round-bottomed flask (i.e., from the procedure above) is dissolved in about 8 mL of CH_3CN (excess CH_3CN lowers the yield), and the orange-red solution is transferred to a 600-mL beaker using a polyethylene pipette. The round-bottomed flask is rinsed with a small amount (ca. 9 mL) of CH_3CN and the washings are combined with the solution in the beaker. (By this step, a total of ca. 17 mL of CH_3CN have been used, and the resulting orange-red solution is clear and homogeneous.) To this solution, while stirring with a spatula, a total of 400 mL of distilled and degassed EtOAc is added slowly in 100-mL portions. A yellow precipitate is formed. After addition of the final portion, the stirring is continued for 15 min using a stirring bar and magnetic stirrer. The yellow precipitate is collected on a 30-mL medium-glass frit, washed twice with 70 mL of degassed diethyl ether, and dried for 30 min under vacuum. (The filtrate, which is discarded, is pale yellow and slightly cloudy.)

For complete removal of any contaminating $[(C_4H_9)_4N]BF_4$, the re-precipitation with EtOAc is repeated. The yellow powdered product is redissolved in approximately 17 mL of CH_3CN in a 140-mL beaker and the solution is filtered through a folded filter paper into a 600-mL beaker. The 140-mL beaker and the filter paper are washed with a few milliliters of CH_3CN, and the washings are added to the filtrate. To the combined filtrate in the 600-mL beaker, a second reprecipitation is performed using 400 mL of EtOAc in 100-mL portions, as above. The suspension containing the yellow precipitate is stirred again for 15–20 min. The mother liquor is almost colorless and slightly cloudy. The yellow precipitate is collected on a 30-mL medium-glass frit and washed three times with 70-mL portions of diethyl ether. The bright yellow powder is dried for 4 h at room temperature under vacuum (drying for 1 wk at 60°C under vacuum is required to remove the last trace of solvates which, otherwise, *interfere with the* ^1H *spectra*, for example[6]). Yield: 8.0 g (1.41 mmol, 78%).

Anal. Calcd. for $Na_3[(C_4H_9)_4N]_5[Ir(\alpha\text{-}Nb_3P_2W_{15}O_{62})\{\eta^4\text{-}C_8H_{12}\}]$: C, 18.63; H, 3.41; N, 1.23; Na, 1.22; Ir, 3.39; P, 1.09; W, 48.6; Nb, 4.91; O, 17.5. Found (and repeat analysis on an independent preparation): C, 18.95 (18.99); H, 3.54 (3.52); N, 1.49 (1.40); Na, 1.43 (1.20); Ir, 2.74 (3.24); P, 1.09 (1.07); W,

48.4 (48.5); Nb, 4.80 (4.66); O, 17.0 (17.6); total, 99.4% (100.18). This compound is stored in a screw-capped polyethylene bottle in the drybox, preferably double-bottled to protect it from trace oxygen. [In addition, the product may be stored over a small layer of the compound in the outer bottle topped by a layer of tissues (i.e., using some of the complex as its own, sacrificial, ideal-affinity oxygen scavenger).]

Properties

The product $Na_3[(C_4H_9)_4N]_5[Ir(\alpha-Nb_3P_2W_{15}O_{62})\{\eta^4-C_8H_{12}\}]$ is very air sensitive, very soluble in acetonitrile, N,N-dimethylformamide and dimethylsulfoxide, slightly soluble in acetone, and insoluble in ethyl acetate, diethyl ether, and similar nonpolar organic solvents. This product serves as an effective precatalyst to an active cyclohexene oxidation catalyst using oxygen as the oxidant[9,10] and to novel polyoxoanion and tetrabutylammonium stabilized $Ir(0)_{\sim 300}$ nanoclusters that are an active and long-lived catalyst for cyclohexene hydrogenation.[11,12] A solution molecular-weight measurement (1×10^{-5} M of the product, in 0.1 M $[(C_4H_9)_4N][PF_6]/CH_3CN$, under N_2) is consistent with the monomeric formulation given [Calcd. for $Na_3[(C_4H_9)_4N]_5[Ir(\alpha-Nb_3P_2W_{15}O_{62})\{\eta^4-C_8H_{12}\}]$, M_r (calculated) = 5670; M_r (found) = 6000 ± 600]. Additional characterization of the product includes IR and ^{31}P, ^{183}W, and ^{17}O NMR spectroscopy [all under strict air-free technique, such as J. Young valve-equipped NMR tubes (from Wilmad) filled in a dry box].[4-6] A 2.3–2.5 Å $Na^+ \cdots [Nb_3P_2W_{15}O_{62}]^{9-}$ "ion-pairing" interaction is present in the solid-state structure of $Na_9[Nb_3P_2W_{15}O_{62}]$.[8] Such ion-pairing persists in solution, causing the extra peaks and line broadening observed in the solution ^{31}P NMR spectra of $Na_3[(C_4H_9)_4N]_5[Ir(\alpha-Nb_3P_2W_{15}O_{62})\{\eta^4-C_8H_{12}\}]$.[4-6] The ^{31}P NMR spectrum in DMSO-d_6, relative to external 85% H_3PO_4, exhibits primarily two equal intensity lines at -8.2 and -14.1 ppm, but the downfield -8.2 ppm resonance is flanked by two smaller peaks at -7.5 and -9.2 ppm. Addition of three equivalents of the cryptand Kryptofix® 2.2.2 (4,7,13,16,21,24-hexaoxa-1,10-diazabicyclo[8.8.8]hexacosane) in the dry box to the solution in the NMR tube collapses the smaller, extra ^{31}P resonances (-7.5 and -9.2 ppm) into the center, -8.2 ppm resonance; Kryptofix® 2.2.2 addition also reduces the peak half-widths to their normal, 4–5 Hz values. The ^{183}W NMR spectrum in DMSO-d_6 without Kryptofix® 2.2.2, relative to 2 M Na_2WO_4 in 1:1 D_2O/H_2O by the external substitution method, shows primarily three relatively broad peaks at -125.0, -150.3, and -194.0 ppm with a 1:2:2 intensity. The addition of three equivalents of Kryptofix® 2.2.2 increases the signal to noise (S/N) of the ^{183}W NMR spectrum twofold, resulting apparently from the removal of the $Na^+ \cdots [Nb_3P_2W_{15}O_{62}]^{9-}$ ion pairing and, hence, the increased concentration of a single solution form of the complex. ^{1}H and ^{13}C NMR spectra and other characterization data, including the informative ^{17}O NMR spectrum,[4] are available elsewhere.[4-6]

C. Preparation of $Na_3[(C_4H_9)_4N]_5[Rh[\alpha\text{-}Nb_3P_2W_{15}O_{62}]\{\eta^4\text{-}C_8H_{12}\}]$

$$\{H_4[Nb_6P_4W_{30}O_{123}]\}^{12-} + 6OH^- + [Rh(C_8H_{12})Cl]_2 + 2Ag^+$$
$$+ 10[(C_4H_9)_4N]^+ + 6Na^+$$
$$\rightarrow 2Na_3[(C_4H_9)_4N]_5[Rh(\alpha\text{-}Nb_3P_2W_{15}O_{65})\{\eta^4\text{-}C_8H_{12}\}] + 5H_2O + 2AgCl$$

Procedure

All manipulations below must be carried out in an oxygen-free environment (a ≤ 1 ppm O_2 dry box). The necessary items* for the dry box experiments should be prepared and assembled before the synthesis is initiated.

In the dry box, solid $[(C_4H_9)_4N]_9[\alpha\text{-}Nb_3P_2W_{15}O_{62}]$* (10.25 g, 1.630 mmol) is placed into a 100-mL, round-bottomed flask and is dissolved with stirring in 30 mL of distilled, degassed CH_3CN to yield a clear solution. This solution is filtered through a folded filter paper (Whatman No. 2), and the clear filtrate is collected in a 200-mL, round-bottomed flask equipped with a stirring bar. The filter paper is rinsed with a small amount of CH_3CN and the washings are combined with the filtrate above. (By this step, a total of ca. 38 mL of CH_3CN has been used.) The combined filtrate is well stirred for the next step.

---

* Solvents (all degassed by distillation under nitrogen): EtOAc (500 mL × 2, distilled over K_2CO_3 under nitrogen), Et_2O (500 mL; distilled under nitrogen by the manufacturer), and CH_3CN (500 mL, distilled over CaH_2 under nitrogen). Reagents: $[Rh(1,5\text{-}COD)Cl]_2$ (0.4007 g, 0.813 mmol), $Ag[BF_4]$ (0.3183 g, 1.635 mmol), and $Na[BF_4]$ (0.5354 g, 4.876 mmol).
* The required $[(C_4H_9)_4N]_9[\alpha\text{-}Nb_3P_2W_{15}O_{62}]$ is prepared from $[(C_4H_9)_4N]_{12}H_4$-$[Nb_6P_4W_{30}O_{123}]$ (9.025 g, 0.815 mmol) and six equivalents of $[(C_4H_9)_4N]OH$ (0.415 M in water, 11.78 mL, 4.890 mmol) as described in the preceding section.

Next, about 15 mL of CH_3CN is placed in a graduated cylinder. About 5–6 mL of the 15 mL CH_3CN is put with the $[Rh(1,5-COD)Cl]_2$[†] (0.4007 g, 0.813 mmol) into a test tube; a yellowish suspension results. Into this well-stirred suspension, a homogeneous solution of $Ag[BF_4]$ (0.3183 g, 1.635 mmol; dissolved in a few milliliters of additional CH_3CN) is added using a pipette. The $[Rh(1,5-COD)Cl]_2$ dissolves completely as it reacts with the Ag^+, a white precipitate of AgCl deposits, and the supernatant solution is yellow. The test tube that contained the $Ag[BF_4]$ is now rinsed three times with a few milliliters of CH_3CN (to accomplish a quantitative transfer; these rinses use up the remaining 9–10 mL of the CH_3CN in the graduated cylinder), and the washings are also added to the well-stirred suspension which now contains all the $[Rh(1,5-COD)(CH_3CN)_2]BF_4$ and 2 equiv of AgCl.

The white precipitate of AgCl is removed by filtration with a folded filter paper, and the filtrate is introduced directly into the clear, well-stirred solution of $[(C_4H_9)_4N]_9[\alpha-Nb_3P_2W_{15}O_{62}]$. Again, to accomplish a quantitative transfer, the precipitate on the filter paper is rinsed several times with a total of about 15 mL of CH_3CN, and the washings are added to the well-stirred solution above which is now red. The stirring is continued for 15 min. (The total volume of the solution in the 200-mL, round-bottomed flask is ca. 68 mL at this point.) To this solution, solid $Na[BF_4]$ (0.5354 g, 4.876 mmol, 3 equiv) is added quantitatively (i.e., the $Na[BF_4]$ left on a wall of the test tube is rinsed into the red solution above using a few milliliters of

---

[†] This compound is prepared by a slight modification of the directions in the literature.[25] In particular, the solvent used for the reaction (water/ethanol vs. just ethanol), the reaction conditions (refluxing under nitrogen), and the isolation procedure are all different. Moreover, commercial $[Rh(1,5-COD)Cl]_2$ sometimes contains a small amount of unreacted $RhCl_3$ in our experience.

Into a 50-mL, round-bottomed flask equipped with a glass-blown stopcock sidearm are placed in water (ca. 40 mL), EtOH (120 mL), and 1,5-cyclooctadiene (6 mL, 0.083 mmol, excess; treated for 1–2 h over Al_2O_3 before use). This mixture is purged with nitrogen gas for 1 h to remove oxygen, and then 12.1 mmol of $RhCl_3 \cdot nH_2O$ is added (ca. 3 g; calculated exactly depending upon the precise nH_2O value). This mixture is refluxed under nitrogen for 5 h. The suspension is cooled to room temperature, and then the bright-orange precipitate is collected by filtration under a stream of nitrogen on a medium-glass frit. The precipitate is washed with 30 mL (3 × 10 mL) of cold (0°C) water and then 45 mL (3 × 15 mL) of cold (0°C) EtOH, both of which had previously been well purged with nitrogen. (If the washing steps are omitted, the product contains a small amount of black powder, presumably unreacted $RhCl_3 \cdot xH_2O$.) The orange precipitate is then dried for at least 1 h under a nitrogen stream on the glass frit, followed by further drying for several hours (to overnight) under vacuum. A bright yellow powder is obtained, yield 2.3 g (4.66 mmol, 77%), which is stored under nitrogen (e.g., in a dry box).^{1}H NMR ($CDCl_3$, 21°C): δ 1.73–1.75, 2.47–2.50, and 4.21 ppm; ^{13}C NMR ($CDCl_3$, 21°C): δ 30.86, 78.55, and 78.74 ppm.

CH_3CN). The stirring is continued at room temperature until all of the $Na[BF_4]$ dissolves (ca. 20 min). The resulting orange-red solution is filtered using a folded filter paper and the filtrate is collected in a 200-mL, round-bottomed flask. Next, the filtrate is rotary-evaporated to dryness and then dried further under vacuum for 5 h (all while still within the dry box). The residual solid, which contains air-sensitive $Na_3[(C_4H_9)_4]N]_5[Rh(\alpha\text{-}Nb_3P_2W_{15}O_{62})\{\eta^4\text{-}C_8H_{12}\}]$ and 4 equiv of $[(C_4H_9)_4N]BF_4$, assumes a darker yellow-orange color.

Isolation and purification are performed using the EtOAc reprecipitation as described in the preceding section (again, all while in the drybox). The dark-orange solid in the 200-mL, round-bottomed flask (i.e., from the step above) is dissolved in about 8 mL of CH_3CN (excess CH_3CN lowers the yield), and the orange-red solution is transferred to a 600-mL beaker using a polyethylene pipette. The round-bottomed flask is rinsed out with a small amount (ca. 7 mL) of CH_3CN and the washings are combined with the solution in the beaker. (By this step, a total of ca. 15 mL of CH_3CN has been used, and the resulting orange-red solution is clear and homogeneous.) While stirring with a spatula, a total of 400 mL of distilled and degassed EtOAc is added slowly to this solution, in 100-mL portions. A yellow precipitate forms. After the addition of the final portion of EtOAc, the stirring is continued for 15 min using a stirring bar and a magnetic stirrer. The orange-yellow precipitate is collected on a 30-mL, medium-glass frit, washed twice with 60-mL portions of degassed Et_2O, and dried for 30 min under a vacuum. (The filtrate, which is discarded, is pale orange and slightly cloudy.)

For complete removal of the contaminating $[(C_4H_9)_4N]BF_4$, the reprecipitation with EtOAc is repeated twice more from about 15 mL of CH_3CN solution (again, use of more CH_3CN than this lowers the yield) and using 400 mL of EtOAc in four 100-mL portions. The final product is collected on a 30-mL medium frit, is washed twice with 60-mL portions of Et_2O, and is dried for 4 h at room temperature under vacuum (drying for 1 wk at 60°C is required to remove the last trace of solvates which, otherwise, *interfere with the* ^1H *spectra*, for example[6]). The final product is an orange-yellow powder, yield 7.0 g (1.25 mmol, 77%).

Anal. Calcd. for $Na_3[(C_4H_9)_4N]_5[Rh(\alpha\text{-}Nb_3P_2W_{15}O_{62})\{\eta^4\text{-}C_8H_{12}\}]$: C, 18.94; H, 3.47; N, 1.26; Na, 1.24; Rh, 1.84; P, 1.11; W, 49.4; Nb, 4.95; O, 17.8. Found: C, 18.50; H, 3.50; N, 1.52; Na, 1.25; Rh, 1.70; P, 1.11; W, 49.6; Nb, 5.09; O, 16.6; total, 98.9%. This compound is stored in a screw-capped polyethylene bottle in the dry box, preferably double-bottled to protect it from trace oxygen. [In addition, the product may be stored over a small layer of the compound in the outer bottle topped by a layer of tissues (i.e., using some of the complex as its own, sacrificial, ideal-affinity oxygen scavenger).]

Properties

$Na_3[(C_4H_9)_4N]_5[Rh(\alpha-Nb_3P_2W_{15}O_{62})\{\eta^4-C_8H_{12}\}]$ is very air-sensitive, quite soluble in acetonitrile, *N,N*-dimethylformamide, and dimethyl sulfoxide, slightly soluble in acetone, and insoluble in ethyl acetate, diethyl ether, and similar less polar organic solvents. The ^{31}P, ^{183}W, ^1H, and ^{13}C NMR spectral data for the product are available and generally resemble those of the corresponding Ir complex[6] (the NMR spectra were again obtained under strict air-free conditions using J. Young valve-equipped NMR tubes filled in the dry box).[6] The polyoxoanion component of the product is characterized by its three, somewhat broad lines in the ^{183}W NMR spectrum (DMSO-d_6 without Kryptofix® 2.2.2) δ -133.9, -159.2, -202.4 ppm (relative to 2 M Na_2WO_4 in 1:1 D_2O/H_2O by the external substitution method), and its two-line ^{31}P NMR spectrum in DMSO-d_6 (without Kryptofix® 2.2.2) (two major signals with primarily equal intensity at δ -8.2 and -14.1 ppm, and smaller extra peaks due to the Na^+ ion pairing at δ -7.4 and -9.2 ppm, relative to external 85% H_3PO_4). If three equivalents of Kryptofix® 2.2.2 are added, similar changes (in the S/N of the ^{183}W NMR spectra and in the number of peaks and the half widths of the ^{31}P NMR spectra) to those seen for the iridium complex are observed. These and other characterization data are available elsewhere.[6]

References

1. R. G. Finke, C. A. Green, and B. Rapko, *Inorg. Synth.*, **27**, 128 (1990) and references therein.
2. R. G. Finke, K. Nomiya, C. A. Green, and M. W. Droege, *Inorg. Synth.*, **29**, 239 (1992).
3. Y. Lin, K. Nomiya, and R. G. Finke, *Inorg. Chem.*, **32**, 6040 (1993).
4. R. G. Finke, D. K. Lyon, K. Nomiya, S. Sur, and N. Mizuno, *Inorg. Chem.*, **29**, 1784 (1990).
5. M. Pohl and R. G. Finke, *Organometallics*, **12**, 1453 (1993).
6. M. Pohl, D. K. Lyon, N. Mizuno, K. Nomiya, and R. G. Finke, *Inorg. Chem.*, **34**, 1413 (1995). This paper should be consulted for additional details of, for example, the spectroscopic characterization of the precursor complexes and final product described herein.
7. (a) D. J. Edlund, R. J. Saxton, D. K. Lyon, and R. G. Finke, *Organometallics*, **7**, 1692 (1988).
7. (b) More recently, since the March 1994 acceptance of the present synthesis, we have even further improved the synthesis of $[(C_4H_9)_4N]_9[Nb_3P_2W_{15}O_{62}]$ to a ca. 4 day (30%) shorter, 116 g total (60% increased) yield procedure, one that omits the need to go through the $[(CH_3)_4N]^+$ salt steps: "Polyoxometalate Catalyst Precursors. An Improved Synthesis, H^+-Titration Procedure, and Evidence for ^{31}P NMR as a Highly Sensitive Support-Site Indicator for the Prototype Polyoxoanion Organometallic-Support System, $[(n-C_4H_9)_4N]_9P_2W_{15}Nb_3O_{62}]$", H. Weiner, J. D. Aiken III, and R. G. Finke, *Inorg. Chem.* submitted.
8. R. G. Finke, D. K. Lyon, K. Nomiya, and T. J. R. Weakley, *Acta Crystallogr.*, **C46**, 1592 (1990).
9. N. Mizuno, D. K. Lyon, and R. G. Finke, *U.S. Patent* 5,250,739, Oct. 5, 1993.
10. N. Mizuno, D. K. Lyon, and R. G. Finke, *J. Catal.*, **128**, 84 (1991).
11. D. J. Edlund, R. G. Finke, and R. J. Saxton, *U.S. Patent* 5,116,796, May 26, 1992.

12. (a) Y. Lin and R. G. Finke, *J. Am. Chem. Soc.*, **116**, 8335 (1994). (b) Y. Lin and R. G. Finke, *J. Inorg. Chem.*, **33**, 4891 (1994).

13. J. D. Aiken III, M. Pohl and R. Finke, unpublished results.

14. R. G. Finke in *Polyoxometalates: From Platonic Solids to Anti-Retroviral Activity*, M. T. Pope and A. Müller, Eds., Kluwer, Dodrecht, 1994, pp. 267–280. (This monograph is Vol. 10 of *Topics in Molecular Engineering*, J. Maruani, Ed.)

15. The use of $[Nb_6O_{19}]^{8-}$ and H_2O_2 as a reagent in polyoxoanion synthesis was first reported by M. Dabbabi and M. Boyer, *J. Inorg. Nucl. Chem.*, **38**, 1011 (1976).

16. W. J. Randall, M. W. Droege, N. Mizuno, K. Nomiya, T. J. R. Weakley, and R. G. Finke, *Inorg. Synth.*, **31**, 167–185 (1996).

17. (a) The α, β-isomerism refers to a $\pi/3$ rotation of a W_3 triad of edge-sharing octahedra (see elsewhere for additional discussion[17b]). (b) M. T. Pope, *Heteropoly and Isopoly Oxometalates*, Springer-Verlag, Berlin, 1983, p. 69.

18. (a) A separate *Inorganic Syntheses* preparation[18b] of $Na_{12}[\alpha\text{-}P_2W_{15}O_{56}]\cdot18H_2O$ from the authors of the original synthesis[18c] has recently appeared; that is, we wish to specifically point out that the original synthesis is not ours, but rather is due to the efforts and insights of others.[18b,c] However, it should be noted that the synthesis reported herein was carried out using our scaled-up version of the synthesis of $Na_{12}[\alpha\text{-}P_2W_{15}O_{56}]\cdot18H_2O$,[16] a preparation that we developed and independently refined over the years since our 1983 paper which first used $Na_{12}[\alpha\text{-}P_2W_{15}O_{56}]$.[18d] Moreover, the synthesis of $Na_{12}[\alpha\text{-}P_2W_{15}O_{56}]\cdot18H_2O$ employed herein (and thus which should be used when duplicating the present synthesis)[16] was independently checked *twice* (i.e., was checked both herein and as part of another, cosubmitted *Inorganic Syntheses*[16]). (b) R. Contant and P. J. Ciabrini, *J. Chem. Res. Synop.*, **1977**, 222; *J. Chem. Res.*, Miniprint 1977, 2601. (c) R. Contant, *Inorg. Synth.*, **27**, 108 (1990). (d) R. G. Finke and M. W. Droege, *Inorg. Chem.*, **22**, 1006 (1983). R. G. Finke, M. W. Droege, and P. J. Domaille, *Inorg. Chem.*, **26**, 3886 (1987). See also reference 7.

19. (a) M. Filowitz, R. K. C. Ho, W. G. Klemperer, and W. Shum, *Inorg. Chem.*, **18**, 93 (1979). (b) Klemperer has reported a preparation of crystalline material. C. J. Besecker, V. W. Day, W. G. Klemperer, and M. R. Thompson, *J. Am. Chem. Soc.*, **106**, 4125 (1984).

20. C. M. Flynn and G. D. Stucky, *Inorg. Chem.*, **8**, 178 (1969).

21. (a) R. H. Cundiff and P. C. Markunas, *Anal. Chem.*, **30**, 1447 (1958). (b) R. H. Cundiff and P. C. Markunas, *Anal. Chem.*, **30**, 1450 (1958). (c) R. H. Cundiff, and P. C. Markunas, *Anal. Chem.*, **28**, 792 (1956).

22. J. L. Herdé, J. C. Lambert, and C. V. Senoff, *Inorg. Synth.*, **15**, 18 (1974).

23. J. L. Herdé and C. V. Senoff, *Inorg. Nucl. Chem. Lett.*, **7**, 1029 (1971).

24. R. H. Crabtree, J. M. Quirk, H. Felkin, and T. Fillebun-Khan, *Syn. React. Inorg. Met.-Org. Chem.*, **12**, 407 (1982).

25. J. Chatt and L. M. Venanzi, *J. Chem. Soc.*, 4735 (1957).

Chapter Three

ORGANOMETALLIC COMPOUNDS

30. ONE-POT SYNTHESIS OF DICARBONYLTRIS(PHOSPHANE)IRON(0) COMPLEXES FROM PENTACARBONYLIRON

SUBMITTED BY J.-J. BRUNET,* F. B. KINDELA,*
and D. NEIBECKER*
CHECKED BY RICHARD L. KEITER[†] and ELLEN A. KEITER[†]

$$Fe(CO)_5 + 2KOH \rightarrow K[FeH(CO)_4] + KHCO_3$$

$$K[FeH(CO)_4] + 3PR_3 + EtOH \rightarrow Fe(CO)_2(PR_3)_3 + H_2 + 2CO + KOEt$$

$$K[FeH(CO)_4] + 3P(OR)_3 + ROH(H_2O) \rightarrow$$

$$Fe(CO)_2[P(OR)_3]_3 + H_2 + 2CO + KOR(KOH)$$

■ **Caution.** *Due to the toxic nature of pentacarbonyliron and to the evolution of highly toxic carbon monoxide (a colorless and odorless gas) and of highly flammable hydrogen, and also because of the toxic nature, bad odor, and risk of ignition of liquid phosphanes, these reactions should be performed in a well-ventilated hood and gloves should be worn.*

Dicarbonyltris(phosphane)iron complexes have been known for more than 30 years. In 1960, Manuel and Stone prepared the complex $Fe(CO)_2[P(C_6H_5)_3]_3$ in low yields (4–13%) by the reaction of a (diene)- or (triene)tricarbonyliron complex with an excess of triphenylphosphine in refluxing ethylcyclohexane.[1]

---

* Laboratoire de Chimie de Coordination du CNRS, unité No 8241 liée par conventions à l'Université Paul Sabatier et à l'Institut National Polytechnique, 205 route de Narbonne, 31077 Toulouse Cedex, France.
† Department of Chemistry, Eastern Illinois University, Charleston, IL 61920.

Today several methods are known, which can be classified into two types, depending on whether they involve the reduction of iron(II) compounds or ligand substitutions on iron(0) complexes.

According to the first method, the dicarbonyltris(phosphane)iron complexes are available by reduction of $FeCl_2$ with sodium amalgam, sodium sand, manganese, or magnesium, under a carbon monoxide or a carbon dioxide atmosphere, in the presence of an excess of a phosphane. Yields are from moderate [$P(CH_3)_3$, 14%[2] and 46%;[3] $P(OCH_3)_3$, $P(OC_2H_5)_3$, 35–40%[4]] to excellent [$P(CH_3)_3$, 92%[5]]. However, the corresponding tributyl and triphenyl phosphite complexes are obtained as a mixture with the tricarbonylbis(phosphite)iron complexes.[4] Similar reduction of $Fe(CO)_2[P(C_6H_5)_3]_2Br_2$ by sodium amalgam in the presence of a fivefold excess of $P(C_6H_5)_3$ affords the complex $Fe(CO)_2[P(C_6H_5)_3]_3$ in 80–90% yield.[6] The same derivative can be obtained by reducing diazonium complexes $\{Fe(ArN_2)(CO)_2[P(C_6H_5)_3]_2\}^{1+}[BF_4]^{1-}$ with $NaOCH_3$ (44% yield)[7] or with $LiOC_2H_5$ in the presence of excess triphenylphosphine (70–80% yield).[8]

According to the second method, the $Fe(CO)_2[P(OCH_3)_3]_3$ complex may be prepared by the thermal reaction (100–140°C) of a large excess of $P(OCH_3)_3$ with either the tricarbonyl(1,3-butadiene)- or the tricarbonyl-(1,3,5-cycloheptatriene)iron complex (quantitative yield).[9] It is also available from the reaction of $Fe[P(OCH_3)_3]_5$ with CO (57%)[2] and, as a mixture with $Fe(CO)_3[P(OCH_3)_3]_2$, by displacement of nitrogen by $P(OCH_3)_3$ from $\{Fe(CO)_2[P(OCH_3)_3]_2\}_2(\mu\text{-}N_2)$ (unreported yield).[10] The $Fe(CO)_2[P(CH_3)_3]_3$ complex can also be prepared by the reaction of CO with the $Fe[P(CH_3)_3]_4$/ $Fe(H)[CH_2P(CH_3)_2][P(CH_3)_3]_3$ equilibrium mixture, but is often contaminated with $Fe(CO)_3[P(CH_3)_3]_2$.[5, 11]

Several other dicarbonyltris(phosphane)iron complexes are mentioned as coproducts in the reaction of tricarbonyl(1,2,3,6,-η-cyclooctenediyl)iron with phosphanes, but only the $P(OC_2H_5)_3$ derivative could be isolated (60%).[12]

All the reported methods suffer from drawbacks, being either multistep processes that are time-consuming or affording difficult-to-separate mixtures of bis- and tris(phosphane) complexes.[13] Moreover, they usually require an excess of phosphane or the cumbersome preparation of sophisticated starting materials, ultimately leading to low overall yields. It is noteworthy that the tris(phosphine) complexes cannot be prepared by substitution of CO by phosphine in the well-known and easily prepared bis(phosphine) complexes.[14]

The procedure reported here involves a one-pot reaction of the inexpensive reagent $K[FeH(CO)_4]$, which is readily available from $Fe(CO)_5$, with a stoichiometric quantity of a phosphane in a protic medium. It affords the complexes $Fe(CO)_2(PR_3)_3$ or $Fe(CO)_2[P(OR)_3]_3$ in fair to high yields using tributylphosphine, dimethylphenylphosphine, and trimethyl, triethyl, and triphenyl phosphite as the phosphorus ligand, but fails in the case of triphenylphosphine and tricyclohexylphosphine.

A. PREPARATION OF ALCOHOLIC OR WATER/THF SOLUTIONS OF POTASSIUM TETRACARBONYLHYDRIDOFERRATE(1−)

$$Fe(CO)_5 + 2KOH \rightarrow K[FeH(CO)_4] + KHCO_3$$

■ **Caution.** *Fe(CO)$_5$ is a toxic liquid and should be handled with hands protected by gloves in a well-ventilated hood.*
Alcoholic (CH_3OH or C_2H_5OH) or H_2O/THF solutions of potassium tetra-carbonylhydridoferrate(1 −) salts are readily prepared by the rapid and quantitative reaction of $Fe(CO)_5$ with 2 equiv of KOH following a procedure described in *Inorganic Syntheses.*[15]

B. DICARBONYLTRIS(TRIBUTYLPHOSPHINE)IRON(0)

$$K[FeH(CO)_4] + 3P(n\text{-}C_4H_9)_3 + EtOH \rightarrow$$
$$Fe(CO)_2[P(n\text{-}C_4H_9)_3]_3 + H_2 + 2CO + KOEt$$

Procedure

All manipulations are carried out under argon using solvents thoroughly degassed by purging with argon for 0.5 h.

A 60-mL ethanolic solution of $K[FeH(CO)_4]$ (11 mmol) is prepared in a 100-mL Schlenk flask under argon as described in Section 30.A. The Schlenk flask is fitted with a reflux condenser equipped with a two-way adaptater sealed with a serum cap and connected to an oil bubbler. Tributyl-phosphine (Aldrich 95%, distilled under argon and stored under argon) (6.68 g, 33 mmol) is syringed through the serum cap into the reaction medium under a slow stream of argon. The stream of argon is stopped and the mixture is stirred vigorously at 10–15°C. Gas evolves slowly during a 24- to 30-h period and the reaction medium turns yellow. It is then brought to reflux for 2 h by means of a preheated oil bath. After cooling to room temperature under argon, the solvent is removed under reduced pressure. The resulting yellow slurry is treated with pentane (60 mL) to give a yellow suspension which is filtered under argon to remove $KHCO_3$ coproduced in the prepara-tion of $K[FeH(CO)_4]$. The $KHCO_3$ is washed with pentane (3 × 20 mL) and the combined filtrates are concentrated under reduced pressure to a volume of 5 mL. The remaining yellow solution is further filtered under argon through a column (3.5-cm diameter) made (from bottom to top) of silica gel

(230–400 mesh, 10 g), alumina (activated, basic, Brockman I, 150 mesh, 10 g) and $CoCl_2 \cdot 6H_2O$ (Prolabo, 96%, 2.5 g), thoroughly degassed by heating under vacuum and cooling under argon cycles. Elution with pentane (200 mL) under argon and evaporation of the resulting yellow solution up to a constant weight affords a yellow syrup requiring no further purification. Yield: 6.62 g, 84% (Note 1).

Anal. Calcd. for $C_{38}H_{81}FeO_2P_3$: C, 63.49; H, 11.36. Found: C, 63.31; H, 11.40.

C. DICARBONYLTRIS(DIMETHYLPHENYLPHOSPHINE)IRON(0)

$$K[FeH(CO)_4] + 3P(CH_3)_2C_6H_5 + EtOH \rightarrow$$
$$Fe(CO)_2[P(CH_3)_2C_6H_5]_3 + H_2 + 2CO + KOEt$$

Procedure

All manipulations are carried out under argon using solvents thoroughly degassed by purging with argon for 0.5 h.

A 30-mL ethanolic solution of $K[FeH(CO)_4]$ (5.5 mmol) is prepared in a 100-mL Schlenk flask under argon as described in Section 30.A. The Schlenk flask is equipped as described in Section 30.B and the reaction medium is stirred vigorously at reflux under a stream of argon during 0.25 h by means of a preheated oil bath. Dimethylphenylphosphine (Aldrich 99%, distilled under argon and stored under argon) (2.28 g, 16.5 mmol) is syringed through the serum cap and falls into the refluxing reaction medium under a slow stream of argon. The stream of argon is then stopped. The vigorous evolution of gas is observed and the reflux is continued for 3 h. After cooling the reaction mixture to room temperature under argon, the solvent is removed under reduced pressure to leave a yellow slurry. This slurry is treated with pentane (60 mL) and the resulting yellow suspension is filtered under argon to remove $KHCO_3$, which is washed with pentane (3 × 20 mL). The combined filtrates are evaporated under reduced pressure and the remaining solid is dissolved in a minimum amount of a $CH_2Cl_2/MeOH$ (1/2) solvent mixture. Cooling to $-20°C$ overnight affords a solid and a yellow liquid phase. The latter is removed with a syringe and the solid is washed with cold ($-20°C$) methanol (3 × 30 mL) and dried at room temperature under reduced pressure to a constant weight. $Fe(CO)_2[P(CH_3)_2C_6H_5]_3$ is a yellow, sometimes green oil, at room temperature. Yield: 2.7 g, 93% (Note 2).

Anal. Calcd. for $C_{26}H_{33}FeO_2P_3$: C, 59.33; H, 6.32. Found: C, 59.38; H, 6.61.

D. DICARBONYLTRIS(TRIMETHYL PHOSPHITE)IRON(0)

$$K[FeH(CO)_4] + 3P(OCH_3)_3 + MeOH \rightarrow$$
$$Fe(CO)_2[P(OCH_3)_3]_3 + H_2 + 2CO + KOMe$$

Procedure

All manipulations are carried out under argon using solvents thoroughly degassed by purging with argon for 0.5 h.

A 30-mL methanolic solution of $K[FeH(CO)_4]$ (11 mmol) is prepared in a 100-mL Schlenk flask under argon as described in Section 30.A. The same equipment and experimental procedure as those described in the preceding section are used, except that trimethyl phosphite (Aldrich 97%, distilled under argon and stored under argon) (4.10 g, 33 mmol) is added dropwise through the serum cap. Vigorous stirring is maintained for 48 h at reflux. During this time, gas evolution is observed and the reaction medium turns light green and then white. After cooling to room temperature, the solvent is removed under reduced pressure to leave a white slurry. This is treated with pentane (60 mL) and the resulting white-green suspension is filtered to remove $KHCO_3$. The latter is washed with pentane (3×20 mL) and the combined filtrates are concentrated under reduced pressure to a 30-mL volume. Cooling to $-78°C$ during 6 h results in the precipitation of white crystals which are collected, washed with cold ($-78°C$) pentane (2×20 mL), and dried at room temperature under reduced pressure to a constant weight. $Fe(CO)_2[P(OCH_3)_3]_3$ is an off-white solid which melts at 46–47°C. Yield: 4.9 g, 91% (Note 3).

Anal. Calcd. for $C_{11}H_{27}FeO_{11}P_3$: C, 27.29; H, 5.62. Found: C, 27.56; H, 5.82.

E. DICARBONYLTRIS(TRIETHYL PHOSPHITE)IRON(0)

$$K[FeH(CO)_4] + 3P(OC_2H_5)_3 + EtOH \rightarrow$$
$$Fe(CO)_2[P(OC_2H_5)_3]_3 + H_2 + 2CO + KOEt$$

Procedure

All manipulations are carried out under argon using solvents thoroughly degassed by purging with argon for 0.5 h.

A 30-mL ethanolic solution of K[FeH(CO)$_4$] (11 mmol) is prepared in a 100-mL Schlenk flask under argon as described in Section 30.A. The equipment and the experimental procedure are the same as those described in Section 30.D, using triethyl phosphite (Fluka > 97%, distilled under argon and stored under argon) (5.48 g, 33 mmol). The reflux time is 8 h. The isolated Fe(CO)$_2$[P(OC$_2$H$_5$)$_3$]$_3$ is an off-white solid which melts at 51–52°C. Yield: 6.5 g, 96% (Note 4).

Anal. Calcd. for C$_{20}$H$_{45}$FeO$_{11}$P$_3$: C, 40.03; H, 7.51. Found: C, 40.31; H, 7.63.

F. DICARBONYLTRIS(TRIPHENYL PHOSPHITE)IRON(0)

$$K[FeH(CO)_4] + 3P(OC_6H_5)_3 + H_2O \rightarrow$$
$$Fe(CO)_2[P(OC_6H_5)_3]_3 + H_2 + 2CO + KOH$$

Procedure

All manipulations are carried out under argon using solvents thoroughly degassed by purging with argon for 0.5 h.

A 30-mL H$_2$O/THF (20/10 mL) solution of K[FeH(CO)$_4$] (11 mmol) is prepared in a 100-mL Schlenk flask under argon as described in Section 30.A. The Schlenk flask is equipped as described in Section 30.B and the reaction medium is stirred and heated under a stream of argon for 0.25 h by means of a preheated oil bath (bath temperature: 80°C). The stream of argon is stopped and triphenyl phosphite (Fluka > 97%, distilled under argon and stored under argon) (10.3 g, 33 mmol) is syringed dropwise through the serum cap. Stirring is maintained for 72 h at reflux during which time the reaction medium turns light green and then yellow. After cooling to room temperature, the reaction medium consists of two phases. The upper yellow phase is syringed into another Schlenk flask and the aqueous phase is extracted under argon with THF until the latter is no longer colored (3 × 40 mL). The combined THF phases are evaporated under reduced pressure to give a yellow oily residue which is dissolved in dichloromethane (40 mL). The CH$_2$Cl$_2$ solution is filtered under argon and concentrated to a volume of 5 mL. Addition of pentane (100 mL) affords a white solid and a yellow liquid phase. The latter is syringed off and the white solid is dried under reduced pressure to a constant weight (6 g). The liquid phase is concentrated to half of its volume and cooled at − 20°C for 12 h. Additional white solid precipitates.

This is separated from the mother liquor and dried under reduced pressure to a constant weight (2.6 g). $Fe(CO)_2[P(OC_6H_5)_3]_3$ is a white solid which melts at 85–86°C. Yield: 8.6 g, 75% (Note 5).

Anal. Calcd. for $C_{56}H_{45}FeO_{11}P_3$: C, 64.51; H, 4.35. Found: C, 64.22; H, 4.47.

Notes

1. The checkers used 8.22 mL of distilled $P(n-C_4H_9)_3$ (Aldrich), 60–200 mesh silica gel, and 80–200 mesh alumina and obtained a 76% yield of $Fe(CO)_2[P(n-C_4H_9)_3]_3$.
2. The checkers used 2.34 mL of $P(CH_3)_2(C_6H_5)$ and obtained a 74% yield of $Fe(CO)_2[P(CH_3)_2(C_6H_5)]_3$.
3. The checkers used 3.90 mL of Aldrich 99 + % NMR grade $P(OCH_3)_3$ and obtained a 69% yield of $Fe(CO)_2[P(OCH_3)_3]_3$ after a 24-h reaction time.
4. The checkers used 5.65 mL of $P(OC_2H_5)_3$ (98% Aldrich) and obtained a 88% yield of $Fe(CO)_2[P(OC_2H_5)_3]_3$.
5. The checkers used 8.70 mL of Aldrich 97% grade $P(OC_6H_5)_3$ and obtained a 42% yield of $Fe(CO)_2[P(OC_6H_5)_3]_3$.

Properties

The compounds $Fe(CO)_2[P(n-C_4H_9)_3]_3$ and $Fe(CO)_2[P(CH_3)_2C_6H_5]_3$ are stable under argon at $-20°C$ in the dark. They are soluble in common organic solvents (ethanol, pentane, hexane, THF, acetone, CH_2Cl_2) at room temperature and less soluble in cold ($-20°C$) methanol and are decomposed by $CHCl_3$. Their solutions gradually turn pale green under argon, without any differences in their IR and NMR spectral characteristics, and are very air-sensitive.

The tris(phosphite) complexes are stable in the solid state under argon at $-20°C$ in the dark. The compounds $Fe(CO)_2[P(OC_2H_5)_3]_3$ and $Fe(CO)_2$-$[P(OC_6H_5)_3]_3$ are also stable at room temperature under argon, but $Fe(CO)_2$-$[P(OCH_3)_3]_3$ gradually turns red after 1 day in the presence of visible light. The compounds $Fe(CO)_2[P(OCH_3)_3]_3$ and $Fe(CO)_2[P(OC_2H_5)_3]_3$ are soluble in methanol, ethanol, THF, pentane, hexane, acetone, CH_2Cl_2, $CHCl_3$, and toluene. The compound $Fe(CO)_2[P(OC_6H_5)_3]_3$ is soluble in THF, acetone, CH_2Cl_2, and toluene, but displays a low solubility in pentane and hexane.

Pertinent spectral characteristics are listed in the following table:

Compound	IR^a v_{CO} (cm^{-1})	^{31}P{^{1}H} NMR^b δ (ppm)	^{13}C{^{1}H} NMR^c δ_{CO} (ppm)	J_{C-P} (Hz) (mult.)
Fe(CO)$_2$[P(n-C$_4$H$_9$)$_3$]$_3$	1820 (vs)^d 1830 (s) 1870 (s)	49.4	222.9^h	23 (q)
Fe(CO)$_2$[P(CH$_3$)$_2$C$_6$H$_5$]$_3$	1825 (vs)^d 1880 (vs)	41.1	220.3^i	22 (q)
Fe(CO)$_2$[P(OCH$_3$)$_3$]$_3$	1875 (vs)^e 1935 (s)	193.7	216.1^j	26 (q)
Fe(CO)$_2$[P(OC$_2$H$_5$)$_3$]$_3$	1865 (vs)^f 1925 (s)	188.0	216.5^k	27 (q)
Fe(CO)$_2$[P(OC$_6$H$_5$)$_3$]$_3$	1900 (vs)^g 1965 (s)	170.7	212.9^l	28 (q)

^{a} Solution IR spectra recorded on a Perkin Elmer 597 spectrophotometer.
^{b} ^{31}P{^{1}H} NMR spectra were recorded on a Bruker WM 250 at 101.26 MHz in CD$_3$COCD$_3$ at 263 K. Chemical shifts are in parts per million downfield from external 85% H$_3$PO$_4$ in D$_2$O.
^{c} ^{13}C{^{1}H} NMR spectra were recorded on a Bruker WM 250 at 62.89 MHz in CD$_3$COCD$_3$ at 263 K. Chemical shifts are in parts per million downfield TMS, assigning the methyl resonance of acetone at 29.20 ppm.
^{d} In hexane.
^{e} In methanol.
^{f} In ethanol.
^{g} In THF.
^{h} Other signals at 31.0, 26.0, 24.7, and 13.7 ppm.
^{i} Other signals at 142.5, 130.3, 128.7, 128.0, and 21.1 ppm.
^{j} Other signal at 50.9 ppm.
^{k} Other signals at 59.7 and 15.9 ppm.
^{l} Other signals at 151.9, 129.5, 124.4, and 121.4 ppm.

References

1. T. A. Manuel and F. G. A. Stone, *J. Am. Chem. Soc.*, **82**, 366 (1960).
2. T. V. Harris, J. W. Rathke, and E. L. Muetterties, *J. Am. Chem. Soc.*, **100**, 6966 (1978).
3. D. L. Allen, M. L. H. Green, and J. A. Bandy, *J. Chem. Soc.*, *Dalton Trans.*, 541 (1990).
4. L. P. Battaglia, T. Boselli, G. P. Chiusoli, M. Nardelli, C. Pelizzi, and G. Predieri, *Gazz. Chim. Ital.*, **115**, 395 (1985).
5. H. H. Karch, H. F. Klein, and H. Schmidbauer, *Chem. Ber.*, **110**, 2200 (1977).
6. W. Hieber and J. Muschi, *Chem. Ber.*, **98**, 3931 (1965).

7. W. E. Carroll, F. A. Deeney, and F. J. Lalor, *J. Chem. Soc., Dalton Trans.*, 1430 (1974).
8. S. Cenini, F. Porta, and M. Pizzotti, *Inorg. Chim. Acta*, **20**, 119 (1976).
9. A. Reckziegel and M. Bigorgne, *J. Organomet. Chem.*, **3**, 341 (1965).
10. H. Berke, W. Bankhardt, G. Huttner, J. von Seyerl, and L. Zsolnai, *Chem. Ber.*, **114**, 2754 (1981).
11. J. Grobe and H. Zimmermann, *Z. Naturforsch.*, **39B**, 808 (1984).
12. B. F. G. Johnson, J. Lewis, and M. V. Twigg, *J. Chem. Soc., Dalton Trans.*, 241 (1974).
13. T. Boselli, A. Mangia, C. Pelizzi, and G. Predieri, *J. Chrom.*, **265**, 347 (1983).
14. J. J. Brunet, F. B. Kindela, and D. Neibecker, *Organometallics*, **11**, 1343 (1992).
15. J. J. Brunet, F. B. Kindela, and D. Neibecker, *Inorg. Synth.*, **29**, 151 and 156 (1992).

31. TRICARBONYLBIS(PHOSPHINE)IRON(0) COMPLEXES

SUBMITTED BY RICHARD L. KEITER,* ELLEN A. KEITER,*
CAROL A. BOECKER,* DAVID R. MILLER,* and KARL H. HECKER*
CHECKED BY GEORGE B. KAUFFMAN,† JOE D. TONEY,†
KIN C. NG,† ERIC ALVIN HAYNIE,† MALRUBIO CABRERA II,†
ROCKY DEAN GIPSON,† BRIAN AMPÈRE SMITH,†
and CHING KIN YIM†

Iron complexes, *trans*-$Fe(CO)_3(PR_3)_2$, continue to be useful starting materials for a variety of reactions.[1] Thermal methods of preparation, utilizing $Fe(CO)_5$ and $Fe_3(CO)_{12}$, have been known for some time to give low-yield mixtures of $Fe(CO)_4PR_3$ and *trans*-$Fe(CO)_3(PR_3)_2$.[2] Photolytic procedures have been optimized to give good yields of disubstituted derivatives, but this approach requires photolysis equipment, excess phosphine, and rather long reaction times.[3] We have found that the reaction of $Fe(CO)_5$ with PR_3 and $Na[BH_4]$ in refluxing 1-butanol gives *trans*-$Fe(CO)_3(PR_3)_2$ selectively. The reaction avoids specialized equipment as well as large excesses of phosphine and is complete within 2 h.[4]

■ **Caution.** *Pentacarbonyliron(0) is a toxic, volatile liquid. It is sensitive to light and heat and can be pyrophoric in air. Carbon monoxide is evolved during the course of the synthesis. Diphenylphosphine, methyldiphenylphosphine, and tributylphosphine are malodorous, air-sensitive, and toxic liquids. All manipulations must be carried out in a well-ventilated fume hood.*

Reagents

All reagents were obtained from Aldrich and used without further purification.

* Department of Chemistry, Eastern Illinois University Charleston, IL 61920.
† Department of Chemistry, California State, Fresno, CA 93740.

Procedure

The reaction is carried out in a side-armed (septum inlet), 200-mL, one-necked, round-bottomed flask equipped with a large magnetic stirring bar and a condenser that is attached to a nitrogen inlet tube.* The assembly is placed in an oil bath that rests on a magnetic stirrer. A mole ratio, $Fe(CO)_5 : PR_3 : Na[BH_4] = 1.0 : 2.1 : 2.0$ was employed for each reaction. The $Na[BH_4]$ is first placed in the reaction flask, followed by 100 mL of 1-butanol. Commercial reagent grade 1-butanol gave good results, so purification is not necessary. The solution is purged with nitrogen gas for 20 min and the phosphine ligand is added with a syringe. Pentacarbonyliron(0) is then added dropwise by syringe through the sidearm. After a short induction period, a vigorous reaction results in the evolution of hydrogen gas:

$$Fe(CO)_5 + BH_4^- \rightarrow Fe(CO)_4CHO^- + BH_3$$

$$BH_3 + 2C_4H_9OH \rightarrow BH(OC_4H_9)_2 + 2H_2$$

To contain the reaction, it is recommended that addition be done *very carefully*, while the solution is vigorously stirred, because a large amount of hydrogen gas is rapidly liberated. The oil bath surrounding the reaction mixture is heated to 120°C and held constant for 2 h after which the solution is allowed to cool to room temperature under nitrogen gas. The *trans*-disubstituted complexes precipitate from solution when PR_3 is $P(C_6H_5)_3$, $P(C_6H_5)_2H$, or $P(C_6H_5)_2CH_3$.

Sodium hydroxide, used in place of $Na[BH_4]$, gives comparable results.[5] Potassium hydroxide in ethanol also gives good results, but a reaction time of 24 h is required.[6]

A. *trans*-TRICARBONYLBIS(TRIPHENYLPHOSPHINE)IRON(0)

$$Fe(CO)_5 + 2P(C_6H_5)_3 \rightarrow \textit{trans-}Fe(CO)_3[P(C_6H_5)_3]_2 + 2CO$$

The reaction is carried out with 1.00 mL (1.49 g, 7.60 mmol) of $Fe(CO)_5$, 4.19 g (16.0 mmol) of $P(C_6H_5)_3$, and 0.58 g (15.3 mmol) of Na $[BH_4]$. Precipitation begins during the course of the reaction. After cooling, the reaction mixture is placed in a freezer (5°C) for 12 h. The precipitate is collected by filtration and washed with three 10-mL aliquots of CH_3OH. Yield: 4.75 g, 94% (pure as shown by IR). An analytically pure sample is prepared by dissolving the original product in 30 mL of CH_2Cl_2 and filtering it into

---

* The checkers used a three-necked flask.

a filter flask containing 50 mL of ice-cold CH_3OH. The precipitate and solution are cooled to 5°C for 12 h. The product is collected by suction filtration, washed with three 10-mL portions of ice-cold CH_3OH, and air dried. Yield: 3.38 g (67%).

Anal. Calcd. for $C_{39}H_{30}O_3P_2Fe$: C, 70.50; H, 4.52; P, 9.34. Found: C, 70.31; H, 4.60; P, 9.11.

B. *trans*-TRICARBONYLBIS(METHYLDIPHENYLPHOSPHINE)-IRON(0)

$$Fe(CO)_5 + 2P(C_6H_5)_2CH_3 \rightarrow trans\text{-}Fe(CO)_3[P(C_6H_5)_2CH_3]_2 + 2CO$$

The reaction is carried out with 1.00 mL (1.49 g, 7.60 mmol) of $Fe(CO)_5$, 2.96 mL (3.18 g, 15.9 mmol) of $P(C_6H_5)_2CH_3$, and 0.58 g (15.3 mmol) of $Na[BH_4]$. Precipitation occurs upon cooling the solution to room temperature. The crude reaction mixture is placed in the freezer (5°C) for 12 h. The precipitate is collected by filtration and washed with three 10-mL aliquots of CH_3OH. Yield: 3.69 g, 90% (pure as shown by IR). An analytically pure sample is prepared by dissolving the original product in 15 mL of CH_2Cl_2 and filtering it into 50 mL of ice-cold CH_3OH. The precipitate and solution are cooled to 5°C for 12 h. The product is collected by suction filtration, washed with three 10-mL portions of ice-cold CH_3OH, and air dried. Yield: 2.87 g (70%).

Anal. Calcd. for $C_{29}H_{26}O_3P_2Fe$: C, 64.46; H, 4.85; P, 11.47. Found: C, 64.01; H, 4.72; P, 11.33.

C. *trans*-TRICARBONYLBIS(DIPHENYLPHOSPHINE)IRON(0)

$$Fe(CO)_5 + 2P(C_6H_5)_2H \rightarrow trans\text{-}Fe(CO)_3[P(C_6H_5)_2H]_2 + 2CO$$

The reaction is carried out with 1.00 mL (1.49 g, 7.60 mmol) of $Fe(CO)_5$, 2.77 mL (2.96 g, 15.9 mmol) of $P(C_6H_5)_2H$, and 0.58 g (15.3 mmol) of $Na[BH_4]$. Precipitation occurs upon cooling the reaction mixture to room temperature. The crude reaction mixture is placed in a freezer (5°C) for 12 h. The precipitate is collected by filtration and washed with three 10-mL aliquots of CH_3OH. Yield: 3.11 g, 79.9% (pure by IR). This sample is dissolved in 15 mL of ice-cold CH_2Cl_2 and filtered into 50 mL of ice-cold CH_3OH. The precipitate and solution are cooled to 5°C for 12 h. The product is collected by suction filtration, washed with three 10-mL portions of ice-cold CH_3OH, and air dried. Yield: 2.57 g (66%).

Anal. Calcd. for $C_{27}H_{22}O_3P_2Fe$: C, 63.31; H, 4.33; P, 12.09. Found: C, 63.11; H, 4.42; P, 11.80.

D. *trans*-TRICARBONYLBIS(TRIBUTYLPHOSPHINE)IRON(0)

$$Fe(CO)_5 + 2P(C_4H_9)_3 \rightarrow \textit{trans-}Fe(CO)_3[P(C_4H_9)_3]_2 + 2CO$$

The reaction is carried out with 1.00 mL (1.49 g, 7.60 mmol) of $Fe(CO)_5$, 3.98 mL (3.23 g, 16.0 mmol) of $P(C_4H_9)_3$, and 0.58 g (15.3 mmol) of $Na[BH_4]$. The solution becomes orange in color during the course of the reaction. After refrigeration of the crude reaction mixture for 12 h at 5°C, some precipitation occurs. The solution is transferred by cannula to another flask and the solvent is removed. To the residue is added 50 mL of CH_3OH. The mixture is refrigerated at 5°C for 12 h and the crystals that result are collected by suction filtration. Yield: 2.23 g (54%).

Anal. Calcd. for $C_{27}H_{54}O_3P_2Fe$: C, 59.58; H, 9.93; P, 11.40. Found: C, 59.63; H, 9.95; P, 11.34.

Properties

The properties of *trans*-$Fe(CO)_3(PR_3)_2$ have been described in previous volumes *Inorganic Syntheses.*[3] The physical properties are listed in the following table:

Compound	mp (°C)	CO $(cm^{-1})^a$	^{31}P $(ppm)^b$
trans-$Fe(CO)_3[P(C_6H_5)_3]_2$	271–273	1881(s), 1886(s), 1944(w)	82.5
trans-$Fe(CO)_3[P(C_6H_5)_2CH_3]_2$	168–170	1872(s), 1882(s), 1966(w)	64.7
trans-$Fe(CO)_3[P(C_6H_5)_2H]_2$	171–173	1880(s), 1886(s), 1978(w)	53.9
trans-$Fe(CO)_3[P(C_4H_9)_3]_2$	54–55	1854(s), 1856(s), 1955(w)	62.4

^{a} Solution IR spectra from $CHCl_3$ solutions were obtained with a Nicolet 20 DX-B spectrometer.
^{b} NMR spectra were obtained with a home-built 250-MHz NMR spectrometer.
^{c} Chemical shifts are parts per million downfield from 85% phosphoric acid.

References

1. S. J. Sherlock, D. C. Boyd, B. Moasser, and W. L. Gladfelter, *Inorg. Chem.*, **30**, 3626 (1991). M. Janicke, H.-U. Hund, and H. Berke, *Chem. Ber.*, **124**, 719 (1991). P. Kölbener, H.-U. Hund, H. W. Bosch, C. Sontag, and H. Berke, *Helv. Chim. Acta*, **73**, 2251 (1990). G. Cardaci, G. Bellachioma, and P. Zanazzi, *Organometallics*, **7**, 172 (1988).

2. A. F. Clifford and A. K. Mukherjee, *Inorg. Synth.*, **8**, 184 (1966).
3. M. J. Therien and W. C. Trogler, *Inorg. Synth.*, **28**, 173 (1990).
4. R. L. Keiter, E. A. Keiter, K. H. Hecker, and C. A. Boecker, *Organometallics*, **7**, 2466 (1988).
5. R. L. Keiter, E. A. Keiter, C. A. Boecker, and D. R. Miller, *Synth. React. Inorg. Met.-Org. Chem.*, **21**(3), 473 (1991).
6. J.-J. Brunet, G. Commenges, F.-B. Kindela, and D. Neibecker, *Organometallics*, **11**, 1343 (1992). J.-J. Brunet, F. B. Kindela, and D. Neibecker, *Inorg. Synth.*, **29**, 151 (1992).

32. (η^5-PENTAMETHYLCYCLOPENTADIENYL) (η^5-CYCLOPENTADIENYL)- IRON [1,2,3,4,5-PENTAMETHYLFERROCENE]

SUBMITTED BY J. M. MANRÍQUEZ,* E. E. BUNEL,† and B. OELCKERS*
CHECKED BY C. VÁSQUEZ§ and JOEL S. MILLER§

The organometallic chemistry of the first-row transition metals generally starts with the binary metal carbonyl organometallic complexes. Noncarbonyl organometallic complexes starting with other easily accessible binary compounds provide entries to a broader spectrum of complexes. In this context, we describe the synthesis of the mixed "sandwich" complex (η^5-pentamethylcyclopentadienyl) (η^5-cyclopentadienyl) iron as an example of the synthetic utility of the solution-stable derivative (η^5-pentamethylcyclopentadienyl) (2,4-pentanedionate) iron.

(η^5-Pentamethylcyclopentadienyl) (η^5-cyclopentadienyl) iron has been prepared by the alkylation of ferrocene under Friedel-Crafts conditions with methylchloride,[1] and also by reaction of ferrous chloride or $FeBr_2 \cdot DME$ (DME = 1,2-dimethoxyethane) with equimolar quantities of pentamethylcyclopentadienyl sodium and cyclopentadienyl lithium.[2, 3]

The procedure outlined here is similar to that reported by Bunel and co-workers,[4] although more experimental details have been included to facilitate the synthesis.

More examples of the usefulness of the (η^5-pentamethylcyclopentadienyl) (2,4-pentanedionate) iron as a source of the (η^5-pentamethylcyclopentadienyl) iron fragment can be found in several references.[4–9]

* Universidad Técnica Federico Santa María, Casilla 110-V, Valparaíso, Chile. Research supported by FONDECYT 799/90.
† Present address: E.I. du Pont de Nemours and Company, Wilmington, DE 19880.
§ E.I. du Pont de Nemours and Company, Wilmington, DE 19880.

A. (η^5-PENTAMETHYLCYCLOPENTADIENYL) (2,4-PENTANEDIONATE)IRON

$$Fe(acac)_2 + (\eta^5\text{-}C_5(CH_3)_5)Li \rightarrow (\eta^5\text{-}C_5(CH_3)_5)Fe(acac) + Li(acac)$$
$$(acac = 2,4\text{-pentanedionate})$$

*Procedure**

A 100-mL, two-necked, round-bottomed flask with a stirring bar, a nitrogen inlet, and a rubber septum is evacuated and filled with nitrogen three times. Iron(II) bis(acetylacetonate) (2.01 g, 7.9 mmol) is added to the flask in a nitrogen bag. Then 20 mL of anhydrous THF, freshly distilled from Na/benzophenone under nitrogen, is added to the flask, the stirring is started, and the dark yellowish-brown solution is cooled to $-78°C$ with a dry ice/acetone bath. A suspension of pentamethylcyclopentadienyl lithium** (1.125 g, 7.9 mmol) in 20 mL of anhydrous THF is added to the reaction flask via syringe. The purity and the stoichiometric amount of pentamethylcyclopentadienyl lithium are crucial to avoid side reactions.

■ **Caution.** *Pentamethylcyclopentadienyl lithium is pyrophoric and must be handled with care under an inert atmosphere.*

The cooling bath is removed and the mixture is left to reach room temperature in an air bath. At this point, a slight color change to a more reddish solution with suspended Li(acac) is observed. $(\eta^5\text{-}C_5(CH_3)_5)Fe(acac)$ is stable only in solution [attempts to isolate the complex result in decomposition to a mixture of decamethylferrocene and iron(II) bis(acetylacetonate)]. The $\eta^5\text{-}C_5(CH_3)_5Fe(acac)$ should be used within 2 h of preparation.

B. (η^5-PENTAMETHYLCYCLOPENTADIENYL) (η^5-CYCLOPENTADIENYL)IRON

$$(\eta^5\text{-}C_5(CH_3)_5)Fe(acac) + (\eta^5\text{-}C_5H_5)Li \rightarrow$$
$$(\eta^5\text{-}C_5(CH_3)_5)Fe(\eta^5\text{-}C_5H_5) + Li(acac)$$

*Procedure**

The apparatus for the preparation of this compound is illustrated in Fig. 1.

---

* Conduct all operations under argon or nitrogen.
** This salt can be prepared in essentially quantitative yield by the careful addition of a butyllithium solution in hexanes to a stirred solution of pentamethylcyclopentadiene in dry pentane under N_2. After 5 h of stirring at room temperature, the resulting white precipitate is filtered, washed with pentane, and dried under vacuum.

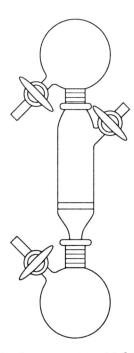

Figure 1. Apparatus for the preparation of $[\eta^5\text{-}C_5(CH_3)_5]Fe(\eta^5\text{-}C_5H_5)$.

A freshly prepared solution of $(\eta^5\text{-}C_5(CH_3)_5)Fe(acac)$ (7.9 mmol) prepared as described above, is cooled to $-78°C$ with a dry ice/acetone bath. A solution of cyclopentadienyllithium (0.566 g, 7.9 mmol; Aldrich Chemical Co.) in 10 mL of anhydrous THF is added to the $\eta^5\text{-}C_5(CH_3)_5Fe(acac)$ solution via syringe. After completion of the addition, the mixture has a yellowish-green color.

■ **Caution.** *Cyclopentadienyllithium is corrosive and moisture sensitive and must be handled with care under an inert atmosphere.*

The cooling bath is removed, and the mixture is left to reach room temperature with an air bath and stirred for 30 min. The solvent is removed under vacuum to afford a brown solid and then 30 mL of petroleum ether (40–60°C), freshly distilled from Na/benzophenone under nitrogen, is added to the flask. The dark yellow suspension is filtered to remove the lithium acetylacetonate, and the solid is washed twice with 7 mL of petroleum ether. The solvent is removed from the filtrate under vacuum to yield a dark yellow solid (1.95 g), which is sublimed at 50–60°C (10^{-3} torr). The product is a yellow crystalline solid. Yield: 1.84 g (86%).*

* The checkers report a 72% yield.

Anal. Calcd. for $C_{15}H_{20}Fe$: C, 70.33; H, 7.87. Found: C, 70.51; H, 7.79. [^{1}H-NMR C_6D_6, 100 MHz ppm: 1.87 (15H, s), 3.52 (5H, s).]

Properties

This mixed metallocene is air-stable as a dry solid and melts at 89–90°C. It oxidizes slowly in solutions exposed to the air. It is very soluble in pentane, benzene, THF, and diethyl ether. It can be recrystallized from methanol. It sublimes without appreciable decomposition.

References

1. A. N. Nesmeyanov and N. S. Kochetkova, *Otdel. Khim. Nauk.*, 242 (1958).
2. M. S. Wrighton, J. L. Robbins, and S. Chao, *J. Am. Chem. Soc.*, **105**, 181 (1983).
3. U. Koelle, B. Fuss, F. Khouzami, and J. Gersdorf, *J. Organomet. Chem.*, **290**(1), 77 (1985).
4. E. E. Bunel, L. Valle, and J. M. Manríquez, *Organometallics*, **4**, 1680 (1985).
5. J. Morrow, D. Catheline, M. H. Desbois, J. M. Manríquez, J. Ruiz, and D. Astruc, *Organometallics*, **6**, 2605 (1987).
6. E. E. Bunel, L. Valle, N. L. Jones, P. J. Carrol, M. González, N. Muñoz, and J. M. Manríquez, *Organometallics*, **7**, 789 (1988).
7. E. E. Bunel, L. Valle, N. L. Jones, P. J. Carrol, M. González, N. Muñoz, J. M. Manríquez, C. Barra, G. Visconti, and A. Aizman, *J. Am. Chem. Soc.*, **110**, 6596 (1988).
8. R. A. Paciello, J. M. Manríquez, and J. E. Bercaw, *Organometallics*, **9**, 260 (1990).
9. S. Rittinger, D. Bucholz, M. H. Desbois, J. Linares, F. Varret, R. Boese, L. Zsolnai, G. Huttner, and D. Astruc, *Organometallics*, **11**, 1454 (1992).

33. PYRAZOLATE-BRIDGED RUTHENIUM(I) CARBONYL COMPLEXES

SUBMITTED BY JAVIER A. CABEZA* and LUIS A. ORO[†]
CHECKED BY RICHARD A. JONES[‡]

The number of known ruthenium(I) carbonyl complexes is relatively small in comparison to those known for ruthenium(0) or ruthenium(II).[1] Moreover, apart from $[Ru_2(C_5H_5)_2(CO)_4]^2$ and carboxylato complexes of the type $[Ru_2(\mu\text{-}RCO_2)_2(CO)_4L_2]$,[3] whose syntheses are well established, most of the

* Instituto de Química Organometálica, Universidad de Oviedo, 33071 Oviedo, Spain.
† Instituto de Ciencia de Materiales de Aragón, Facultad de Ciencias, Universidad de Zaragoza-CSIC, 50009 Zaragoza, Spain.
‡ Department of Chemistry and Biochemistry, The University of Texas at Austin, Austin, TX 78712.

other ruthenium(I) carbonyl compounds have been obtained in low yields, generally as by-products of reactions of $[Ru_3(CO)_{12}]$ or other polynuclear clusters.[1,4] We describe here a simple synthesis of the complexes $[Ru_2(\mu\text{-}L)_2(CO)_6]$ [L = 3,5-dimethylpyrazolato (Me$_2$pz) and pyrazolato (pz)], starting from $RuCl_3 \cdot nH_2O$,[5] the cheapest starting material for ruthenium compounds, in higher yields than those reported starting from the more expensive $[Ru_3(CO)_{12}]$.[6] Several ruthenium(I) carbonyl complexes are efficient catalyst precursors for the homogeneous hydroformylation of olefins[7] and for the addition of carboxylic acids to alkynes.[8]

■ **Caution.** *All the manipulations must be carried out with the use of gloves in an efficient fume hood due to the toxicity of RuCl$_3 \cdot n$H$_2$O and carbon monoxide. The 2-methoxyethanol used should be free of peroxides.[9]*

A. BIS(μ-3,5-DIMETHYLPYRAZOLATO-N^1: N^2)-BIS[TRICARBONYLRUTHENIUM(I)] (*Ru-Ru*)

$$RuCl_3 \cdot nH_2O + CO + CH_3OCH_2CH_2OH \rightarrow \text{"yellow solution"}$$

"yellow solution" + Me$_2$pzH + Zn + CO $\rightarrow$

$$[Ru_2(\mu\text{-}Me_2pz)_2(CO)_6] + H_2 + ZnCl_2$$

Procedure

A 100-mL reaction flask equipped with a gas inlet and a reflux condenser connected to a mineral oil bubbler is charged with $RuCl_3 \cdot nH_2O$* (3 g, 10.74 mmol) and 2-methoxyethanol (60 mL). Carbon monoxide is bubbled through the initially brown solution at reflux temperature until the color is pale yellow[†] (2–3 h). Then, 3,5-dimethylpyrazole[‡] (1.102 g, 10.74 mmol) and granular zinc[‡] (5 g) are added (the use of zinc dust reduces the yield of the final product). The mixture is stirred at reflux temperature maintaining a slow stream of carbon monoxide. After 2 h, the mixture is cooled to room temperature and the pale brown-grey suspension is decanted from the unreacted zinc and poured under air into a beaker containing 200 mL of distilled water. The yellowish precipitate is collected by filtration with a sintered-glass filter, washed with water (five 10-mL portions), and dried by suction. This solid is dissolved in the filter in the minimum volume of dichloromethane and the solution is chromatographed on a silica gel (Merck, 35–70 mesh) column

---

* Obtained from Johnson Matthey, 38.7% Ru.
[†] The composition of this yellow solution is uncertain; however, it has proven to be a useful source of mononuclear ruthenium(II) carbonyl complexes (ref. 1, p. 693).
[‡] Obtained from Aldrich.

(15 × 3 cm). Elution with hexane affords two bands. The first one is yellow and contains a small amount of a mixture of $[Ru_3(CO)_{12}]$ and $[Ru_4H_4(CO)_{12}]$. The second one is colorless and moves very slowly with hexane and the eluant is changed to hexane–dichloromethane (1 : 1).* Vacuum removal of the solvent gives the product as white crystals. Yield 2.65 g (82%).

Anal. Calcd. for $C_{16}H_{14}N_4O_6Ru_2$: C, 34.3; H, 2.5; N, 10.0. Found: C, 34.5; H, 2.6; N, 9.8.

B. BIS(μ-PYRAZOLATO-N^1: N^2) BIS[TRICARBONYLRUTHENIUM(I)] (*Ru-Ru*)

Procedure

This compound is prepared by the same procedure as that described for $[Ru_2(\mu\text{-}Me_2pz)_2(CO)_6]$, in 65% yield.

Anal. Calcd. for $C_{12}H_6N_4O_6Ru_2$: C, 28.9; H, 1.2; N, 11.1. Found: C, 28.9; H, 1.1; N, 11.1.

Properties

Both compounds are white air-stable solids, which are very soluble in acetone, ethers, chlorinated solvents, and aromatic hydrocarbons, slightly soluble in alcohols and aliphatic hydrocarbons, and insoluble in water. On heating, they decompose without melting. Spectroscopic data for $[Ru_2(\mu\text{-}Me_2pz)_2(CO)_6]$: v(CO) (hexane): 2089(m), 2057(s), 2013(vs), 1998(m), 1971(w), 1964(w) cm^{-1}. ^{1}H NMR (CDCl$_3$): 5.60 (s, 2H), 2.10 (s, 12H) ppm. Fast atom bombardment (FAB) mass spectrum (m/z): 560 (M^+) and the successive loss of six CO groups. Spectroscopic data for $[Ru_2(\mu\text{-}pz)_2(CO)_6]$: v(CO) (hexane): 2094(m), 2062(s), 2018(vs), 2005(m), 1982(w), 1969(w) cm^{-1}. ^{1}H NMR (CDCl$_3$): 7.29 (d, J = 2.1 Hz, 4H), 6.07 (t, J = 2.1 Hz, 2H) ppm. FAB mass spectrum (m/z): 504 (M^+) and the successive loss of six CO groups. The X-ray structure of the Me$_2$pz complex has been reported.[5] Both compounds present a rich and somewhat different derivative chemistry; that is, both can be oxidized with iodine to give $[Ru_2(\mu\text{-}I)(\mu\text{-}L)_2(CO)_6]I_3$,[5] but upon reaction with phosphine ligands the pz complex gives the disubstituted derivatives $[Ru_2(\mu\text{-}pz)_2(CO)_4(PR_3)_2]$[6] while the Me$_2$pz complex only affords the monosubstituted ones $[Ru_2(\mu\text{-}Me_2pz)_2(CO)_5(PR_3)]$.[5] The complex

---

* The elution of the colorless band can be monitored by qualitative TLC, observing the plates under UV light, or by running periodical IR spectra of the eluted solution.

$[Ru_2(\mu\text{-}pz)_2(CO)_4(PPh_3)_2]$ can also be prepared by reaction of $[Ru_3(CO)_9\text{-}(PPh_3)_3]$ with pyrazole in refluxing toluene.[10]

References

1. M. I. Bruce in *Comprehensive Organometallic Chemistry*, Vol. 4, G. Wilkinson, F. G. A. Stone, and E. W. Abel (eds.), Pergamon, Oxford, 1984; E. A. Seddon and K. R. Seddon, *The Chemistry of Ruthenium*, Elsevier, Amsterdam, 1984.
2. A. P. Humphries and S. A. R. Knox, *J. Chem. Soc. Dalton Trans.*, 1710 (1975).
3. G. R. Cooks, B. F. G. Johnson, J. Lewis, I. G. Williams, and G. Gamlen, *J. Chem. Soc. (A)*, 2761 (1969).
4. J. A. Cabeza and J. M. Fernández-Colinas, *Coord. Chem. Rev.*, **126**, 319 (1993).
5. J. A. Cabeza, C. Landázuri, L. A. Oro, D. Belletti, A. Tiripicchio, and M. Tiripicchio-Camellini, *J. Chem. Soc., Dalton Trans.*, 1093 (1989).
6. F. Neumann and G. Süss-Fink, *J. Organomet. Chem.*, **367**, 175 (1989).
7. P. Kalck, M. Siani, J. Jenck, B. Peyrille, and Y. Peres, *J. Mol. Catal.*, **67**, 19 (1991). A. Béguin, H.-C. Böttcher, G. Süss-Fink, and B. Walther, *J. Chem. Soc., Dalton Trans.*, 2133 (1992).
8. M. Rotem and Y. Shvo, *J. Organomet. Chem.*, **448**, 189 (1993).
9. D. D. Perrin, W. L. F. Armarego, and D. R. Perrin, *Purification of Laboratory Chemicals*, 2nd ed., Pergamon, Oxford, 1980.
10. P. L. Andreu, J. A. Cabeza, V. Riera, F. Robert, and Y. Jeannin, *J. Organomet. Chem.*, **372**, C15 (1989).

34. MAIN GROUP-TRANSITION METAL CARBONYL COMPLEXES

SUBMITTED BY KENTON H. WHITMIRE,* JOHN C. HUTCHISON,*
MICHAEL D. BURKART,* JACK LEE,* SYLVIA EZENWA,*
ANDREW L. MCKNIGHT,* CAROLYN M. JONES*,
and ROBERT E. BACHMAN*
CHECKED BY N. C. NORMAN,† G. A. FISHER,† and N. L. PICKETT†

Metal carbonyl anions react with main group halides and oxides to yield a number of main-group transition-metal carbonyl complexes in good yields. These complexes serve as starting materials for a number of higher nuclearity cluster complexes.

General Procedure

All operations are carried out under an inert atmosphere using standard Schlenk line techniques.[1] Either deionized or distilled water which has been

* Department of Chemistry, Rice University, 6100 Main Street, Houston, TX 77005-1892.
† Department of Chemistry, University of Bristol, Bristol, UK.

deoxygenated may be used. Organic solvents are dried, distilled, and sparged with nitrogen prior to use. Eye protection must be worn when handling methanolic KOH solutions.

■ **Caution.** *Metal carbonyls are toxic and should only be used in a well-ventilated fume hood. Care should be taken when handling the main-group element reagents as some may also be very toxic, especially thallium and lead salts.*

A. $[Et_4N]_3[Bi\{Fe(CO)_4\}_4]^2$

$$Fe(CO)_5 \xrightarrow[\text{2. NaBiO}_3]{\text{1. KOH}} Na_3[Bi\{Fe(CO)_4\}_4] \xrightarrow{\text{[Et}_4\text{N]Br/MeOH}} [Et_4N]_3[Bi\{Fe(CO)_4\}_4] \downarrow$$

Procedure

Reagent grade KOH (5 g) is dissolved in MeOH (150 mL) in a 500-mL Schlenk flask. After the solution is cooled in an ice bath, pentacarbonyliron (5.0 mL, 0.037 mol, 7.3 g) is added by syringe. A 200-mL pressure equalizing addition funnel containing a slurry of $NaBiO_3$ (4.2 g, 0.014 mol) in MeOH (70 mL) is attached to the Schlenk flask. While bubbling with N_2, the slurry is added slowly to the flask with care being taken to avoid clogging of the funnel. The reaction mixtue changes color from dirty yellow-brown to dark green within about 15 min. After 1 h of stirring, the mixture is filtered through a medium frit. A solution of $[Et_4N]Br$ (6.0 g, 0.21 mol) dissolved in MeOH is then added by syringe to the filtrate. The microcrystalline, deep-green precipitate which forms is collected by filtration and washed with MeOH (4 × 30 mL) until the wash is no longer brown. The crude product is recrystallized by dissolving the crystals in a minimum of MeCN, filtering, and precipitating the pure product with at least twice the volume of CH_2Cl_2. The product is collected by filtration, washed with CH_2Cl_2, and dried under vacuum. Yield: 8.2 g (69% based on Fe).

Anal. Calcd. for $C_{40}H_{60}N_3O_{16}BiFe_4$: C, 37.79; H, 4.75; N, 3.30; Fe, 17.57; Bi, 16.43. Found: C, 37.14; H, 4.85; N, 3.2; Fe, 16.76, Bi, 16.96.

Properties

Tetraethylammonium *tetrakis*(tetracarbonylferrio)bismuthate(3 −), $[Et_4N]_3$-$[BiFe_4(CO)_{16}]$, is a green, oxygen-sensitive, crystalline solid whose crystal structure has been reported.[2, 3] It is soluble in acetonitrile, but is insoluble in most other organic solvents. The bis(triphenylphosphine)iminium ([PPN]$^+$) salt can be prepared by using [PPN]Cl in place of [Et$_4$N]Br. The [PPN]$^+$ salt is more soluble in organic solvents, but is more difficult to purify. The characteristic IR bands (MeCN, cm^{-1}) are 1962s, 1906m, and 1867m. Visible spectra are obtained on 0.11-mM solutions of the [Et$_4$N]$^+$ salt in acetonitrile with $\lambda_{max} = 617$ nm; ε ca. 5000. ^{13}C NMR Spectrum ($-85°$C, Me-d_3CN/ EtCN): 225 and 218 ppm (ca. 1:3 intensity ratio).

B. [Et$_4$N]$_3$[Sb{Fe(CO)$_4$}$_4$]^4

$$SbCl_5 + 4Na_2[Fe(CO)_4] \cdot 3/2 \text{ dioxane} \rightarrow Na_3[SbFe_4(CO)_{16}] + 5NaCl$$

$$Na_3[SbFe_4(CO)_{16}] + [Et_4N]Br \rightarrow [Et_4N]_3[SbFe_4(CO)_{16}] + 3NaBr$$

Procedure

Na$_2$[Fe(CO)$_4$] · 3/2 dioxane (10.8 g, 31.3 mmol), which can be obtained commerically (Aldrich Chemical Co.) or prepared after the literature method,[5] is dissolved in MeCN (100 mL) in a Schlenk flask (with stirbar) connected to a gas-bubbler outlet and fitted with a rubber septum. Antimony pentachloride (1.0 mL) is added dropwise by syringe to the solution (■ **Caution.** *Antimony pentachloride is extremely reactive and hydrolyses exothermically*). After slight warming, there is vigorous gas evolution (*CARE!*). The resulting mixture is stirred for 24 h and then filtered through celite. The filtrate is collected and the solvent is removed under vacuum at room temperature to give a red, oily residue. Excess [Et$_4$N]Br (10 g) is dissolved in methanol (30 mL), and the solution is added by syringe to the residue. This solution is allowed to stir for 24 h, after which time the red precipitate is collected by filtration, washed with methanol (2 × 15 mL), and dried under vacuum. Yield: 4.2 g (46% based on Sb or Fe).

Anal. Calcd. for C$_{40}$H$_{60}$N$_3$O$_{16}$SbFe$_4$: Sb, 10.28; Fe, 18.87. Found: Sb, 9.76; Fe, 18.39.

Properties

Tetraethylammonium *tetrakis*(tetracarbonylferrio)antimonate(3 −), [Et$_4$N]$_3$- [SbFe$_4$(CO)$_{16}$], is a red, powdery solid that may be stored indefinitely under nitrogen. The crystal structure has been reported and is isomorphous to the bismuth analogue.[4] The characteristic IR bands are found at (MeCN, cm^{-1}) 1971, 1910, and 1882. A maximum is observed in the visible region of the absorption spectrum (MeCN) at 480 nm (ε, 2800).

C. $BiCo_3(CO)_{12}$[6]

$$3 NaCo(CO)_4 + BiCl_3 \rightarrow BiCo_3(CO)_{12} + 3 NaCl$$

Procedure

A solution of $Na[Co(CO)_4]$ is prepared as follows. To a solution of octa-carbonyldicobalt (1.67 g, 19.5 mmol) dissolved in tetrahydrofuran (THF, 40 mL) is added Na/Hg amalgam (5 mL, 1%) via syringe and the resultant solution is stirred for several hours (■ **Caution.** *Care should be taken in the handling of Na/Hg because a vigorous evolution of H_2 gas could result from reaction with moisture or acid and Hg is a cumulative poison, see ref. 1*). The reaction is monitored by IR spectroscopy and considered to be complete when the only carbonyl stretching frequency observable is the band at 1890 cm^{-1} due to the tetracarbonylcobalt(-1) anion. The amalgam is then removed carefully through a sidearm of the Schlenk flask. After filtering the solution through celite to remove traces of amalgam, the solvent is removed under reduced pressure and the residue is redissolved in water (30 mL).

An aqueous solution of $BiCl_3$ is prepared as follows. Anhydrous $BiCl_3$ (1.0 g, 3.2 mmol) is added to water (30 mL) with stirring, resulting in the immediate precipitation of BiOCl. The resultant slurry is acidified by dropwise addition of concentrated hydrochloric acid until the BiOCl redissolves, giving a clear solution containing complex chlorobismuth anions. An equivalent amount of BiOCl, Bi_2O_3, or $Bi_2O_2(CO_3)$ may be substituted for the $BiCl_3$. The aqueous solution of $Na[Co(CO)_4]$ is added via cannula to the bismuth chloride solution, giving rise to the formation of a dark black precipitate. The resulting slurry is magnetically stirred for approximately 1 h, after which time the product is isolated by filtration. The solid is washed with three portions of distilled water (15 mL) and recrystallized by dissolving in CH_2Cl_2, filtering, and subsequently removing the solvent under reduced pressure. Yield: 1.7–1.9 g (70–80% based on Bi).

Anal. Calcd. for $C_{12}BiCo_3O_{12}$: C, 19.96; Bi, 28.95; Co, 24.49. Found: C, 20.13; Bi, 28.77; Co, 23.38.

Properties

Tris(tetracarbonylcobaltio)bismuth, $BiCo_3(CO)_{12}$, is a purplish-black air-sensitive crystalline solid which must be stored under nitrogen. The crystal structure is known.[6] On heating $BiCo_3(CO)_{12}$ in solution, $BiCo_3(CO)_9$ is formed.[7] In reactions with diphenylacetylene $BiCo_3(CO)_{12}$ functions as a source of $Co(CO)_4$ radicals. It is soluble in hexane, toluene, CH_2Cl_2, THF,

MeOH, and Et_2O. IR spectrum (CH_2Cl_2, cm^{-1}): 2115(w), 2083(vs), 2060(w), 2033(vs), 2020(vw), 2015(vs), and 1995(sh).

D. $BiCo_3(CO)_9$[7]

$$BiCo_3(CO)_{12} \rightarrow BiCo_3(CO)_9 + 3CO$$

Procedure

Tris(tetracarbonylcobaltio)bismuth, $BiCo_3(CO)_{12}$, (0.25 g, 0.35 mmol) is dissolved in THF (30 mL) in a 100-mL Schlenk flask. A reflux condenser with a bubbler outlet is then attached to the mouth of the Schlenk flask. The solution is refluxed and magnetically stirred for approximately 30 min while being monitored by IR spectroscopy. Longer heating promotes decomposition. After cooling, the solvent is removed under reduced pressure. The crude product is extracted with hexane (50 mL). The extraction solvent is removed carefully under reduced pressure, yielding long needle-like crystals of the product. Yield: 0.13–0.16 g (60–70%).

Anal. Calcd. for $C_9BiCo_3O_9$: Bi, 32.76; Co, 27.72. Found: Bi, 32.65; Co, 27.68.

Properties

(μ_3-Bismuthido)nonacarbonyltricobalt, $BiCo_3(CO)_9$, crystallizes as long shiny purple needles that may be stored under nitrogen. This compound was prepared independently by two groups who have also reported its crystallographic characterization.[7] Upon further heating, the cluster decomposes to metallic bismuth and cobalt. The reverse reaction back to $BiCo_3(CO)_{12}$ can be achieved in about 6 days under 750 psi of CO. IR (hexane, cm^{-1}): 2083(m), 2041(s), 2022(ms), 2012(m), 1981(w), and 1885(m). The compound is soluble in hexane, toluene, Et_2O, CH_2Cl_2, MeOH, acetone, MeCN, and THF.

E. $[Cp_2Co][Bi\{Co(CO)_4\}_4]$[8]

$$[Cp_2Co][Co(CO)_4] + BiCo_3(CO)_{12} \rightarrow [Cp_2Co][Bi\{Co(CO)_4\}_4]$$

Procedure

Cobaltocene (0.066 g, 0.35 mmol) and octacarbonyldicobalt (0.059 g, 0.17 mmol) are placed in a Schlenk flask (50 mL). Dichloromethane (30 mL) is syringed into the flask and the mixture is stirred for 1.5 h. The solution is filtered into a Schlenk flask containing $BiCo_3(CO)_{12}$ (0.25 g, 0.35 mmol) and the resulting mixture is stirred for approximately 1 h after which time it is filtered through a medium-porosity frit. The solvent is removed from the filtrate under reduced pressure, yielding approximately 0.32 g of a reddish-brown solid (0.30 mmol, 86%). The product forms cleanly if nearly exact stoichiometric amounts of the reactants are used. Stoichiometric variations may cause the formation of small amounts of neutral by-products, which can be removed by washing the product with toluene.

Anal. Calcd. for $C_{26}H_{10}BiCo_5O_{16}$: Bi, 19.31; Co, 27.23. Found: Bi, 20.34; Co, 27.56.

Properties

Cobaltocenium *tetrakis*(tetracarbonylcobaltio)bismuthate(− 1), $[Cp_2Co]$-$[Bi\{Co(CO)_4\}_4]$, is a reddish-brown solid which must be stored under nitrogen.[8] The $[Me_4N]^+$ salt has been isolated and structurally characterized.[9] The Cp_2Co^+ salt is insoluble in hexane, toluene, and only partially soluble in diethyl ether. It dissolves in CH_2Cl_2, MeOH, acetone, MeCN, and THF. IR (CH_2Cl_2, cm^{-1}): 2066(m), 2028(vs), 1969(s), and 1890(w).

F. $[Et_4N]_2[Pb\{Fe_2(CO)_8\}\{Fe(CO)_4\}_2]$[10]

$$4Fe(CO)_5 + 2PbCl_2 + 2KOH \rightarrow Pb^0 + K_2[Pb\{Fe(CO)_4\}_2\{Fe_2(CO)_8\}]$$

$$K_2[Pb\{Fe(CO)_4\}_2\{Fe_2(CO)_8\}] + [Et_4N]Br \rightarrow$$
$$[Et_4N]_2[Pb\{Fe(CO)_4\}_2\{Fe_2(CO)_8\}] + 2KBr$$

Procedure

Pentacarbonyliron (1.0 mL, 1.46 g, 7.4 mmol) is added by syringe to a 100-mL Schlenk flask containing a solution of deoxygenated methanolic KOH (1.0 g KOH, 20 mL MeOH) in an ice bath as described in Section 32.A. A slurry of $PbCl_2$ (1.72 g, 0.00618 mol) in MeOH (16 mL) is added slowly to the flask through a pressure-equalizing addition funnel after being sparged for 5 min with an inert gas. There is an immediate color change from orange-yellow to brown-purple. The reaction mixture is stirred for 20 h, during which time a lead mirror forms on the flask. The solution is then filtered through Celite and treated with a concentrated solution of $[Et_4N]Br$ (3.0 g, 14 mmol) in MeOH, followed by addition of an excess of H_2O. A precipitate results which is collected by filtration and washed with H_2O (2 × 10 mL). The fine black crystals are dried under vacuum overnight.

Purification is accomplished by dissolving the product in CH_2Cl_2 (30 mL), followed by filtering through a frit with Celite and precipitation by addition of excess hexane. Dark blackish-purple crystals are obtained. Yield: 1.10 g (54% based on Fe).

Anal. Calcd. for $C_{32}H_{40}N_2O_{16}PbFe_4$: CO, 0.0140 mol/g. Found: CO, 0.0140 mol/g.

Properties

$[Et_4N]_2[Pb\{Fe_2(CO)_8\}\{Fe(CO)_4\}_2]$[10] decomposes rapidly upon exposure to air and is soluble in MeOH, CH_2Cl_2, THF, and MeCN, but is insoluble in hexane, toluene, and H_2O. The crystal structure shows a central lead atom ligated by two trigonal bipyramidal $Fe(CO)_4$ groups and an $Fe_2(CO)_6$-$(\mu$-CO$)_2$ fragment. IR (CH_2Cl_2, cm^{-1}): 2042(m), 1999(s), 1990(s), 1975(s), 1900(s), and 1777(m).

G. $[Sn\{Co(CO)_4\}_4]$[11]

$$SnCl_4 \cdot 5H_2O + 4Na[Co(CO)_4] \rightarrow [Sn\{Co(CO)_4\}_4] + 4NaCl + 5H_2O$$

Procedure

An aqueous solution of $Na[Co(CO)_4]$ is prepared as described in Section 32.C using 2.10 g $Co_2(CO)_8$ (12.3 mmol), 100 mL THF, and 5 mL sodium amalgam (1%). The volume of H_2O employed is 60 mL. An aqueous solution of $SnCl_4 \cdot 5H_2O$ (1.08 g, 3.08 mmol in 100 mL deionized water) is added dropwise to the $Na[Co(CO)_4]$ solution. After stirring for several hours, the burnt-orange precipitate of $[Sn\{Co(CO)_4\}_4]$ is isolated by filtration, washed with water, and dried in vacuo. Yield 53% (1.30 g, 1.62 mmol). *Tetrakis*-(tetracarbonylcobaltio)tin is air-sensitive. The compound is purified by redissolving it in CH_2Cl_2 (50 mL) followed by filtration and removal of the solvent under reduced pressure.

Anal. Calcd. for $C_{16}O_{16}Co_4Sn$: CO, 0.0199 mol/g. Found: CO, 0.0194 mol/g.

Properties

Tetrakis(tetracarbonylcobaltio)tin, $[Sn\{Co(CO)_4\}_4]$, has been structurally characterized and is isoelectronic and isostructural to the anion $[Bi\{Fe(CO)_4\}_4]^{3-}$.[12] IR (CH_2Cl_2, cm^{-1}): 2080(s), 2016(s), and 1994(sh). It is sparingly soluble in water, but is very soluble in CH_2Cl_2. It has been used as a source of $Co(CO)_4$ in some catalysis reactions.[13]

H. [Et$_4$N]$_2$[Tl$_2$Fe$_4$(CO)$_{16}$][14]

$$4Fe(CO)_5 + KOH + 2TlCl_3 \Rightarrow K_2[Tl_2Fe_4(CO)_{16}] + 2[Et_4N]Br \rightarrow$$

$$[Et_4N]_2[Tl_2Fe_4(CO)_{16}]$$

Procedure

Pentacarbonyliron (1.0 mL, 5.1 mmol) is added to a deoxygenated solution of KOH (0.99 g, 18 mmol) in MeOH (30 mL) in a standard Schlenk flask as described in Section 32.A. To this solution is added solid $TlCl_3 \cdot 4H_2O$ (1.35 g, 3.52 mmol). To the dark brown solution which results immediately is added a solution of [Et$_4$N]Br (0.73 g, 3.47 mmol) in H_2O (30 mL) by way of a pressure-equalizing addition funnel. A reddish-brown precipitate settles out of the slightly yellow solution. If the solution remains highly colored, additional [Et$_4$N]Br in H_2O may be added to precipitate the remainder of the product and leave the solution colorless. After filtering and washing with H_2O, the precipitate is left to dry overnight under vacuum. The product is recrystallized by dissolving it in CH_2Cl_2, filtering it, and removing the solvent under reduced pressure.

Anal. Calcd. for $C_{32}G_{40}N_2O_{16}Fe_4Tl_2$: N, 2.09; Fe, 16.65; Tl, 30.47. Found: N, 1.86; Fe, 15.10; Tl, 30.50.

Properties

Tetraethylammonium*bis*(tetracarbonyliron)*bis*(μ-tetracarbonyliron)dithallate(2 $-$), [Et$_4$N]$_2$[Tl$_2$Fe$_4$(CO)$_{16}$], decomposes rapidly upon exposure to air and is soluble in MeOH, CH_2Cl_2, and most polar organic solvents. Its crystal structure shows it to be a weak dimer of [Tl{Fe(CO)$_4$}$_2$]$^-$ and it probably exists as the monomer in solution.[15] IR (CH_2Cl_2, cm^{-1}): 1985(m), and 1908(s). It may be oxidized or irradiated to yield higher nuclearity thallium-iron carbonyl compounds.[14, 15]

References

1. D. F. Shriver and M. A. Drezdon, *The Manipulation of Air-Sensitive Compounds*, 2nd *Ed.* Wiley-Interscience, New York, 1986.
2. K. H. Whitmire, C. B. Lagrone, M. R. Churchill, J. C. Fettinger, and L. V. Biondi, *Inorg. Chem.*, 23, 4227 (1990).

3. M. R. Churchill, J. C. Fettinger, K. H. Whitmire, and C. B. Lagrone, *J. Organomet. Chem.*, **303**, 99 (1986).
4. S. Luo and K. H. Whitmire, *Inorg. Chem.*, **28**, 1424 (1989).
5. R. G. Finke and T. N. Sorrell, *Org. Synth.*, **59**, 102 (1980).
6. G. Etzrodt, R. Boese, and G. Schmid, *Chem. Ber.*, **112**, 2574 (1979).
7. (a) K. H. Whitmire, J. S. Leigh, and M. E. Gross, *J. Chem. Soc., Chem. Commun.*, 926 (1987). (b) S. Martinengo and G. Ciani, *J. Chem. Soc., Chem. Commun.*, 1589 (1987).
8. J. S. Leigh and K. H. Whitmire, *Angew. Chem.*, **100**, 399 (1988); *Angew. Chem. Int. Ed. Eng.*, **27**, 396 (1988).
9. S. Martinengo, A. Fumagalli, G. Ciani, and M. Moret, *J. Organomet. Chem.*, **347**, 413 (1988).
10. C. B. Lagrone, K. H. Whitmire, M. R. Churchhill, and J. C. Fettinger. *Inorg. Chem.*, **25**, 2080 (1986).
11. (a) G. Schmid and G. Etzrodt, *J. Organomet. Chem.*, **131**, 477 (1977). (b) P. J. Patmore and W. A. G. Graham, *Inorg. Chem.*, **7**, 771 (1968). (c) M. Bigorgne and A. Quintin, *C. R. Acad. Sci.*, **264**, 2055 (1967).
12. J. S. Leigh and K. H. Whitmire, *Acta Cryst.*, **C46**, 732 (1990).
13. A. Cabrera, H. Samain, A. Mortreux, F. Petit, and A. J. Welch, *Organometallics*, **9**, 959 (1990).
14. K. H. Whitmire, J. M. Cassidy, A. L. Rheingold, and R. R. Ryan. *Inorg. Chem.*, **27**, 1347 (1988).
15. J. M. Cassidy and K. H. Whitmire, *Inorg. Chem.*, **28**, 1432 (1989).

35. $M_4^{II}(\mu_4$-O$)[(CO)_9Co_3(\mu_3$-CCO$_2)]_6$, M = Co, Zn

SUBMITTED BY WEI CEN* and THOMAS P. FEHLNER*
CHECKED BY B. H. ROBINSON† and N. DUFFY†

(1) $(CO)_9Co_3(\mu_3$-CCl$) \xrightarrow{\text{1. 3AlCl}_3/2.\text{H}_2\text{O}} (CO)_9Co_3[\mu_3$-CCO$_2H](I) + HCl$

(2) $22\,(CO)_9Co_3(\mu_3$-CCO$_2$H$) + 7\,O_2 \rightarrow$

$$3\,Co_4(\mu_4\text{-O})[(CO)_9Co_3(\mu_3\text{-CCO}_2)]_6(II) \downarrow + 44CO \uparrow + 11H_2O$$

(3) $6\,(CO)_9Co_3(\mu_3$-CCO$_2$H$) + 6ZnEt_2 \rightarrow$

$$Zn_4(\mu_4\text{-O})[(CO)_9Co_3(\mu_3\text{-CCO}_2)]_6(III) \downarrow + 8C_2H_6 \uparrow$$

$$+ [(CO)_9Co_3(\mu_3\text{-CCO})_2O \cdot (ZnEt_2)_2$$

In the procedure described herein, a cluster carboxylic acid $[(CO)_9Co_3$-$[\mu_3$-CCO$_2$H$]]$ (I) is chosen as a ligand to bridge $M_4^{II}O^{6+}$ (M = Zn, Co) cores in order to chemically assemble two clusters of $Co_4(\mu_4$-O$)[(CO)_9Co_3$-$(\mu_3$-CCO$)_2]_6$ (II) and $Zn_4(\mu_4$-O$)[(CO)_9Co_3(\mu_3$-CCO$_2)]_6$ (III).[1,2] The prep-

---

* Department of Chemistry and Biochemistry, University of Notre Dame, Notre Dame, IN 46556.
† Chemistry Department, University of Otago, P.O. Box 56, Dunedin, New Zealand.

aration of **I** from an ester of the acid has been reported previously.[3,4] A convenient and clean preparation of **II** is aerial oxidation of **I**.[2,5] As a heteronuclear analogue of **II**, the second compound (**III**) is formed by a direct reaction of **I** with $ZnEt_2$.[1]

Procedure

All procedures are carried out under a dry nitrogen atmosphere using conventional Schlenk techniques. Dichloromethane is dried over and distilled from P_2O_5. All other solvents are dried on molecular sieves, deoxygenated with dry nitrogen, and distilled from sodium metal and benzophenone under nitrogen. The $AlCl_3$ powder and $ZnEt_2$ (1 M solution in hexane) are purchased from Aldrich and used without further processing. The $(CO)_9Co(\mu_3\text{-}CCl)$ is prepared from CCl_4 and $Co_2(CO)_8$ using a published procedure and purified on a silica gel column.[6]

■ **Caution.** *Decomposition of the cobalt carbonyl clusters produces CO, which is an odorless, toxic gas. The hexane solution of $ZnEt_2$ is convenient to handle, but is a source of $ZnEt_2$, which is a pyrophoric and volatile liquid. Handling these compounds in a well-ventilated hood and wearing eye protection are recommended.*

A. $(CO)_9Co_3(\mu_3\text{-}CCO_2H)$

■ **Caution.** *Always add HCl very slowly in an open system during the protonation step in the following procedure, otherwise an explosion may result from the reaction heat and gas evolution. Adjustment of the reaction time is necessary according to the activity of the $AlCl_3$. A well-ground and fresh powdery $AlCl_3$ has higher activity and therefore requires a shorter reaction time. Handling in a well-ventilated hood and weaving gloves and eye-protecting glasses are recommended.*

In a typical reaction $(CO)_9Co_3(\mu_3\text{-}CCl)$ (12.3 g, 25.2 mmol) is dissolved in 200 mL of distilled CH_2Cl_2 in a 800-mL two-necked Schlenk flask equipped with a bubbler. Powdered $AlCl_3$ (10.3 g, 77.3 mmol) is added rapidly and stirred under N_2 for ≈ 12 min. Control of timing is critical and monitoring the reaction mixture by TLC (IB2-F flexible sheets, J. T. Baker, CH_2Cl_2 as elutant) shows the disappearance of the purple band due to $Co_3(CO)_9(\mu_3\text{-}CCl)$ at high R_f position and the appearance of a new major dark-brown band ($R_f = 0$) covered by a purple-brown band ($R_f = 0.3$). The color changes from purple to yellow brown and the reaction mixture thickens with some precipitation. To this mixture is added another 250 mL of CH_2Cl_2 followed immediately by addition dropwise of 50 mL of 38% HCl while the reaction flask is cooled with an ice-water bath. A vigorous gas evolution results from this acid addition. As the mixture stops bubbling and cools down, 38% HCl and water are used alternately to wash it in a 800-mL separation funnel until the blue color in the aqueous phase disappears. The organic layer is then separated after washing and 1 mL of acetic acid or an equal amount of hydrocinnamic (3-phenylproanoic) acid is added. Following a further wash with 50 mL of water, this solution is dried over 15 g of $MgSO_4$ for several hours and filtered. At this stage, TLC (IB2-F flexible sheets, J. T. Baker, CH_2Cl_2 as elutant) shows acid **I** as the major component (purple-brown band; $R_f = 0.3$) with a very pale brown impurity band at higher position ($R_f = 0.8$). Evaporation of the filtrate yields dark, truncated triangular, thick plate-like crystals. The total amount of crystalline material obtained is 7.20 g, corresponding to a yield of 60%.

B. $Co_4(\mu_4\text{-}O)[(CO)_9Co_3(\mu_3\text{-}CCO_2)]_6$

A quantity of **I** (0.200 g, 0.412 mmol) is dissolved in 20 mL of acetone in a 100-mL Schlenk flask in the air. After 48 h at room temperature in the air, the deep purple color of the solution fades to a pale brown with precipitation of a crystalline material. The reaction is complete when a TLC (IB2-F flexible sheets, J. T. Baker, CH_2Cl_2 as elutant) shows the disappearance of the band due to **I** ($R_f = 0.3$) and the appearance of the band due to **II** ($R_f = 0$). The reaction mixture is dried under vacuum. The solid product is washed with 5 mL of diethyl ether and 20 mL of hexane and dried again to give shining black crystals. The total weight of crystals is 0.171 g (0.0541 mmol), corresponding a 96% yield based on total cobalt when optimized. The checkers observed variable and generally lower yields.

C. $Zn_4(\mu_4\text{-}O)[(CO)_9Co_3(\mu_3\text{-}CCO_2)]_6$

A typical synthesis is carried out in a two-necked, 250-mL Schlenk flask equipped with a 50-mL filtration funnel and a magnetic stirring bar with the side neck sealed by rubber septum. This setup effectively prevents contamination from traces of **II** on the surface of crystalline **I**. A quantity of freshly prepared **I** (2.70 g, 5.56 mmol) is loaded into the filtration funnel, which is then sealed by a rubber septum under N_2. To this funnel is added 125 mL of Et_2O by syringe in three aliquots. The intense purple-brown filtrate is collected in the reaction flask. When the filtrate becomes pale and only a very minor amount of black impurity is left on the funnel, the funnel is removed and replaced with a rubber septum. A 1.0 M solution of $ZnEt_2$ in hexane (5.56 mL, 5.56 mmol) is added dropwise into the stirring filtrate with a 10 mL disposable syringe over a period of 10 min. Black microcrystals begin precipitating out of the solution immediately and precipitation is complete after another 20 min. The microcrystals are filtered, collected, and washed with cool Et_2O using the second neck of the flask under N_2. Note that before washing the microcrystals are pyrophoric. At this point both the solid and solution IR spectra show that these microcrystals are pure **III**. The microcrystals are redissolved in 50 mL of warm tetrahydrofuran (THF) (50°C), and diffusion of an equal amount of Et_2O overnight results in well-formed crystals with ideal or truncated rhombohedral shapes. The total weight of recrystallized material is 1.84 g (0.564 mmol) corresponding to a 81% yield based on the total cobalt and the formula $Zn_4(\mu_4$-$O)[(CO)_9Co_3$-$(\mu_3$-$CCO_2)]_6 \cdot (C_2H_5)_2O$.

Properties

Crystalline **I** is purple in solution. However, it is air-sensitive and becomes brown on spontaneous conversion to **II**. FT-IR shows a characteristic pattern from the C=O stretching vibrations of the $(CO)_9Co_3C$ fragment [KBr pellet: 2111(w), 2058(vs), 2048(s), 2032(s), 2018(ms), and 2005(ms) cm^{-1}; in CH_2Cl_2: 2109(w), 2066(vs), and 2047(s) cm^{-1}] and carboxyl [KBr pellet: 1640 cm^{-1}; in CH_2Cl_2: 1686(w), 1644(m), and 1606(vw) cm^{-1}]. Mass spectrum (-FAB) shows a peak at m/z 458 (M-28), but no parent ion. The NMR spectra exhibit the following features: ^1H (CD_2Cl_2, 20°C, δ) 11.5(br, w); ^{13}C (CD_2Cl_2, 20°C, δ) 202.5. Note that the acid proton shift reported previously[4] is incorrect and indicates a wet sample. Elemental analysis confirms the composition of **I** (Calc. for $Co_3O_{11}C_{11}H$: Co 36.38, C 27.19, H 0.21; Found: Co 37.25, C 27.31, H < 0.5).

Crystalline **II** and **III** are soluble in THF and barely soluble in CH_2Cl_2 to form a brown solution with similar UV/visible spectra ($\varepsilon_{370\,nm} = 1.5 \times 10^4$ $M^{-1}cm^{-1}$, $\varepsilon_{515\,nm} = 5.4 \times 10^3$ $M^{-1}cm^{-1}$). In the FT-IR spectrum, except for distinguishable antisymmetric carboxylate stretching modes [1539(1) for **II** and 1557(1) for **III** cm^{-1}], the rest of the vibrational bands are similar [carbonyl stretching modes 2109(w), 2070 $\pm$ 1(s), 2042 $\pm$ 1(vs), COO^- symmetric stretching 1386 $\pm$ 1 cm^{-1}] in CH_2Cl_2. Note that the previously reported IR spectrum of **II** is actually that of **I**.[5] Elemental analyses of the two confirm the formulas of **II** [Calc. for $Co_{22}O_{67}C_{67}Cl_2H_2$ (CH_2Cl_2 solvate): C 24.79, H 0.06; Found: C 24.80, H 0.21] and **III** (Calc. for $Zn_4Co_{18}O_{67}C_{66}$: Zn 8.21, Co 33.29, C 24.87, O 33.64; Found: Zn 7.68, Co 30.52, C 24.77, O 33.41).

References

1. W. Cen, K. J. Haller, and T. P. Fehlner, *Inorg. Chem.*, **30**, 3120 (1991).
2. W. Cen, K. J. Haller, and T. P. Fehlner, *Organometallics*, **11**, 3499 (1991).

3. J. E. Hallgren, C. S. Eschbach, and D. Seyferth, *J. Am. Chem. Soc.*, **94**, 2547 (1972).
4. D. Seyferth, J. E. Hallgren, and C. S. Eschbach, *J. Am. Chem. Soc.*, **96**, 1730 (1974).
5. R. L. Sturgeon, M. M. Olmstead, and N. E. Schore, *Organometallics*, **10**, 1649 (1991).
6. C. L. Nivert, G. H. Williams, and D. Seyferth, *Inorg. Syn.*, **20**, 234 (1980).

36. 1,2,3,4-TETRAMETHYL-5-(TRIFLUOROMETHYL)CYCLO-PENTADIENE (Cp‡H) AND DI-μ-CHLORODICHLOROBIS-[η^5-TETRAMETHYL(TRIFLUOROMETHYL)-CYCLOPENTADIENYL]DIRHODIUM(III)

SUBMITTED BY PAUL G. GASSMAN,*[†] JOHN W. MICKELSON,*[§]
and JOHN R. SOWA, JR.*[§]
CHECKED BY LUIS P. BARTHEL-ROSA[‡] and JOHN H. NELSON[‡]

Recently it has been demonstrated that 1,2,3,4-tetramethyl-5-(trifluoro-methyl)cyclopentadienide (Cp‡) combines the electronic properties of cyclo-pentadienide (Cp) with the steric properties of pentamethylcyclopentadienide (Cp*).[1] Electronically, the trifluoromethyl group is four times as electron-withdrawing[2] as one methyl group is electron-donating.[3] Thus, one tri-fluoromethyl group electronically balances four methyl groups around the periphery of a cyclopentadienide ring. The availability of this new ligand provides a vehicle for the separation of steric versus electronic effects of the five methyl groups of Cp*. In addition, it opens up new realms of investiga-tion for the synthesis of transition metal complexes where the steric effects of Cp* are desired, but the electron-donating properties of Cp* are not advant-ageous.

In large part, the synthesis of Cp‡H is modeled on the Nazarov cyclization[4] developed by Threlkel and Bercaw[5, 6] for the synthesis of Cp*H. The syn-thesis of [Cp‡RhCl$_2$]$_2$ is given as an example of an organotransition metal complex which can be prepared from Cp‡H.

* Department of Chemistry, University of Minnesota, Minneapolis, MN 55455.
§ Current addresses: JRS, Department of Chemistry, Seton Hall University, South Orange, NJ 07079; JWM, 3M Co., 3M Center, St. Paul, MN 55144.
† Deceased April 21, 1993.
‡ Department of Chemistry, University of Nevada-Reno, Reno, NV 89557.

A. 1,2,3,4-TETRAMETHYL-5-(TRIFLUOROMETHYL)-CYCLOPENTADIENE (Cp‡H)

Procedure

■ **Caution.** *Although the toxicity of Cp‡H has not been investigated, due care is recommended when handling this compound. Bromine, 2-bromo-cis-2-butene, and methanesulfonic acid are hazardous compounds and contact with skin or inhalation should be avoided.*

Preparation of meso-2,3-dibromobutane.[7a] A 500-mL, three-necked, round-bottomed flask, marked at the 250-mL volume level with a glass marking pencil, is equipped with a magnetic stir bar, a dry ice/isopropyl alcohol (IPA) condenser, a 250-mL addition funnel, and a 24/40 rubber septum. The flask is cooled to − 78°C (dry ice/IPA) and 250 mL of *trans*-2-butene (bp = 1°C, Matheson Gas Co., East Rutherford, NJ) is condensed into the flask (to the 250-mL volume level) through a 12-in. syringe needle in the rubber septum. (The rate of gas flow may be monitored with a mineral oil bubbler attached to the top of the dry ice/IPA condenser. To counteract Joule–Thompson cooling, the checkers indicated that it was necessary to warm the cylinder of *trans*-2-butene to affect complete transfer.) The addition funnel is then carefully charged with bromine (150 mL, 2.9 mol) which is added to the *trans*-2-butene dropwise at − 78°C over a period of 45 min to give a clear liquid with a red tinge. Hydrogen bromide, which is generated as a by-product, is removed by washing the pale red liquid with saturated $NaHCO_3$ (50 mL) and water (50 mL). (The checkers also washed the reaction mixture with saturated 50 mL of sodium bisulfite to remove excess Br_2.) After drying over anhydrous magnesium sulfate and filtration, the clear, colorless liquid is distilled at atmospheric pressure through a 12-in. Vigreux column to give 523 g (83% yield[8]) of *meso*-2,3-dibromobutane, bp 155–157°C. As some *d,l*-2,3-dibromobutane is also produced in this reaction, the isomeric purity of the distilled product is checked by ^{1}H NMR spectroscopy.[7b] The compound is stored in a brown glass bottle.

Preparation of 2-bromo-cis-2-butene.[7a, 9] A 1-L, three-necked, round-bottomed flask is equipped with a 250-mL addition funnel, an overhead stirrer, and a simple distillation apparatus. The flask is charged with 200 g (0.93 mol) of *meso*-2,3-dibromobutane and 150 mL of ethylene glycol. The reaction mixture is then heated to 115°C in an oil bath, and, with stirring, a solution of 68.5 g (1.0 mol, 85% purity) of potassium hydroxide in 250 mL of ethylene glycol is added dropwise over a period of 45 min. During this time, the oil bath temperature is maintained between 115 and 120°C and the 2-bromo-*cis*-2-butene that is generated is distilled over as an azeotrope with water at 75–93°C. After the addition of base is complete, the oil bath is maintained at 125–130°C for an additional 1-h period to obtain additional product. The distillate is then dried over anhydrous magnesium sulfate, filtered, and distilled through a 12-in. column packed with glass helicies (bp = 92.5–94.0°C). Taking the middle fraction gives 80 g (64% yield[8]) of 2-bromo-*cis*-2-butene as a clear, colorless liquid which shows ~15% of the 2-bromo-*trans*-2-butene as a contaminant. The 2-bromo-*cis*-2-butene is stored over ~5 g of Davison 4-Å molecular sieves at − 20°C. However, it should be used within a few days of preparation.

Preparation of 3,5-dimethyl-4-(trifluoromethyl)-2,5-heptadien-4-ol. A 1-L, three-necked, round-bottomed flask, equipped with a reflux condenser, a magnetic stir bar and an addition funnel is flame dried under an atmosphere of argon. After the apparatus has cooled, 350 mL of anhydrous diethyl ether (freshly distilled from sodium benzophenone[10] under argon) is added. Lithium wire (6.9 g, 1.0 mol, 3.2-mm diameter, 0.01% Na content, Aldrich Chem. Co., Milwaukee, WI), which is cut into 5- to 10-mm pieces and washed with hexanes, is added to the flask under a counterstream of argon gas. The reaction flask is cooled to 0°C in an ice bath and 68.9 g (0.51 mol) of 2-bromo-*cis*-2-butene (prepared in the previous step[9]) in 50 mL of anhydrous diethyl ether is added dropwise over a 45-min period while the reaction mixture is stirred. The reaction solution becomes cloudy due to the formation of lithium bromide. Stirring is continued for an additional 1.5–2 h at 0°C.

(■ **Caution.** *Some unreacted lithium metal may remain.*) Next, the reaction flask is cooled to − 40°C in a dry ice/acetonitrile bath and 32.7 g (0.23 mol) of ethyl trifluoroacetate (PCR, Inc., Gainesville, FL) in 50 mL of anhydrous diethyl ether is added dropwise over a 45-min period while stirring. The reaction mixture is stirred for an additional 1.5 h while allowing the cold bath to warm slowly to 0°C. The reaction flask is then immersed in an ice bath, and the reaction is quenched by dropwise addition of 250 mL of 2 *N* aqueous HCl. After the unreacted lithium is destroyed, the biphasic mixture is transferred to a 2-L separatory funnel along with an additional 250 mL of 2 *N* HCl, and the layers are gently mixed. After separation, the aqueous layer is extracted with diethyl ether (3 × 75 mL), and the combined organic extracts are washed successively with saturated aqueous NaHCO₃ (50 mL), water (50 mL), and saturated brine (50 mL) and dried over anhydrous magnesium sulfate. After filtration, the solvent is removed on a rotary evaporator and the clear, yellow liquid is distilled under reduced pressure to give 36.6 g (76% yield[8] based on ethyl trifluoroacetate) of 3,5-dimethyl-4-(trifluoromethyl)-2,5-heptadien-4-ol as a pale, yellow liquid, bp 78–83°C (20 mm). Gas chromatographic analysis (Hewlett Packard 5890A Capillary GC, isothermal elution at 80°C) indicates the presence of two isomers in a 60/40 ratio.

Preparation of the Mixture of Double Bond Isomers of 1,2,3,4-Tetramethyl-5-(trifluoromethyl)cyclopentadiene (Cp‡H). The sample (36.6 g, 0.176 mol) of 3,5-dimethyl-4-(trifluoromethyl)-2,5-heptadien-4-ol prepared in the previous step is dissolved in 915 mL of dichloromethane (4% w/v solution) in a 2-L, three-necked, round-bottomed flask fitted with a magnetic stirrer and an addition funnel. (*Note: The next step is done as quickly as possible as prolonged reaction times reduce the yield and purity of Cp‡H*). The solution is then stirred vigorously with a magnetic stir bar and 91.5 mL (1.41 mol) of neat methanesulfonic acid (i.e., 8.0 equiv of methanesulfonic acid are used per equivalent of dienol) is added rapidly (∼1 min) through an addition funnel. The deep red solution is stirred at ambient temperature for an additional 20 s and the reaction mixture is poured into a 2-L separatory funnel containing 500 mL of a crushed ice/water mixture. The layers are thoroughly mixed to quench the reaction and the organic layer is separated and thoroughly washed with an additional 500 mL of cold water. The combined aqueous layers are extracted with dichloromethane (3 × 100 mL) and the combined organic extracts are washed with saturated NaHCO₃ (2 × 100 mL), water (2 × 100 mL), and saturated brine (2 × 100 mL), and dried over anhydrous magnesium sulfate. After filtration, the solvent is removed on a rotary evaporator to give a clear, yellow liquid which is distilled under vacuum through a 12-in. column of glass helices to give 24.7 g (74% yield[8] or 56% overall yield based on ethyl trifluoroacetate) of Cp‡H as a clear, colorless liquid, bp 38–40°C (4 mm). Gas chromatographic analysis (Hewlett Packard 5890A Capillary GC, isothermal elution at 80°C) indicates a purity of 91%. The chromatogram shows two closely spaced peaks due to the presence of at least two of the three possible isomers of Cp‡H. If desired, the compound can be obtained in ∼99% purity by careful fractional vacuum distillation at 20 mm through a 12-in. column packed with glass helicies, bp 64–65°C (20 mm).

Properties

The mixture of double bond isomers of 1,2,3,4-tetramethyl-5-(trifluoro-methyl)cyclopentadiene (Cp‡H) is a clear, colorless liquid with a sweet ole-finic odor similar to that of pentamethylcyclopentadiene.[6] Cp‡H is an air-stable compound but it should be stored in the refrigerator. Although gas chromatographic analysis shows two peaks, the presence of all three of the possible cyclopentadiene isomers is indicated by ^{1}H NMR spectroscopy as three quartets for the ring protons are observed at 3.26 (J_{HF} = 10.0 Hz), 2.99 (J_{HH} = 6.7 Hz), and 2.66 ppm (J_{HH} = 6.8 Hz) with a relative ratio of approx-imately 1:16:2.4 for isomers I, II, and III, respectively. Resonances between 4.8 and 5.0 ppm are attributed to products that result from an elimination mechanism rather than cyclization in the last step of the synthesis.[4b] These impurities boil very close to that of Cp‡H, but they are removed by careful fractional vacuum distillation as part of the forerun. The IR spectrum (neat liquid) exhibits significant bands at 2977(vs), 2937(vs), 2884(s), 2864(s), 2751(w), 1658(m), 1599(s), 1446(s), 1384(vs), 1357(vs), 1280(s), 1258(vs), 1164(vs), 1099(vs), 1012(vs), and 686(m) cm^{-1}. The syntheses of organotransi-tion metal complexes are usually carried out with the 91% purity Cp‡H product. Because alkali metal and thallium salts of Cp‡ are unstable, organo-transition metal complexes are prepared from Cp‡H itself,[1] as are many complexes of pentamethylcyclopentadiene and cyclopentadiene.[11]

B. DI-μ-CHLORODICHLOROBIS[η^5-TETRAMETHYL-(TRIFLUOROMETHYL)CYCLOPENTADIENYL]-DIRHODIUM(III), [Cp‡RhCl$_2$]$_2$

$$2RhCl_3 + 2Cp‡H \xrightarrow{MeOH} [Cp‡RhCl_2]_2 + 2HCl$$

Procedure

A 200-mL, round-bottomed flask, equipped with a magnetic stir bar, is charged with 1.00 g (3.80 mmol) of rhodium(III) chloride trihydrate (Johnson

Matthey, Ltd.), 0.86 g (4.5 mmol, 91% purity) of Cp‡H,[9b] and 100 mL of reagent-grade methanol. The flask is fitted with a reflux condenser and an inert gas-bubbler adapter, and the reaction mixture is refluxed under an argon atmosphere for 72 h. An orange-red solid precipitates from solution. The reaction mixture is then cooled to 0°C and the air-stable precipitate (1.04 g) is collected by filtration on a medium-porosity glass frit and is washed with 25 mL each of cold methanol and hexanes. A second crop of product (0.05 g) is obtained by evaporation of the mother liquor on a rotary evaporator to a volume of ~25 mL. The combined orange-red solid is then dissolved (in air) in ~150 mL of warm dichloromethane, and the solution is added to a 13 × 5 cm column of silica gel (Baker, 60–200 mesh) with a 1-cm top layer of sand. After washing the absorbed compound with 50 mL of dichloromethane, the compound is eluted with a mixture of 10% anhydrous diethyl ether in dichloromethane. A leading, thick (~3–5 cm), red band is eluted with about 550 mL of the solvent mixture. The trailing end of the first band is eluted with an additional 450 mL of the solvent mixture and is collected in 150-mL fractions. The solvent is removed separately from each of the four fractions to about 1/4 of the original volume on a rotary evaporator and the remaining solutions are diluted with a twofold excess of hexanes to precipitate the [Cp‡RhCl$_2$]$_2$ fractions. After cooling in ice for 10 min, each fraction is collected by filtration on a medium-porosity glass frit and washed with 2 × 10 mL portions of hexanes to give 0.83, 0.10, 0.02, and 0.01 g of [Cp‡RhCl$_2$]$_2$ from the first through the fourth fractions, respectively. The purity of each fraction is assessed by ^{1}H NMR spectroscopy and the pure fractions (1, 2) are combined to give 0.93 g (67% yield[8]) of [Cp‡RhCl$_2$]$_2$.

Properties

The complex [Cp‡RhCl$_2$]$_2$ is an air-stable, orange-red solid which shows remarkable thermal stability (mp > 300°C decomposes). The compound also appears to be relatively stable in aerated solvents. The identity and purity of [Cp‡RhCl$_2$]$_2$ are readily assessed by ^{1}H NMR spectroscopy as only two resonances in CDCl$_3$ are observed [1.71(s) and 1.89 (q, J_{HF} = 1 Hz)] for the methyl groups on the Cp‡ ring. The trifluoromethyl group in the Cp‡ ligand appears at − 55.4 ppm in CDCl$_3$ (vs. CFCl$_3$) in the ^{19}F NMR spectrum. Fractions 3 and 4 from the chromatography (0.03 g, 2% yield) contain common impurities which are indicated by two resonances at ~3.9 ppm, and as shoulders on the 1.71- and 1.89-ppm resonances in the ^{1}H NMR spectrum. As these impurities are sometimes carried over in further reactions of [Cp‡RhCl$_2$]$_2$, fractions 3 and 4 should not be used without additional column chromatography. [Note: The use of high-purity Cp‡H (99%) in the procedure gives approximately the same amount of impurity.] The complexes derived from $[Cp^{\ddagger}RhCl_2]_2$ may be prepared,[1, 14] using procedures similar to those developed for the synthesis of the related pentamethylcyclopentadienide (Cp*) complexes from the $[Cp*RhCl_2]_2$ derivative.[11g, 12, 13]

Acknowledgments

JRS and JWM dedicate this work to the memory of Professor Paul G. Gassman. This research was supported by the National Science Foundation. We are grateful to Johnson Matthey for a loan of rhodium trichloride.

References and Notes

1. (a) P. G. Gassman, J. W. Mickelson, and J. R. Sowa, Jr., *J. Am. Chem. Soc.*, **114**, 6942 (1992); (b) P. G. Gassman, J. R. Sowa, Jr., and J. W. Mickelson, *U.S. Patent No.* 5,245,064, Sept. 14, 1993; (c) P. G. Gassman, J. R. Sowa, Jr., M. G. Hill, and K. R. Mann, *Organometallics*, **14**, 4879 (1995); (d) M. C. Baschky, J. R. Sowa, Jr., P. G. Gassman, and S. R. Kass, *J. Chem. Soc.*, *Perkin, Trans. 2*, 213 (1996).

2. P. G. Gassman and C. H. Winter, *J. Am. Chem. Soc.*, **108**, 4228 (1986).

3. P. G. Gassman, D. W. Macomber, and J. W. Hershberger, *Organometallics*, **2**, 1470 (1983).

4. (a) C. Santelli-Rouvier and M. Santelli, *Synthesis*, 429 (1983); (b) P. H. Campbell, N. W. K. Chiu, K. Deugau, I. J. Miller, and T. J. Sorensen, *J. Am. Chem. Soc.*, **91**, 6404 (1969).

5. R. S. Threlkel and J. E. Bercaw, *J. Organomet. Chem.*, **136**, 1 (1977).

6. (a) J. M. Manriquez, P. J. Fagan, L. D. Schertz, and T. J. Marks, *Inorg. Synth.*, **28**, 317 (1990); (b) J. M. Manriquez, P. J. Fagan, L. D. Schertz, and T. J. Marks, *Inorg. Synth.*, **21**, 181 (1982); (c) C. M. Fendrick, L. D. Schertz, E. A. Mintz, and T. J. Marks, *Inorg. Synth.*, **29**, 193 (1992).

7. (a) F. G. Bordwell and P. S. Landis, *J. Am. Chem. Soc.*, **79**, 1593 (1957); (b) A. A. Bothner-By and C. Naar-Colin, *J. Am. Chem. Soc.*, **84**, 743 (1962).

8. The checkers report the following yields: *meso*-2,3-dibromobutane (67%), 2-bromo-*cis*-2-butene (59%), 3,5-dimethyl-4-(trifluoromethyl)-2,5-heptadien-4-ol (51%), $Cp^{\ddagger}H$ (60%, $\sim 80\%$ purity), $[Cp^{\ddagger}RhCl_2]_2$ (88% of crude product).

9. (a) Use of commercially available 2-bromo-2-butene (Aldrich Chem. Co., Milwaukee, WI) which consists of 80–90% of 2-bromo-*trans*-2-butene ultimately gives $Cp^{\ddagger}H$ in lower yield (44%) and purity (62%). (b) The $[Cp^{\ddagger}RhCl_2]_2$ complex was obtained in 67% yield from the 62% purity $Cp^{\ddagger}H$ above with the use of a 2.5-fold excess of the $Cp^{\ddagger}H$ product.

10. D. D. Perrin, W. L. F. Armarego, and D. R. Perrin, *Purification of Laboratory Chemicals*, 2nd ed., Pergamon, New York, 1980.

11. (a) J. J. Eisch and R. B. King, *Organomet. Synth.*, **1**, (1965); (b) R. B. King, W. M. Douglas, and A. Efraty, *J. Organomet. Chem.*, **69**, 131 (1974); (c) R. B. King and M. B. Bisnette, *J. Organomet. Chem.*, **282**, 357 (1985); (d) M. O. Albers, D. C. Liles, D. J. Robinson, A. Shaver, E. Singleton, M. B. Wiege, J. C. A. Boeyens, and D. C. Levendis, *Organometallics*, **5**, 2321 (1986); (e) S. A. Frith and J. L. Spencer, *Inorg. Synth.*, **28**, 273 (1990); (f) T. S. Piper, F. A. Cotton, and G. Wilkinson, *J. Inorg. Nucl. Chem.*, **1**, 165 (1955); (g) J. W. Kang, K. Moseley, and P. M. Maitlis, *J. Am. Chem. Soc.*, **91**, 5970 (1969); (h) U. Koelle and J. Kossakowski, *Inorg. Synth.*, **29**, 225 (1992); (i) S. P. Nolan, C. D. Hoff, and J. T. Landrum, *J. Organomet. Chem.*, **282**, 357 (1985).

12. C. White, A. Yates, and P. M. Maitlis, *Inorg. Synth.*, **29**, 228 (1992).

13. (a) B. L. Booth, R. N. Haszeldine, and M. Hill, *J. Organomet. Chem.*, **16**, 491 (1969); (b) J. W. Kang and P. M. Maitlis, *J. Organomet. Chem.*, **26**, 393 (1971).
14. E. Hauptman, S. Sabo-Etienne, P. S. White, M. Brookhart, J. M. Garner, P. J. Fagan, and J. C. Calabrese, *J. Am. Chem. Soc.*, **116**, 8038 (1994).

37. ACETONITRILE-SUBSTITUTED DERIVATIVES OF $Rh_6(CO)_{16}$: $Rh_6(CO)_{16-x}(NCMe)_x$ $(x = 1, 2)$

SUBMITTED BY S. P. TUNIK,* A. V. VLASOV,* and V. V. KRIVYKH[†]

CHECKED BY JOSEF TAKATS,[‡] JASON COOKE,[‡]

and JOHN WASHINGTON[‡]

The reactivity of carbonyl clusters can be enhanced substantially by the use of their activated derivatives which contain such labile ligands as acetonitrile, ethene, or cyclooctene. For example, it was the synthesis of acetonitrile derivatives $Os_3(CO)_{12-x}(NCMe)_x$ $(x = 1, 2)$[1] that spurred the development of the chemistry of this trinuclear cluster.[2-4] In general, the reactions of the activated clusters occur under mild conditions and result in high selectivities and yields.

Here we report the simple and convenient syntheses of the mono- and disubstituted clusters, $Rh_6(CO)_{16-x}(NCMe)_x$ $(x = 1, 2)$. These compounds are useful synthons[5-7] and also have some catalytic applications.[8]

■ **Caution.** *Metal carbonyls, carbon oxide, and trimethylamine N-oxide are highly toxic and should be handled in a well-ventilated hood.*

The starting material, $Rh_6(CO)_{16}$, is synthesized by refluxing $Rh_4(CO)_{12}$ (0.5 g) in hexane solution (150 mL) for 4–5 h. The initially dark red solution of the tetranuclear cluster turns light brown and a black crystalline powder of $Rh_6(CO)_{16}$ precipitates in approximately quantitative yield. IR spectrum of $Rh_6(CO)_{16}$: CHCl₃, $v(CO)$, cm⁻¹, 2078ᵥₛ, 2044ᵥ, 1810ₘ,ᵦᵣ. The cluster $Rh_4(CO)_{12}$ is synthesized by the literature procedure.[9] All solvents are purified by distillation. Commercial grade $Me_3NO \cdot 2H_2O$ is used without further purification. All operations are carried out in air, unless otherwise stated.

* Department of Chemistry, St. Petersburg University, Universitetskii pr., 2, St. Petersburg, 198904, Russia.
† A. N. Nesmeyanov Institute of Organoelement Compounds, 28 Vavilov Street, Moscow, 117813, Russia.
‡ Department of Chemistry, University of Alberta, Edmonton, Alberta T6G 2G2, Canada.

A. SYNTHESIS OF $Rh_6(CO)_{15}(NCMe)$

(Acetonitrile)tetra-μ_3-carbonyl-undecacarbonyl-octahedrohexarhodium

The monosubstituted cluster $Rh_6(CO)_{15}(NCMe)$ can be synthesized in two ways.

Photochemical synthesis

$$Rh_6(CO)_{16} + MeCN \xrightarrow[\text{CHCl}_3]{hv(\lambda_{irr} > 300\,nm)} Rh_6(CO)_{15}(NCMe) + CO \qquad (1)$$

$Rh_6(CO)_{16}$ (206 mg, 0.193 mmol), trichloromethane (150 mL), and aceto-nitrile (10 mL) are placed in an Erlenmeyer Pyrex flask (200 mL capacity) equipped with a reflux condenser. Pyrex glass is not transparent for the short-wavelength irradiation ($\lambda_{irr} < 300$ nm) which causes decomposition of the starting cluster in the $CHCl_3$ solution. The reaction mixture is irradiated under vigorous stirring with a 400-W Hg lamp placed approximately 15 cm away from the flask. Heating of the solution above 50°C during the irradiation should be avoided in order to avoid side reactions. In the course of the irradiation the starting cluster dissolves completely and the initial light-brown solution turns dark brown. The reaction is monitored by the IR spectroscopy and TLC spot tests and should be stopped (after approximately 1 h) after the complete consumption of the parent $Rh_6(CO)_{16}$ cluster (disap-pearance of the most intense absorption band at 2078 cm^{-1}). The solution is then reduced in volume to about 100 mL under vacuum, diluted with heptane (15 mL) and concentrated again such that a narrow starting band is observed when the solution is loaded to a silica column (total final volume is ca. 20 mL). This condition can be checked by a spot test as well. The solution is applied to a TLC plate with a capillary and the diameter of the light solvent spot should be at least twice that of the dark brown spot of the rhodium clusters components. The solution thus obtained is loaded on to a chromato-graphic column (3.5 × 9 cm, silica 5–40 mesh). If some residue of the clusters is formed while concentrating the reaction mixture, it should be dissolved in a minimal amount of $CHCl_3$ (ca. 1 mL) diluted with heptane, and applied to the column as well. Elution under the pressure of Ar with hexane/trichloromethane/acetonitrile (20/10/1) mixture gives the following bands; (1) a light brown band containing traces of unreacted $Rh_6(CO)_{16}$; (2) the main dark-brown band containing $Rh_6(CO)_{15}NCMe$; (3) trace amounts of the disubstituted cluster $Rh_6(CO)_{14}(NCMe)_2$. The main fraction is diluted with heptane (5–10 mL) and concentrated under reduced pressure until near complete precipitation of the product and the initial dark brown solution turns transparent (the solvent over the precipitated product is only slightly colored). The precipitate is washed with pentane and dried in an air flow. The Rh$_6$(CO)$_{15}$NCMe is obtained as a dark-brown crystalline powder. Yield 145–165 mg, (70 –80%). Found: C, 18.98%; H, 0.45%. Calculated for C$_{17}$H$_3$N$_1$O$_{15}$Rh$_6$: C, 18.93%; H, 0.28%.

NOTE. Reaction (1) is quite clean and the reaction mixture obtained after irradiation contains up to 90–95% of the monosubstituted product. If Rh$_6$(CO)$_{15}$NCMe is to be used for a further synthesis involving separation or purification of products, the crude Rh$_6$(CO)$_{15}$NCMe product can be obtained without chromatographic separation in the following manner. Heptane (20 mL) is added to the final reaction mixture, which is concentrated under reduced pressure until complete precipitation of the product as described above.

Synthesis Using Trimethylamine N-oxide

$$Rh_6(CO)_{16} + MeCN + Me_3NO \xrightarrow{CHCl_3} Rh_6(CO)_{15}NCMe + CO_2 + Me_3N$$

$$(2)$$

Rh$_6$(CO)$_{16}$ (107 mg, 0.1 mmol) is dissolved completely in a mixture of trichloromethane (150 mL) and acetonitrile (10 mL). A solution of trimethylamine N-oxide (Me$_3$NO·2H$_2$O) (12 mg, 0.11 mmol) in MeOH/CHCl$_3$ (0.5/5.0 mL) is added dropwise under vigorous stirring to the cluster solution. The reaction mixture is allowed to stand for an additional 15 min. The product is isolated by the procedure described above. Yield of 70–75 mg of Rh$_6$(CO)$_{15}$NCMe: 64–70%.

Properties

The cluster is insoluble in hydrocarbons, but is very soluble in polar solvents such as trichloromethane and dichloromethane. The cluster is not air sensitive and can be stored at 0°C for a few months. However, it decomposes on prolonged standing in solution in the absence of added acetonitrile.

The IR frequencies (cm^{-1}) in CHCl$_3$ are: 2103 w, 2068 s, 2040 m, 2018 sh, 1788 m, br (ν CO), 2360 vvw (ν CN).

The ^{13}C NMR spectrum of Rh$_6$(CO)$_{15}$NCMe in CDCl$_3$: CD$_3$CN (10 : 1) solution at 25°C displays the following signals: δ (multiplicity, J$_{Rh-C}$ Hz; assignment) 123.9 (s; NCMe), 178.6 (d, 70; 2 ∗ CO), 179.8 (d, 69; (1 + 2) ∗ CO), 180.9 (d, 68; 2 ∗ CO), 181.0 (d, 68; 2 ∗ CO), 182.5 (d, 70; 2 ∗ CO), 230.6 (q, 27; 1 ∗ CO), 233.7 (q, 28; 1 ∗ CO), 240.2 (q, 27; 2 ∗ CO).

These spectroscopic data are very similar to those obtained for other monosubstituted derivatives of $Rh_6(CO)_{16}$,[5,6] in particular, to that of $Rh_6(CO)_{15}PPh_3$, whose structure has been determined by single-crystal X-ray analysis.[6] Hence, the structure of $Rh_6(CO)_{15}NCMe$ can be shown schematically as

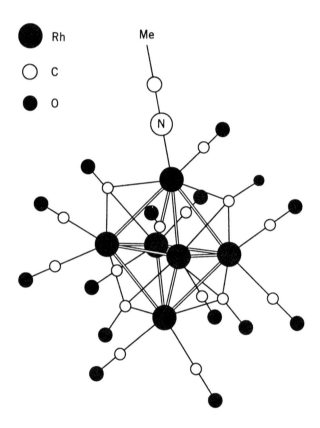

where acetonitrile occupies one of the 12 terminal positions in the structure of the parent $Rh_6(CO)_{16}$ cluster.

The acetonitrile ligand in $Rh_6(CO)_{15}NCMe$ is very labile and can be readily substituted by other two-electron ligands.[5-8]

$$Rh_6(CO)_{15}NCMe + L \longrightarrow Rh_6(CO)_{15}L + NCMe \qquad (3)$$

$$L = PPh_3, P(OPh)_3, \text{pyridine}, I^-, \text{ethene, cyclooctene}$$

B. SYNTHESIS OF $Rh_6(CO)_{14}(NCMe)_2$

$$Rh_6(CO)_{16} + 2MeCN + 2Me_3NO \xrightarrow[CHCl_3]{} Rh_6(CO)_{14}(NCMe)_2$$

$$+ 2CO_2 + 2Me_3N \tag{4}$$

$Rh_6(CO)_{16}$ (100 mg, 0.094 mmol) is dissolved in a mixture of trichloromethane (120 mL) and acetonitrile (10 mL). A solution of trimethylamine N-oxide ($Me_3NO \cdot 2H_2O$) (22 mg, 0.2 mmol) in $MeOH/CHCl_3$ (0.5/5.0 mL) is added dropwise under vigorous stirring to the cluster solution. The mixture is allowed to stand for an additional 15 min and then prepared for column separation (3.5×8 cm, silica 5–40 mesh) as described above for the synthesis of the monosubstituted cluster. Elution under pressure of Ar with a hexane/trichloromethane/acetonitrile (15/15/1) mixture gives the following bands: (1) a light-brown band containing $Rh_6(CO)_{15}NCMe$: (2) the main dark-brown band containing $Rh_6(CO)_{14}(NCMe)_2$: (3) trace amounts of polysubstituted derivatives. Heptane (5 mL) is added to the $Rh_6(CO)_{14}(NCMe)_2$ containing fraction and the solvents are removed under reduced pressure until complete precipitation of the product occurs. The precipitate is washed with pentane and dried in an air flow. The $Rh_6(CO)_{14}(NCMe)_2$ is obtained as a dark-brown crystalline powder. Yield 60–68 mg, 57–69%. Found: C, 19.51%: H, 1.00%: N, 2.26%. Calculated for $C_{18}H_6N_2O_{14}Rh_6$: C, 19.91%: H, 0.56%; N, 2.58%.

Properties

The $Rh_6(CO)_{14}(NCMe)_2$ cluster is insoluble in hydrocarbons, but is very soluble in polar solvents such as trichloromethane and dichloromethane. The cluster is not air sensitive and can be stored at 0°C for a few months. However, it decomposes on standing in solution in the absence of added acetonitrile. The IR frequencies (cm^{-1}) in $CHCl_3$ are: 2092 w, 2056 s, 2028 m, 2000 sh, 1776 m, br (v CO), 2360 vvw (v CN).

References

1. B. F. G. Johnson, J. Lewis, and D. Pippard, *J. Organomet. Chem.*, **160**, 263 (1978), **213**, 249 (1981).
2. K. Burgess, *Polyhedron*, **3**, 1175 (1984).
3. R. D. Adams, I. T. Horvath, and P. Mathur, *Organometallics*, **3**, 623 (1984).
4. A. J. Arce, P. A. Bates, S. P. Best, R. J. H. Clark, A. J. Deeming, M. B. Hursthous, R. C. S. McQueen, and N. I. Powell, *J. Chem. Soc., Chem. Comm.*, 478 (1988).
5. S. P. Tunik, A. V. Vlasov, A. B. Nikol'skii, V. V. Krivykh, and M. I. Rybinskaya, *Metallorg. Khim.* (Russ. J.), **3**, 387 (1990).

6. S. P. Tunik, A. V. Vlasov, N. I. Gorshkov, G. L. Starova, A. B. Nikol'skii, M. I. Rybinskaya, A. S. Batsanov, and Yu. T. Struchkov, *J. Organomet. Chem.*, **433**, 189 (1992).
7. S. P. Tunik, M. V. Osipov, and A. B. Nikol'skii, *J. Organomet. Chem.*, **426**, 105 (1992).
8. S. P. Tunik, A. V. Vlasov, A. B. Nikol'skii, V. V. Krivykh, and M. I. Rybinskaya, *Metallorg. Khim.* (Russ. J.), **4**, 586 (1991).
9. S. Martinengo, P. Chini, and G. Giordano, *J. Organomet. Chem.*, **27**, 389 (1971).

38. TETRAPHENYLARSONIUM CARBONYLTRICHLOROPLATINATE(II)

SUBMITTED BY MITSURU KUBOTA* and GREGORY F. PARKS*
CHECKED BY DANIELA BELLI DELL'AMICO† and
FAUSTO CALDERAZZO†

The anionic $[Pt(CO)Cl_3]^-$ complex has been synthesized in low yields by elaborate, time-consuming processes, including the reaction of formic acid and $Na_2[PtCl_6]$,[1] the reaction of $K_2[PtCl_4]$, NO, CO, and $CuCl_2$,[2] and the reaction of $(n\text{-}Bu_4N)_2[PtCl_4]$ and CO in $SOCl_2$.[3] The $[Pt(CO)Cl_3]^-$ anion may be conveniently prepared by the reaction of CO with Zeise's anion $[Pt(C_2H_4)Cl_3]^-$.

$$[Pt(C_2H_4)Cl_3]^- + CO \rightarrow [Pt(CO)Cl_3]^- + C_2H_4$$

Procedure

■ **Caution.** *Carbon monoxide is a colorless, odorless gas of extreme toxicity. The reaction of carbon monoxide and ethene should be carried out in a well-ventilated hood.* A solution of $K[Pt(C_2H_4)Cl_3]$ is prepared by an adaptation of the method of Chock, Halpern, and Paulik.[4] The starting reagents and the products were handled without any special precautions to avoid contact with air. Potassium tetrachloroplatinate(II) (2.25 g, 5.4 mmol) (Engelhard), 20 mg tin(II) chloride hydrate, $SnCl_2 \cdot 2H_2O$ (0.1 mmol), and a magnetic stirring bar are placed in a 200-mL Schlenk flask with gas inlet and outlet connections. The outlet is connected to 5-mm glass tubing placed in a test tube containing water to monitor the pressure and rate of gas flow. Aqueous hydrochloric acid (25 mL of 5 M HCl) is added to the flask and

* Department of Chemistry, Harvey Mudd College, Claremont, CA 91711.
† Dipartimento di Chimica e Chimica Industriale, Via Risorgimento 35, University of Pisa, 56126 Pisa, Italy.

ethene is introduced to the flask while the sample mixture is stirred vigorously. After the flask is flushed with ethene, a positive pressure of ethene is maintained while the stirring is continued until the solution, initially red, turns orange and eventually yellow and all the $K_2[PtCl_4]$ is dissolved. This process should take about 2 h. A positive pressure of carbon monoxide is then maintained over the gently stirred solution for 5 min with care being taken to avoid splashing the solution on the sides of the flask. Treatment with carbon monoxide should not be prolonged beyond 5 min to prevent contamination of the product by $[Pt_2Cl_4(CO)_2]^{2-}$ which has IR bands at 2046 and 2027 cm^{-1}.[5] A solution of 2.4 g (5.5 mmol) of tetraphenylarsonium chloride hydrate in 100 mL of water is added slowly to the reaction mixture to precipitate a pale yellow solid. The product is filtered on a glass frit and washed, first with 35 mL of cold 0.5 M hydrochloric acid, and then with 15 mL cold ethanol. The product is dried in vacuo. The yellow product is recrystallized by dissolving in 30 mL of dichloromethane, filtering, and adding 50 mL of diethyl ether. The yield is 3.54 g (91%).*

Anal. Calcd. for $AsC_{25}Cl_3H_{20}OPt$: C, 42.13; H, 2.83; Cl, 14.92. Found: C, 42.11; H, 3.10; Cl, 15.13.

Properties

The compound $[Ph_4As][Pt(CO)Cl_3]$ is characterized by IR absorption at 2077 cm^{-1} (Nujol) and 2095 cm^{-1} in dichloromethane. Yellow crystals of the compound decompose in air at 179°C. The compound is unreactive in air, but decomposes slowly in water. It is soluble in dichloromethane and dimethylformamide, slightly soluble in ethanol, and insoluble in water, diethyl ether, and hexane. The chlorine ligand *trans* to the CO is readily displaced by other halogens and other ligands.[6] The compound reacts with chlorine to form $[Pt(CO)Cl_5]^-$.[7]

References

1. M. J. Cleare and W. P. Griffith, *J. Chem. Soc. A.*, 372 (1969).
2. K. S. Sun, K. C. Kong, and C. H. Cheng, *Inorg. Chem.*, **30**, 1998 (1991).
3. D. B. Dell Amico and F. Calderazzo, *Inorg. Chem.*, **23**, 3030 (1984).
4. P. B. Chock, J. Halpern, and F. E. Paulik, *Inorg. Synth.*, **28**, 349 (1990).
5. P. L. Goggin and R. J. Goodfellow, *J. Chem. Soc. Dalton Trans.*, 2355 (1973).
6. L. Spaulding, B. A. Reinhard, and M. Orchin, *Inorg. Chem.*, **11**, 2092 (1972).
7. C. Crocker, P. L. Goggin, and R. J. Goodfellow, *J. Chem. Soc. Chem. Commun.*, 1056 (1978).

* The checkers obtained yields of 72 and 76%.

Chapter Four

TRANSITION METAL, LANTHANIDE, AND ACTINIDE COMPLEXES

39. DICHLORODIOXOBIS(DIMETHYLSULPHOXIDE)-MOLYBDENUM(VI)

SUBMITTED BY FRANCISCO J. ARNAIZ*
CHECKED BY GEORGE B. KAUFFMAN[†] and
SCOTT D. PENNINGTON[†]

Dichlorodioxomolybdenum(VI), MoO_2Cl_2, is a valuable starting material for preparing numerous molybdenum compounds, but because it is moisture-sensitive, it must be prepared immediately before use, by chlorinating molybdenum(IV) oxide.[1] Inasmuch as other related compounds, such as $MoO_2Br_2(DMSO)_2$ (DMSO = dimethyl sulphoxide), can be used to prepare new dioxomolybdenum(VI) compounds,[2] a rapid, inexpensive method for preparing $MoO_2Cl_2(DMSO)_2$, requiring only 3–4 h, is given here. This compound was isolated by Horner and Tyree[3] by the reaction of MoO_2Cl_2 with DMSO in methylene chloride. The synthesis described here is based on the existence of $MoO_2Cl_2(H_2O)_2$ in aqueous solution[4] and the ability of DMSO to displace water from this substance:

$$MoO_3 + 2HCl + H_2O \rightarrow MoO_2Cl_2(H_2O)_2$$

$$MoO_2Cl_2(H_2O)_2 + 2DMSO \rightarrow MoO_2Cl_2(DMSO)_2 \downarrow + 2H_2O$$

Procedure

A mixture of 20.0 g (0.138 mol) of MoO_3 powder, 100 mL of concentrated (12 M) hydrochloric acid, and 100 mL of distilled water is heated to just

---

* Laboratorio de Química Inorgánica, Universidad de Burgos, 09001 Burgos, Spain.
† Department of Chemistry, California State University, Fresno, CA 93740.

below boiling with mechanical stirring until the MoO_3 has dissolved (ca. 1.5–2 h)* (■ **Caution.** *The operation should be carried out in a well-ventilated hood.*) The resulting solution is cooled,[†] and 50.0 mL (0.705 mol) of dimethyl sulphoxide is added with mechanical stirring (■ **Caution.** *Because DMSO readily penetrates skin and other tissues, rubber gloves should be worn. Also, because it is flammable, it should be kept away from open flames.*) After 15 min the white, microcrystalline precipitate is collected by suction filtration, washed with four 30-mL portions of acetone, and dried in vacuo. Yield: 43.0 g (86.4%).[‡]

Anal. Calcd. for $MoO_2Cl_2[(CH_3)_2SO]_2$: Mo, 27.0; Cl, 19.9; C, 13.6; H, 3.4. Found: Mo, 26.9; Cl, 20.1; C, 13.5; H, 3.4.[§]

Properties

$MoO_2Cl_2(DMSO)_2$ melts at 172–173°C with decomposition. The IR spectrum, taken as a Nujol mull, has characteristic bands of *cis*-MoO_2 at 921 and 892 cm^{-1}. It is slightly soluble in dichloromethane, very soluble in hot methanol, soluble in acetonitrile, and less soluble in acetone. It can be recrystallized from the last two solvents. It is very stable at room temperature and it can be manipulated without special care (after storage for 6 months in a dessicator over KOH, it remains unchanged). Several compounds, such as $MoO_2Cl_2(L)_2$ (L = hexamethylphosphoramide, 1/2 bipyridine), $MoO_2(S_2CNEt_2)_2$, and $MoO_2(dpc)(L)$ (dpc = pyridine-2,6-dicarboxylate; L = pyridine, DMSO, hexamethylphosphoramide) are easily prepared in good yields starting from this material.

References

1. R. S. Schrock, J. S. Murdzek, G. C. Bazan, J. Robins, M. DiMare, and M. O'Reagan, *J. Am. Chem. Soc.*, **112**, 3875 (1990).
2. S. A. Roberts, C. G. Young, C. A. Kipke, W. E. Cleland, Jr., K. Yamanouchi, M. D. Carducci, and J. H. Enemark, *Inorg. Chem.*, **29**, 3650 (1990).
3. S. M. Horner and S. Y. Tyree, Jr., *Inorg. Chem.*, **1**, 122 (1962).
4. J. M. Coddington and M. J. Taylor, *J. Chem. Soc., Dalton Trans.*, **1**, 41 (1990).

---

* The volume is not reduced appreciably during heating. The temperature should be kept just below the boiling point, and 125 mL rather than 100 mL of HCl should be used to dissolve the maximum of the MoO_3. Nevertheless, we were never able to get all the MoO_3 to dissolve, thus requiring the subsequent filtration step.

† Room temperature (20°C) rather than ice-bath temperature gives the maximum yield.

‡ The checkers obtained 41.3 g (81.3%).

§ Found: C, 13.74; H, 3.33 (by checkers).

40. METAL-CATALYZED SYNTHESIS OF *cis*-[Re(CO)₄LI] [L = P(OMe)₃, PMe₂Ph, PPh₃]

SUBMITTED BY ANN E. LEINS* and NEIL J. COVILLE*
CHECKED BY JOHN A. GLADYSZ[†]

Attempts to remove only one CO ligand from [Re(CO)₅X] [1] (X = Cl, Br, I) by L (L = group 15 donor ligand) can lead to replacement of X[1] or replacement of more than one CO group.[2] Thus, in nearly all reports in which substitution has been achieved (thermal,[3] photochemical,[4] and [BH₄]⁻[5] induced procedures) [Re(CO)₃L₂X] is the dominant if not exclusive product obtained from the reaction. Hence a major drawback of these strategies entails the lack of product control and the presence of undesired products. The most general synthetic procedures for obtaining the complex [Re(CO)₄LX] in high yield have depended on indirect methods of substitution. These have normally entailed the cleavage of metal dimer complexes by L. For instance, cleavage of the halide bridged dimer [Re(CO)₄X]₂ by L occurs under mild conditions to give [Re(CO)₄LX].[6-8]

The transition-metal-*catalyzed* substitution of [1] by L has, by way of contrast, proved to be a *direct and rapid* synthetic route that gives the product, *cis*-[Re(CO)₄LX], under mild conditions and in high yield. The syntheses of some typical examples are reported here.

General Procedure

Decacarbonyldirhenium is available from Strem Chemicals (7 Mullikan Way, Dexter Industrial Park, P.O. Box 108, Newburyport, MA 01950) and [Re(CO)₅I] is synthesized by published procedures.[9] Palladium oxide (mesh: 200–300 μm) may be purchased from Engelhardt (Valley Road, Cinderford, Glos. GL 142PB). Trimethylphosphite was obtained from BDH Laboratories (Chemical Division, Poole, England), PMe₂Ph from Strem Chemicals, and PPh₃ from Merck (Schuchardt, 8011 Hohenbrunn, Munich). Reagent-grade toluene is distilled over sodium under nitrogen prior to use. Thin-layer chromatography plates are of silica gel and contain a fluorescent indicator (Merck 60F, 0.2 mm thickness). All reactions are routinely carried out under a nitrogen atmosphere in a two-necked, round-bottomed flask and exposed to normal laboratory light (fluorescent). One neck is fitted with a nitrogen

* Department of Chemistry, University of the Witwatersrand, P.O. Wits 2050, Johannesburg 2001, Republic of South Africa.
[†] Department of Chemistry, University of Utah, Salt Lake City, UT 84112.

inlet and the other with a glass stopper. The reaction solution is heated by a paraffin-oil bath preset at 40–45°C and magnetically stirred.

■ **Caution.** *Decacarbonyldirhenium, like all metal carbonyl compounds, should be considered to be toxic. $P(OMe)_3$ and PMe_2Ph have a pungent odor and should be handled in a well-ventilated hood. $P(OMe)_3$ is flammable and corrosive and PPh_3 is particularly hazardous when in contact with the skin. PMe_2Ph is particularly air-sensitive. Hence, all ligands should be handled with care.*

SYNTHESIS OF cis-[Re(CO)LI] [L = P(OMe)₃, PMe₂Ph, PPh₃] USING PALLADIUM(II) OXIDE AS CATALYST

$$[Re(CO)_5I] + L \rightarrow [Re(CO)_4LI] + CO$$

Procedure

Pentacarbonyliodorhenium (3.00 mmol; 1.360 g) and toluene (or benzene) (40 mL) are placed in the reaction flask and allowed to be exposed to normal laboratory light. The stirred mixture is heated to 40–45°C and the appropriate amount of ligand required to achieve monosubstitution [3.30 mmol; L = P(OMe)₃, 0.410 mL; L = PMe₂Ph, 0.456 mL; L = PPh₃, 0.866 g] is added by microsyringe or as a solid to the warmed reaction mixture. The catalyst, PdO (0.030 g), is added last and as a solid using a spatula. After an initial induction period of a few minutes, rapid effervescence is observed for L = PMe₂Ph and P(OMe)₃. For L = PPh₃, the reaction is obvious in the initial stages only by a slow bubbling.

Completion of the reaction (Table I) is indicated by monitoring the reaction using IR spectroscopy or thin-layer chromatography (mobile phase: hexane–toluene, 1:1). The mixture is cooled to room temperature, filtered

TABLE I. Reaction Times, Yields, and Melting-Point Determinations for [Re(CO)₄LI] [L = P(OMe)₃, PMe₂Ph, PPh₃]

Complex	Reaction Time (min)	Yield (%)	mp (°C)
[Re(CO)₄{P(OMe)₃}I]	40	81	35–36
[Re(CO)₄(PMe₂Ph)I]	5	98	82–83[a]
[Re(CO)₄(PPh₃)I]	~ 660	98	139–141[b]

[a] Checkers found a melting point value of 70.5–71°C.
[b] Checkers observe softening of Re(CO)₄(PPh₃)I at 148°C and subsequent slow decomposition.

through a cellulose layer to remove catalyst, and solvent removal is effected under vacuum, giving the crude products as pale yellow solids. Purification is achieved by recrystallization (hot hexane). The products are obtained from these procedures in > 80%* yields as pale yellow crystalline materials (Table II).

Infrared spectroscopy (to detect starting material and disubstituted products) is used as an indication of purity (Table III). If necessary, column chromatography can be used to purify the materials (silica; eluent: hexane–toluene, 1:1).

Anal. Calcd. for $C_7H_9IO_7PRe$: C, 15.31; H, 1.65. Found: C, 14.93; H, 1.50. Calcd. for $C_{12}H_{11}IO_4PRe$: C, 25.59; H, 1.97. Found: C, 25.61; H, 1.88. Calcd. for $C_{22}H_{15}IO_4PRe$: C, 38.44; H, 2.20. Found: C, 40.13; H, 2.24.

Properties

The complexes $[Re(CO)_4LI]$ $[L = P(OMe)_3, PMe_2Ph, PPh_3]$ are air-stable, pale-yellow crystalline solids[†] soluble in most common organic solvents (e.g., hexane, benzene, toluene, tetrahydrofuran, and chloroform). They can be identified readily by their IR and NMR spectra (Table I). The IR spectra in polar solvents show the four absorptions characteristic of a *cis* complex. However, five absorptions are observed for $L = P(OMe)_3$ when the spectrum is recorded in hexane. The IR spectra of some disubstituted complexes, $Re(CO)_3L_2X$, are given in Table III. As will be seen, the presence of mono- and disubstituted complexes can be detected readily by IR spectroscopy. The good separation of the methyl resonances [$P(OMe)_3$ and PMe_2Ph ligands] of the mono- and disubstituted[3,5,10] complexes permits ready assessment of product purity.*

The reaction has been found to be influenced by both laboratory light and traces of O_2. This can result in variable reaction times if standardized procedures are not observed.[11] The reaction is not restricted to the ligands described herein. Thus a wide range of group 15 donor ligands (L) have been used to synthesize $[Re(CO)_4LX]$ in high yield. However, the reaction is limited to ligands with a Tolman cone angle $\theta < 150°$.[12]

---

* Checker reports yields > 89% when using half-scale quantities.
[†] Checkers have found complexes to be white crystalline solids that change color to pale yellow on exposure to laboratory light, at room temperature, over a period of days.
* Checkers have observed that ^{31}P NMR spectroscopy (if available) provides an even better method of establishing product purity, i.e., presence of disubstituted complexes.

TABLE II. Spectroscopic Data for cis-[Re(CO)₄LI] [L = P(OMe)₃, PMe₂Ph, PPh₃]

Complex	IR^a [$\nu(CO)/cm^{-1}$]	NMR^b (^1H/ppm)	$^{31}P^{e,\,f}$/ppm
[Re(CO)₄{P(OMe)₃}I]	2108(m), 2027(sh), 2008(s), 1964(m), 1956(m)^c	3.15 [d, 9H, CH₃, J(PH) 11]	100.52
[Re(CO)₄(PMe₂Ph)I]	2101(m), 2101(sh), 2003(s), 1946(m)	1.49 [d, 6H, CH₃, J(PH) 9]	−44.26
		6.94–7.10^d	
[Re(CO)₄(PPh₃)I]	2103(m), 2018(sh), 2001(s), 1949(m)	6.90–6.99^d	−5.30
		7.55–7.65^d	

^{a} Recorded in C₆H₆.
^{b} Recorded in C₆D₆ relative to TMS.
^{c} Recorded in hexane.
^{d} Complex multiplet.
^{e} Data provided by checkers.
^{f} Recorded in C₆H₆ relative to H₃PO₄.

251

TABLE III. Spectroscopic Data for [Re(CO)$_5$I] and Some Disubstituted Derivatives

Complex	IR [ν(CO)/cm^{-1}]		
mer-[Re(CO)$_3$(PMe$_2$Ph)$_2$I][a]	2000(w)	1945(s)	1895(m)
fac-[Re(CO)$_3$(PMe$_2$Ph)$_2$Br][b]	2034(s)	1955(s)	1899(s)
fac-[Re(CO)$_3$(PPh$_3$)$_2$I][a]	2049	1966	1904
fac[Re(CO)$_3$(PPh$_3$)$_2$Br][c]	2029(s)	1949(s)	1906(s)
[Re(CO)$_5$I][d]	2152(m)	2044(s)	1995(s)

[a] See ref. 10.
[b] See ref. 5.
[c] See ref. 3e.
[d] See ref. 9.

References

1. For example, see *Gmelin Handbook of Inorganic Chemistry*. Inorganic Compounds B11/B12, Springer Verlag, Berlin (1983/1984).
2. J. P. Fawcett and A. J. Poe, *J. Chem. Soc. Dalton Trans.*, 1302 (1977).
3. (a) L. K. Peterson, I. W. Johnson, J. K. Hoyano, S. Au-Yeung, and B. Gour, *J. Inorg. Nucl. Chem.*, **43**, 935 (1981). (b) H. Schumann and H. Neumann, *Z. Naturforsch.*, **36B**, 708 (1981). (c) J. R. Wagner and D. G. Hendricker, *J. Organomet. Chem.*, **98**, 321 (1971). (d) P. W. Jolly and F. G. A. Stone, *J. Chem. Soc.*, 5259 (1965). (e) A. M. Bond, R. Cotton, and M. E. McDonald, *Inorg. Chem.*, **17**, 2842 (1978).
4. (a) M. M. Glezen and A. J. Lees, *J. Chem. Soc., Chem. Commun.*, **23**, 1752 (1987). (b) M. M. Glezen and A. J. Lees, *J. Am. Chem. Soc.*, 3892 (1988). (c) D. R. Tyler and D. P. Petrylak, *Inorg. Chem. Acta*, **53**, L185 (1981).
5. R. H. Riemann and E. Singleton, *J. Organomet. Chem.*, **59**, 309 (1973).
6. J. Hoyano and L. K. Peterson, *Can. J. Chem.*, **54**, 2697 (1976).
7. F. Zingales, U. Sartorelli, F. Canziani, and M. Raveglia, *Inorg. Chem.*, **6**, 154 (1967).
8. E. W. Abel and G. Wilkinson, *J. Chem. Soc.*, 1501 (1959).
9. S. P. Schmidt, W. C. Trogler, and F. Basolo in J. M. Shreeve (ed.), *Inorganic Synthesis*, Vol. 23, Wiley, New York, 1989.
10. J. T. Moelwyn-Hughes, A. W. B. Garner, and N. Gordon, *J. Organometl. Chem.*, **26**, 373 (1971).
11. A. E. Leins and N. J. Coville, *J. Organomet. Chem.*, **407**, 359 (1991).
12. C. A. Tolman, *Chem. Rev.*, **77**, 313 (1977).

41. TRIS(1,2-BIS(DIMETHYLPHOSPHINO)ETHANE)-RHENIUM(I) TRIFLUOROMETHANESULFONATE, [Re(DMPE)₃][CF₃SO₃]*

SUBMITTED BY LIHSUEH (SHERRY) CHANG[†] and EDWARD DEUTSCH[†]
CHECKED BY JOHN R. KIRCHHOFF[‡]

Tris(1,2-bis(dimethylphosphino)ethane)rhenium(I), $[Re(DMPE)_3]^+$ is a simple, symmetrical cation which contains three identical bidentate phosphine ligands. This complex provides a Re(II/I) redox couple with properties that are very convenient for the study of outer-sphere electron transfer reactions.[1] Specifically, this couple is stable in both alkaline and acidic media and it exhibits a reversible, one-equivalent redox potential in an accessible region $[E^O,(II/I) = 565$ mV vs. NHE]. Moreover, this complex has been used to obtain information about the biological mechanism of action of ^{186}Re and ^{188}Re radiopharmaceuticals.[2, 3]

Tris(1,2-bis(dimethylphosphino)ethane)rhenium(I) was first synthesized by the use of excess phosphine ligand [DMPE, 1,2-bis(dimethylphosphino)-ethane] to both reduce and displace chloride ions from the Re(III) complex *trans*-$[Re^{III}(DMPE)_2Cl_2]^+$.[2, 3] This reaction is conducted in a Teflon-lined bomb reactor at 155°C for 12 h and affords the product in 70% yield.[2, 3] An easier and safer procedure which does not require the use of a bomb reactor involves reaction of a Re(V) complex such as $[Re^V(O)_2(Py)_4]^+$ (Py = pyridine) or $Re^VOCl_2(OEt)(PPh_3)_2$ (Et = $-C_2H_5$, Ph = $-C_6H_5$) with excess DMPE in the presence of a thiophenol catalyst. This newer route is also advantageous in that the starting materials are inexpensive and/or easy to synthesize, but it does involve a chromatographic purification to remove impurities such as $[Re^{III}(DMPE)_2(SPh)_2]^+$.

Procedure

■ **Caution.** *Benzenethiol is a highly noxious liquid. It should be used in a well-ventilated hood and gloves should be worn when handling this reagent. Benzenethiol can be replaced by p-thiocresol which is a solid and is somewhat less noxious.*

---

* The systematic name is tris[1,2-ethandiylbis(dimethylphosphine)]rhenium(I).
† Biomedical Chemistry Research Center, University of Cincinnati, Cincinnati, OH 45221-0172.
‡ Department of Chemistry, The University of Toledo, Toledo, OH 43606.

A. PREPARATION OF TRIS(1,2-BIS(DIMETHYLPHOSPHINO)-ETHANE)RHENIUM(I) TRIFLUOROMETHANESULFONATE FROM DIOXOTETRAKIS(PYRIDINE)RHENIUM(V) TRIFLUOROMETHANESULFONATE

$$[Re^VO_2(Py)_4]^+ + 4DMPE \rightarrow [Re^I(DMPE)_3]^+ + DMPE(O)_2$$

$$DMPE = (CH_3)_2P-CH_2CH_2-P(CH_3)_2$$

Dioxotetrakis(pyridine)rhenium(V) chloride can be prepared by one of several literature procedures;[4] a convenient synthesis[4c] starts from dichlorooxoethoxobis(triphenylphosphine)rhenium(V).[5] Five mL of pyridine (62 mmol) are added to a suspension of 1 g (1.19 mmol) of $Re^VOCl_2(OEt)(PPh_3)_2$ in 25 mL of ethanol and the resulting mixture is boiled under reflux for 2 h before being filtered and concentrated to a volume of 10 mL on a rotary evaporator. The yellow-orange product is precipitated by dropwise addition of diethyl ether, removed by filtration, washed with diethyl ether, and dried in air. (Yield 95%, ca. 0.7 g).

Dioxotetrakis(pyridine)rhenium(V) trifluoromethanesulfonate[4c] is prepared by dissolving 0.7 g of $[ReO_2(Py)_4]Cl \cdot 2H_2O$ (1.15 mmol) in 60 mL of a 2/49/49 (v/v/v) pyridine/ethanol/water solution. Any solid impurities are removed by filtration. Then, 1.5 g of $LiCF_3SO_3$ (9.6 mmol), dissolved in a minimum amount of water, is added and the solution is concentrated on a rotary evaporator. The solid product is removed by filtration, washed with cold ethanol/diethyl ether and diethyl ether, and dried in vacuo. (Yield 50%, ca. 0.4 g).

To a 50-mL pear-shaped flask containing $[ReO_2Py_4]CF_3SO_3$ (200 mg, 0.297 mmol) in absolute ethanol (9 mL) is added neat benzenethiol (0.33 mL, 3.25 mmol) and the mixture is stirred and heated at 100°C for 5 min under an argon atmosphere; DMPE (0.4 mL, 2.4 mmol) is then added under argon. The resulting red solution is heated at reflux for at least 8 h, whereupon the solution becomes red or pale purple in color. (The final color of the reaction solution depends upon the relative amounts of $[Re^{II}(DMPE)_3]^{2+}_2$ and $[Re^{III}(DMPE)_2(SPh)_2]^+$ impurities formed during the reaction.[6]) A stream of argon is bubbled through this solution to evaporate the ethanol. The minimum amount of dichloromethane necessary to dissolve the product mixture is added, and the resulting solution is transferred to a Sephadex LH-20 column (Sigma Chemical Company, Lipophilic Sephadex, bead size = 25–100 μ; column = 2.5 cm i.d. × 20 cm length; dichloromethane solvent) which has been prepared by swelling in dichloromethane. Subsequent elution with dichloromethane yields two colored bands–first purple and then pale yellow. FAB-MS analyses show that the yellow band contains $[Re^V(O)_2(DMPE)_2]^+$ while the purple band contains primarily the colorless

Re(I) product $[Re^I(DMPE)_3]^+$ along with small amounts of the red-purple Re(II) analog $[Re^{II}(DMPE)_3]^{2+}$ and the red Re(III) impurity $[Re^{III}(DMPE)_2(SPh)_2]^+$.[6] This purple solution is concentrated by solvent evaporation to provide a semisolid which is redissolved in dichloromethane (or methanol) (1 mL). Diethyl ether is added dropwise until no further white precipitate appears; impurities may coprecipitate if the diethyl ether is added too rapidly. The purple solution can then be removed and the white solid collected by filtration to afford $[Re(DMPE)_3]CF_3SO_3$ in about 50% yield.

B. PREPARATION OF TRIS(1,2-BIS(DIMETHYLPHOSPHINO)-ETHANE)RHENIUM(I) TRIFLUOROMETHANESULFONATE FROM DICHLORO(ETHOXO)OXOBIS-(TRIPHENYLPHOSPHINE)RHENIUM(V)

$$[Re^V OCl_2(OEt)(PPh_3)_2 + 4DMPE + H_2O \rightarrow [Re^I(DMPE)_3]$$

$$+ DMPE(O)_2 + Cl^- + HCl + EtOH$$

$$DMPE = (CH_3)_2P-CH_2CH_2-P(CH_3)_2$$

To a 50 mL pear-shaped flask containing $ReOCl_2(OEt)(PPh_3)_2$ (600 mg, 0.712 mmol)[5] in EtOH (12 mL) is added neat DMPE (0.9 mL, 5.4 mmol). The mixture is stirred and heated at 100°C for 5 min under an argon atmosphere, and then benzenethiol (0.76 mL, 7.4 mmol) is added under argon. The workup is the same as that described above. FAB-MS and TLC (silica gel) analyses show that colorless triphenylphosphine elutes before and with the red-purple band which contains the colorless Re(I) product $[Re(DMPE)_3]^+$. The purple eluant is concentrated by solvent evaporation to provide a semisolid which is redissolved in methanol (3 mL). Eight drops of a methanolic, saturated solution of $Li[CF_3SO_3]$ is added, and to this solution diethyl ether is added dropwise until no further white precipitate appears; impurities may coprecipitate if the diethyl ether is added too rapidly. Filtration affords $[Re(DMPE)_3]CF_3SO_3$ in about 52% yield.

Properties

Tris(1,2-bis(dimethylphosphino)ethane)rhenium(I) trifluoromethanesulfonate is a white solid which can be oxidized to the purple Re(II) analog. This Re(I) salt is soluble in methanol, ethanol, dichloromethane, acetone, dimethyl sulfoxide (DMSO), and acetonitrile, and insoluble in diethyl ether [N.B. This Re(I) complex is unstable in nitromethane, presumably suffering oxidation to Re(II)].

Anal. Calcd. for $ReC_{19}H_{48}P_6S_3O_3F_3$: C, 29.05; H, 6.16; P, 23.65. Found: C, 29.24; H, 6.22; P, 23.69.
For $[Re(DMPE)_3]F_3CSO_3$, UV(CH_3CN): λ_{max}, nm (ε, M^{-1} cm^{-1}): 253 (19, 180); 222 (19, 692). IR(KBr pellet): v, cm^{-1} (intensity): 466(w), 636(s), 681(m), 708(m), 792(w), 829(m), 886(s), 925(s), 1030(s), 1145(s), 1221(m), 1264(s), 1302(m), 1384(m), and 1420(m). FAB-MS (positive ion mode, *p*-nitrobenzyl alcohol matrix): 637/635 (M_+) (corresponding to $^{187}Re/^{185}Re$). The ^{1}H-NMR (δ, DMSO)[7] spectrum exhibits two peaks of equal intensity at 1.427 and 1.544 ppm. The ^{31}P-NMR (δ, CD_2Cl_2) spectrum[7, 8] exhibits a broad peak at 3.280 ppm. E^O,Re(II/I) = 565 mV vs. NHE (0.5 M tetraethylammonium perchlorate in *N,N*-dimethylformamide at a platinum disk working electrode).

References

1. K. Libson, M. Woods, J. C. Sullivan, J. W. Watkins, II, R. C. Elder, and E. Deutsch, *Inorg. Chem.*, **27**, 3614 (1988).
2. E. Deutsch, K. Libson, J.-L. Vanderheyden, A. R. Ketring, and H. R. Maxon, *Nucl. Med. Biol.*, **13**, 465 (1986).
3. J.-L. Vanderheyden, M. J. Heeg, and E. Deutsch, *Inorg. Chem.*, **24**, 1666 (1985).
4. (a) N. P. Johnson, C. J. L. Lock, and G. Wilkinson, *J. Chem. Soc.*, 1054 (1965). (b) J. H. Beard, J. Casey, and R. K. Murmann, *Inorg. Chem.*, **4**, 797 (1965). (c) L. S. Chang, J. Rall, F. Tisato, E. Deutsch, and M. J. Heeg, *Inorg. Chim. Acta*, **205**, 35 (1993).
5. N.P. Johnson, C. J. L. Lock, and G. Wilkinson, *Inorg. Syn.*, **9**, 145 (1967).
6. L. S. Chang, E. Deutsch, and M. J. Heeg, *Transition Met. Chem.*, **18**, 335 (1993).
7. (a) ^{31}P and ^{1}H NMR specra were obtained on an IBM/Bruker AF-250 spectrometer at 101.256 and 250.134 MHz. (b) ^{1}H NMR shifts are relative to solvent peaks (DMSO, 2.55 ppm) as an internal reference. (c) J.-L. Vanderheyden, University of Cincinnati, *Ph.D. Thesis*, pp. 230–231 (1983).
8. (a) ^{31}P chemical shifts are relative to 85% H_3PO_4 at δ 0.0 with shifts downfield of the reference considered positive. (b) The peak becomes sharper at 223 K and even sharper at 200 K.

42. TETRAHALO OXORHENATE ANIONS

SUBMITTED BY J. R. DILWORTH,* W. HUSSAIN,† A. J. HUTSON,*
C. J. JONES,† and F. S. McQUILLAN†
CHECKED BY J. M. MAYER‡ and J. B. ARTERBURN‡

The rhenium(V) complex $[ReOCl_4]^-$ is a useful precursor to a variety of complexes containing the $\{ReO\}^{3+}$ core[1-10] and has the advantage, over the well-known[1] complex $ReOCl_3(PPh_3)_2$, that no phosphine ligands are present to complicate the chemistry of the system. We have found the previously reported synthesis of $[ReOCl_4]^-$ by the reduction of ReO_4^- with zinc in acid[11] to be unreliable and have instead prepared this complex from $[ReOCl_5]^-$.[12] We have now developed a convenient alternative synthetic route to $[ReOX_4]^-$ (X = Cl or Br) which proceeds in good yield. Either tetrabutylammonium or tetraphenylphosphonium salts of these complexes can be prepared. The former counter ion provides a material with superior solubility in organic solvents. For completeness, the preparation of the tetrabutylammonium and tetraphenylphosphonium salts of tetraoxo-rhenate(VII) from $NaReO_4$ are also described.

A. TETRABUTYLAMMONIUM TETRAOXORHENATE(VII)

$$Na[ReO_4] + [n\text{-}Bu_4N]Cl \rightarrow [n\text{-}Bu_4N][ReO_4] + NaCl$$

Procedure

Sodium perrhenate (2.0 g, 7.3 mmol) is dissolved in demineralized water (100 mL) in a 250-mL conical flask. The solution is stirred and heated to 60°C then added to a warm (60°C) solution of tetrabutylammonium chloride (2.6 g, 9.3 mmol) in demineralized water (100 mL) in a 250-mL conical flask. The solution is left to stir for 30 min at room temperature. The white solid which forms is collected on a No. 3 sintered-glass filter tube, by filtration, and washed with distilled water (4 × 20 mL) and diethyl ether (4 × 30 mL) and dried in vacuo for 5 h at 56°C (boiling acetone) (yield 2.86 g, 79%).

* Department of Biological and Chemical Sciences, University of Essex, Colchester, Essex, CO4 3SQ, UK.
† School of Chemistry, The University of Birmingham, Edgbaston, Birmingham B15 2TT, UK.
‡ Department of Chemistry, University of Washington, Seattle, WA 98195.

Properties

Infrared (NaCl disc, Nujol mull) v(Re = 0) 905 cm^{-1}, v(CH) 722 cm^{-1}. Soluble in chloroform and dichloromethane.

B. TETRAPHENYLPHOSPHONIUM TETRAOXORHENATE(VII)

$$Na[ReO_4] + [Ph_4P]Br \rightarrow [Ph_4P][ReO_4] + NaBr$$

Procedure

A warm (60°C) solution of sodium perrhenate (1.5 g, 5.5 mmol) in demineralized water (60 mL) in a 250-mL conical flask is added to a warm (60°C) solution of tetraphenylphosphonium bromide (2.8 g, 6.7 mmol) in demineralized water (60 mL) contained in a 250-mL conical flask. The mixture is then stirred for 30 min at room temperature. The white precipitate which forms is collected by filtration on a No. 3 sintered-glass filter tube, washed with distilled water (4 × 20 mL) and absolute ethanol (4 × 30 mL), and then dried in vacuo for 5 h at 56°C (boiling acetone) (yield 2.80 g, 87%).

Properties

Infrared (NaCl disc, Nujol mull). v(Re = 0) 909 cm^{-1}, v(CH) 852, 724, 690 cm^{-1}, v(C–C) 1584 and 1482 cm^{-1}. Sparingly soluble in chloroform and dichloromethane.

C. TETRABUTYLAMMONIUM TETRACHLOROOXORHENATE(V)

$$[n\text{-}Bu_4N][ReO_4] + 6Me_3SiCl + 6MeOH \rightarrow$$

$$[n\text{-}Bu_4N][ReOCl_4] + 6Me_3SiOMe + 3H_2O + Cl_2$$

■ **Caution.** *Both bromotrimethylsilane and chlorotrimethylsilane are corrosive, flammable liquids. Chloroform is a possible carcinogen. Rubber gloves should be worn and all operations should be carried out in an efficient hood. Dry oxygen-free solvents must be used in this reaction, and all operations should be carried out under dry nitrogen gas. The toxicological properties of sodium perrhenate and the tetrahalooxorhenates have not been investigated, and these compounds should, therefore, be handled with care. Solid tetrahalooxorhenate is hygroscopic and should be stored in a desiccator.*

Procedure

A two-necked, 100-mL, round-bottomed flask, fitted with a nitrogen inlet, stopcock adaptor, and a stopper, is charged with a solution of $[n\text{-Bu}_4\text{N}]$-$[\text{ReO}_4]$ (0.83 g, 1.68 mmol) in chloroform (6.0 mL). The solution is stirred and cooled in an ice bath. Chlorotrimethylsilane (4.9 mL, 39 mmol) is added slowly, over a period of 5 min, to give a dark brown solid. Dry degassed methanol (1.6 mL, 39 mmol) is added and the solution is stirred in the ice bath for 3 h. The volume of the solution is then reduced under vacuum to about 2 mL. Absolute ethanol (10 mL) mixed with chlorotrimethylsilane (10 mL) is added to the round-bottomed flask, and the solution is shaken vigorously for 1 min. The solution is stirred at room temperature for 1 h and then the volume is reduced under vacuum to about 10 mL, after which the concentrated solution is placed in the deep freeze (− 15°C) for 3 h. The yellow/green crystals that form are collected by filtration under nitrogen, washed with cold (0°C) dry deoxygenated hexane (4 × 30 mL), and dried in vacuo for 5 h at 56°C (boiling acetone) (yield 0.93 g, 94%).

Properties

$[n\text{-Bu}_4]$$[\text{ReOCl}_4]$ forms hygroscopic yellow/green crystals which decompose on exposure to moist air to give a black solid. Infrared (NaCl, Nujol mull) $v(\text{Re} = 0)$ 1030 cm^{-1}. $\text{C}_{16}\text{H}_{36}\text{NOCl}_4\text{Re}$ requires C, 32.8; H, 6.18; N, 2.39; Cl, 24.2; found: C, 32.6; H, 6.3; N, 2.4; Cl, 24.7. δ_H (d_6-DMSO) 0.98–1.1 (unresolved multiplet, 12H, $\underline{\text{CH}}_3\text{CH}_2\text{CH}_2\text{CH}_2\text{N}$), 1.3–1.5 (unresolved multiplet, 8H, $\text{CH}_3\text{CH}_2\underline{\text{CH}}_2\text{CH}_2\text{N}$), 1.5–1.7 (unresolved multiplet, 8H, $\text{CH}_3\underline{\text{CH}}_2\text{CH}_2\text{CH}_2\text{N}$), 3.1–3.2 (unresolved multiplet, 8H, $\text{CH}_3\text{CH}_2\text{CH}_2 \underline{\text{CH}}_2\text{N}$). Soluble in dichloromethane, chloroform, dry ethanol, acetone, and THF. Insoluble in hexane.

D. TETRABUTYLAMMONIUM TETRABROMOOXORHENATE(V)

$$[n\text{-Bu}_4\text{N}] [\text{ReO}_4] + 6\text{Me}_3\text{SiBr} + 6\text{MeOH} \rightarrow$$

$$[n\text{-Bu}_4\text{N}] [\text{ReOBr}_4] + 6\text{Me}_3\text{SiOMe} + 3\text{H}_2\text{O} + \text{Br}_2$$

Procedure

■ **Caution.** *See Section C.* A two-necked, 100-mL, round-bottomed flask, fitted with a nitrogen inlet, stopcock adaptor, and a stopper, is charged with a solution of $[n\text{-Bu}_4\text{N}][\text{ReO}_4]$ (1.86 g, 3.77 mmol) in chloroform (10.0 mL). The solution is stirred and cooled in an ice bath. Bromotrimethylsilane (4.0 mL, 30.0 mmol) is added slowly, over a period of 5 min, to give a dark brown solution. Dry degassed methanol (1.25 mL, 30.0 mmol) is added and the solution is stirred in the ice bath for 3 h. The volume of the solution is then reduced under vacuum to about 2 mL. Absolute ethanol (10 mL) mixed with bromotrimethylsilane (10 mL) is added to the round-bottomed flask, and the solution is shaken vigorously for 1 min. The solution is stirred at room temperature for 1 h, then the volume is reduced under vacuum to about 8.0 mL, after which the concentrated solution is placed in the deep freeze ($-15°C$) for 3 h. The brown/purple solid that forms is collected by filtration under nitrogen, washed with cold (0°C) dry, deoxygenated diethyl ether (4×30 mL), and dried in vacuo for 5 h at 56°C (boiling acetone) (yield 2.40 g, 83%).

Properties

$[n\text{-Bu}_4N][ReOBr_4]$ forms a hygroscopic brown/purple solid which decomposes on exposure to moist air to give a black solid. Infrared (NaCl, Nujol mull) $v(Re = O)$ 1025 cm^{-1}. $C_{16}H_{36}NOBr_4Re$ requires C, 25.1; H, 4.75; N, 1.83; Br, 41.8; found: C, 25.1; H, 4.85; N, 1.7; Br, 41.9. δ_H (d_6-Acetone) 0.92–1.15 (unresolved multiplet, 12H, $\underline{CH_3}CH_2CH_2CH_2N$), 1.3–1.5 (unresolved multiplet, 8H, $CH_3\underline{CH_2}CH_2CH_2N$), 1.6–1.7 (unresolved multiplet, 8H, $CH_3\underline{CH_2}CH_2CH_2N$), 3.0–3.25 (unresolved multiplet, 8H, $CH_3CH_2CH_2\underline{CH_2}N$). Soluble in dichloromethane, chloroform, dry ethanol, acetone, and THF. Insoluble in hexane.

E. TETRAPHENYLPHOSPHONIUM TETRACHLOROOXORHENATE(V)

$$[Ph_4P][ReO_4] + 6Me_3SiCl + 6MeOH \rightarrow$$

$$[Ph_4P][ReOCl_4] + 6Me_3SiOMe + 3H_2O + Cl_2$$

Procedure

■ **Caution.** *See Section C.* A two-necked, 100-mL, round-bottomed flask, fitted with a nitrogen inlet, stopcock adaptor, and a stopper is charged with a suspension of $[Ph_4P][ReO_4]$ (2.07 g, 3.51 mmol) in chloroform (6.0 mL). The suspension is stirred and cooled in an ice bath, then chlorotrimethylsilane (4.5 mL, 35.0 mmol) is added slowly, over a period of 5 min, to give a dark brown solution. Dry degassed methanol (1.25 mL, 30.0 mmol) is added and the solution is stirred in the ice bath for 1 h. The volume of the solution is then reduced under vacuum to about 2 mL. Absolute ethanol (10 mL) mixed with chlorotrimethylsilane (10 mL) is added to the round-bottomed flask, and the solution is shaken vigorously for 1 min. The solution is stirred at room temperature for 1 h then the volume is reduced under vacuum to about 10 mL, after which the concentrated solution is placed in the deep freeze ($-15°C$) for 3 h. The yellow/brown solid that forms is collected by filtration under nitrogen, washed with cold ($0°C$), dry, deoxygenated diethyl ether (4×30 mL), and dried in vacuo for 5 h at $56°C$ (boiling acetone) (yield 2.36 g, 98%).

Properties

[Ph$_4$P] [ReOCl$_4$] forms a brown solid which decomposes in moist air to give a black solid. Infrared (NaCl, Nujol mull) v(Re $= 0$) 1025 cm^{-1}, v(CH) 725, 690 cm^{-1}. C$_{24}$H$_{20}$OPCl$_4$Re requires C, 42.2; H, 2.9; Cl, 20.7; found: C, 41.9; H, 3.3; Cl, 19.7. δ_p (d_6-DMSO) 28.00 (s, PPh$_4$). Slightly soluble in dichloro methane, chloroform, dry ethanol, acetone, and THF. Insoluble in hexane.

F. TETRAPHENYLPHOSPHONIUM TETRABROMOOXORHENATE(V)

$$[Ph_4P][ReO_4] + 6Me_3SiBr + 6MeOH \rightarrow$$
$$[Ph_4P][ReOBr_4] + 6Me_3SiOMe + 3H_2O + Br_2$$

Procedure

■ **Caution.** *See Section C.* A two-necked, 100-mL, round-bottomed flask, fitted with a nitrogen inlet, stopcock adaptor, and a stopper, is charged with a suspension of [Ph$_4$P][ReO$_4$] (0.84 g, 1.42 mmol) in chloroform (6.0 mL). The suspension is stirred and cooled in an ice bath and then bromotrimethylsilane (1.5 mL, 11.4 mmol) is added slowly, over a period of 5 min, to give a dark brown solution. Dry degassed methanol (0.46 mL, 11.4 mmol) is added and the solution is stirred in the ice bath for 1 h. The volume of the solution is then reduced under vacuum to about 2 mL. Absolute ethanol (10 mL) mixed with bromotrimethylsilane (10 mL) is then added to the round-bottomed flask and the solution is shaken vigorously for 1 min. The solution is then stirred at room temperature for 1 h and the volume is reduced under vacuum to about 6.0 mL, after which the concentrated solution is placed in the deep freeze ($-15°C$) for 3 h. The purple/brown solid that forms is collected by filtration under nitrogen, washed with cold ($0°C$), dry, deoxygenated diethyl ether (4×30 mL), and dried in vacuo for 5 h at $56°C$ (boiling acetone) (yield 0.65 g, 54%).

Properties

$[Ph_4P][ReOBr_4]$ forms a brown solid which decomposes in moist air to give a black solid. Infrared (NaCl, Nujol mull) $v(Re = 0)$ 1027 cm^{-1}, v(C–C) 1586 cm^{-1}, v(CH) 753, 725, 697 cm^{-1}. $C_{24}H_{20}OPBr_4Re$ requires C, 33.5; H, 2.34; Br, 37.1; found: C, 33.8; 2.4; Br, 37.1. δ_p (d_6-DMSO) 27.80 (s, PPh$_4$). Slightly soluble in dichloromethane, chloroform, dry ethanol, acetone, and THF. Insoluble in hexane.

References

1. K. A. Conner and R. A. Walton, *Comprehensive Coordination Chemistry*, Vol. 4, G. Wilkinson, R. D. Gillard, and J. A. McCleverty, eds., Pergamon, Oxford, 1987, p. 125.
2. F. A. Cotton and S. J. Lippard, *Inorg. Chem.*, **4**, 1621 (1964).
3. U. Mazzi, F. Refosco, G. Bandolini, and M. Nicolini, *Trans. Met. Chem.*, **10**, 121 (1985).
4. U. Mazzi, F. Refosco, G. Bandolini, and M. Nicolini, *J. Chem. Soc., Dalton Trans.*, 611 (1988).
5. H. J. Banberry, T. A. Hamor, F. S. McQuillan, C. J. Jones, and J. A. McCleverty, *Polyhedron*, **8**, 559 (1989).
6. H. J. Banberry, T. A. Hamor, F. S. McQuillan, C. J. Jones, and J. A. McCleverty, *J. Chem. Soc., Dalton Trans.*, 1405 (1989).
7. H. J. Banberry, W. Hussain, T. A. Hamor, C. J. Jones, and J. A. McCleverty, *J. Chem. Soc., Dalton Trans.*, 657 (1990).
8. H. J. Banberry, F. S. McQuillan, T. A. Hamor, C. J. Jones, and J. A. McCleverty, *Inorg. Chim. Acta*, **170**, 23 (1990).
9. H. J. Banberry, F. S. McQuillan, T. A. Hamor, C. J. Jones, and J. A. McCleverty, *Polyhedron*, **9**, 615 (1990).
10. H. J. Banberry, W. Hussain, T. A. Hamor, C. J. Jones, and J. A. McCleverty, *Polyhedron*, **10**, 243 (1991).
11. F. A. Cotton and S. J. Lippard, *Inorg. Chem.*, **5**, 9 (1966).
12. T. Lis and B. Jezowska-Trzeliankowska, *Acta. Cryst. Sec. B*, **33**, 1248 (1977).

43. A RHENIUM(I) DINITROGEN COMPLEX CONTAINING A TERTIARY PHOSPHINE

SUBMITTED BY GREGORY A. NEYHART,* KEVIN SEWARD,*
and B. PATRICK SULLIVAN*

Dinitrogen complexes of rhenium are relatively stable to air and moisture under ambient conditions so that they serve as convenient starting materials for a number of Re(I), Re(II), and Re(III) complexes. The basis of their reactivity is the lability of the N$_2$ ligand in either the Re(I) or Re(II) oxidation states.

* Department of Chemistry, University of Wyoming, Laramie, WY 82071-3838.

The procedure here describes the preparation of $Re(N_2)(PMe_2Ph)_4Cl$ from the precursor "green chelate," $Re(NNC(O)Ph)Cl_2(PPh_3)_2$. This example provides a general preparation of many structurally similar arylalkylphosphine Re(I) dinitrogen complexes.

A. (BENZOYLDIAZENIDO-N^2,O)DICHLOROBIS-(TRIPHENYLPHOSPHINE)RHENIUM(III)

$$ReOCl_3(PPh_3)_2 + [PhC(O)NHNH_3]Cl \rightarrow$$
$$Re(NNC(O)Ph)Cl_2(PPh_3)_2 + 2HCl + 2H_2O$$

The procedure is based on the work of Chatt et al.[1] It involves attack of a protonated hydrazine at the oxo ligand of $ReOCl_3(PPh_3)_2$, although the exact mechanistic details have yet to be unravelled.[2] In our hands, the use of the protonated hydrazine is absolutely necessary for the preparation to work.

Procedure

The preparation of $[PhC(O)NHNH_3]Cl$ is as follows. To 3 g of PhC(O)-NHNH$_2$ (22 mmol) in 12 mL of absolute EtOH is added 3 mL of concentrated HCl (36 mmol). The mixture is stirred for 5 min at which point 100 mL of Et$_2$O is added. The white precipitate that forms is collected by suction filtration and is washed subsequently with 3 × 50 mL portions of Et$_2$O. After drying the solid overnight under vacuum 3.2 g (18.6 mmol; 84%) of product is obtained.

The preparation of $Re(NNC(O)Ph)Cl_2(PPh_3)_2$ is as follows. A mixture of 3.0 g of $ReOCl_3(PPh_3)_2$ (3.6 mmol),[2] 3.0 g of $[PhC(O)NHNH_3]Cl$ (17.4 mmol), 3.0 g of PPh$_3$ (11.4 mmol), 250 mL of absolute ethanol, and 2 mL of concentrated HCl is placed in a 500-mL, round-bottomed flask. The mixture is heated at reflux under N$_2$ with vigorous magnetic stirring for 30 min. During this period, the yellow suspension gradually becomes green. After cooling to room temperature, the product is collected by suction filtration and washed with 3 × 25 mL portions of absolute ethanol followed by 3 × 25 mL portions of diethyl ether. The crude reaction product is then dissolved in 100 mL of chloroform and filtered to remove insoluble impurities. The complex is reprecipitated by layering 100 mL of absolute ethanol on the chloroform solution and stirring slowly in order to mix the two layers over a period of 2 h (for convenience, the checker suggested that the preparation of the crude material be done in a single day and the recrystallization procedure the following day). The green solid is then collected by suction filtration, washed with 3 × 25 mL portions of absolute ethanol followed by

3×25 mL portions of diethyl ether, and air dried by suction for 30 min. Isolated in this manner, the complex contains an ethanol of crystallization. Average yield: 2.5 g (72%). The reaction can be conveniently doubled in scale. Elemental analysis, calculated: C, 56.23%; H, 4.31%; N, 2.91%. Found: C, 56.04%; H, 4.02%; N, 2.78%.

Properties

$Re(NNC(O)Ph)Cl_2(PPh_3)_2$ is an emerald-green, air-stable solid that is soluble in low-polarity solvents such as benzene, toluene, chloroform, and dichloromethane, but is insoluble in high-polarity solvents. Solutions in chloroform are stable for days, but in dichloromethane the complex begins to decompose within minutes. The melting point is reported to be 196°C,[1] but from our preparation the complex begins to decompose at 184–189°C before melting at 196°C. The electronic absorption spectrum shows a maximum at 672 nm ($\varepsilon = 450$ M^{-1}cm^{-1}) which is responsible for the green color. In addition, a shoulder at 355 nm ($\varepsilon = 6000$ M^{-1}cm^{-1}) and a maximum at 300 nm ($\varepsilon = 17,000$ M^{-1}cm^{-1}) are distinguishing features. Cyclic voltammetry in 0.10 M tetrabutylammonium perchlorate/dichloromethane shows an irreversible oxidation wave at $+0.52$ V vs. the ferricenium/ferrocene couple. The ^{1}H NMR spectrum in CDCl$_3$ exhibits several multiplets in the phenyl region (7.1–7.6 ppm δ) as well as resonances corresponding to an ethanol of crystallization. The ^{31}P NMR spectrum shows one resonance at -3.5 ppm vs. 85% H$_3$PO$_4$. The Re–Cl stretch has been reported[1] to occur at 325 cm^{-1}.

Electronic structural considerations dictate two limiting forms for the Re–N bond, either as a Re^{III}[diazenido$(1-)$] or a Re^V[diazanido$(3-)$] linkage. No definitive evidence is available concerning this issue. Although weak bands are present in the $\nu_{C=O}$ and $\nu_{N=N}$ regions of the IR spectrum, the absence of characteristic diazenido modes has been taken as an indication of a Re(V) formulation. This argument is not convincing because the IR spectrum has not been assigned. An X-ray photoelectron spectrum is purportedly in accord with a Re(III) formulation.[3]

The initial reactivity of the complex $Re(NNC(O)Ph)Cl_2(PPh_3)_2$ centers around the lability of the oxygen bound both to Re and the diazenido ligand. For example, reaction with a number of reagents such as chlorine, nitriles, pyridine, and phosphines results in complexes containing a monodentate, N-bound diazenido ligand.[1] Further reaction results in the sequential replacement of the triphenylphosphine ligands.[4] As described here, $Re(NNC(O)Ph)Cl_2(PPh_3)_2$ is also used as a starting material for Re(I) dinitrogen complexes. Reaction with carbon monoxide results in Re(N$_2$)-Cl(CO)$_2$(PPh$_3$)$_2$.[4,5] Reaction with monodentate and bidentate phosphines in methanol results in complexes with the formulation $Re(N_2)Cl(P)_4$.[5, 6] Cleavage of the acyl group from the diazenido ligand is aided by the presence of a Lewis base.

B. CHLOROTETRAKIS(DIMETHYLPHENYLPHOSPHINE) (DINITROGEN)RHENIUM(I)

$$Re(NNC(O)Ph)Cl_2(PPh_3)_2 + 4PMe_2Ph \rightarrow$$

$$Re(N_2)(PMe_2Ph)_4Cl + 2PPh_3 + PhC(O)Cl$$

The procedure is based on that given by Chatt et al.[5] This involves reaction of the diazenido complex $Re(NNC(O)Ph)Cl_2(PPh_3)_2$ described in Section 43.A with the phosphine. During the reaction, base cleavage of the diazenido ligand at the acyl group results in a bound dinitrogen. Care should be taken to not heat the reaction mixture too long since overreaction results in complexes that no longer contain dinitrogen.[5]

Procedure

A mixture of 4.0 g of $Re(NNC(O)Ph)Cl_2(PPh_3)_2$ (4.45 mmol) and 80 mL of 1:1 methanol/toluene in a 500-mL, round-bottomed flask equipped with a reflux condenser and a nitrogen inlet is deaerated for 25 min. Four grams of PMe_2Ph (28.95 mmol) is then added with a minimum exposure to air, after which the mixture is heated to boiling. After several minutes at reflux the green suspension turns into a brown solution. After 2 h at reflux, the resulting brown to yellow solution is cooled to room temperature and 150 mL methanol is added. The volume of the solution then is reduced to about 50 mL by rotary evaporation in a *well-ventilated* hood. At this point, the toluene has been removed and a brown to yellow oil begins to form from the solution. Now, methanol (ca. 100 mL) is added to redissolve the oil, and, upon dissolution, a pale yellow precipitate *may* begin to form. (We wish to thank the checker for noting that the product is exceedingly soluble in many solvents; washing the crude product is not recommended, and the use of a minimum amount of MeOH is a wise idea.) Regardless, the mixture is placed in the freezer (− 25°C) overnight. The pale yellow crystals are collected by filtration. The crude complex is recrystallized by dissolving the solid in 15 mL of toluene, filtered to remove insoluble impurities, and dripped by means of a medium-porosity sintered glass funnel into 150 mL of stirring methanol. Stirring is ceased and crystals of the complex begin to form. The mixture is again placed in the freezer (− 25°C) overnight. The crystals are collected by filtration and washed with 2 × 10 mL portions of cold methanol.

Yield: 1.88 g (53%). Elemental analysis, calculated; C, 47.90%; H, 5.54%; N, 3.49%. Found: C, 47.86%; H, 5.53%; N, 3.50%.

Properties

$Re(N_2)(PMe_2Ph)_4Cl$ is an air-stable, pale yellow solid that is soluble in weakly polar solvents such as toluene, chloroform, and dichloromethane. However, chloroform solutions are stable for only a few minutes under normal laboratory conditions. The melting point is reported to be 165–169°C. The electronic absorption spectrum in dichloromethane exhibits a high-energy transition at λ_{max} of about 230 nm and two absorptions which appear as shoulders at 260 nm ($\varepsilon = 16,000$ $M^{-1}cm^{-1}$) and at 300 nm ($\varepsilon = 8200$ $M^{-1}cm^{-1}$). The $Re^{II/I}$ redox potential has been reported to be + 0.22 (vs. SCE).[7] The ^1H NMR spectrum in CD_2Cl_2 shows three resonances, a broad multiplet at 7.53 ppm δ, a multiplet at 7.19 ppm δ, and a singlet at 1.57 ppm δ. Integration yields area ratios of 2:3:6, respectively. The ^{31}P NMR spectrum exhibits one resonance at − 31.6 ppm vs. 85% H_3PO_4. The $N_2(\nu_{NN})$ stretch occurs at 1925 cm^{-1}.

Reactivity studies have been confined to oxidation processes and Lewis-base reactions at the coordinated dinitrogen. Thus, $Re^I(N_2)(PMe_2Ph)_4Cl$ can be oxidized with halogens, Cu(II) chloride, or Fe(III) chloride to produce $Re^{II}(N_2)(PMe_2Ph)_4Cl^+$, with the associated anion dependent on the choice of oxidant.[5] The complex can interact at the terminal nitrogen with a variety of electron acceptors[8,9] forming dinitrogen bridged complexes.

Other air stable complexes of tertiary phosphines may be prepared from $Re(NNC(O)Ph)Cl_2(PPh_3)_2$ by procedures that are very similar to that presented above for $Re(N_2)(PMe_2Ph)_4Cl$. The original literature in references 10–12 represent appropriate leads to a number of these useful reactions.[13]

References

1. J. Chatt, J. R. Dilworth, G. J. Leigh, and V. D. Gupta, *J. Chem. Soc. (A)*, 2631 (1971).
2. G. Parshall, *Inorg. Synth.*, **17**, 110 (1977).
3. V. I. Nefedov, *Soviet J. Coord. Chem. (Engl. Transl.)*, **4**, 965 (1980).
4. G. J. Leigh, R. H. Morris, C. J. Pickett, D. R. Stanley, and J. Chatt, *J. Chem. Soc., Dalton Trans.*, 800 (1981).
5. J. Chatt, J. R. Dilworth, and G. J. Leigh, *J. Chem. Soc., Dalton Trans.*, 612 (1973).
6. J. Chatt, W. Hussain, G. J. Leigh, H. M. Ali, C. J. Pickett, and D. A. Rankin, *J. Chem. Soc., Dalton Trans.*, 1131 (1985).
7. J. Chatt, W. Hussain, G. J. Leigh, H. M. Ali, C. J. Pickett, and D. A. Rankin, *J. Chem. Soc., Dalton Trans.*, 1131 (1985).
8. J. Chatt, J. R. Dilworth, G. J. Leigh, and R. L. Richards, *J. Chem. Soc., Chem. Commun.*, 955 (1970).
9. J. Chatt, R. C. Fay, and R. L. Richards, *J. Chem. Soc. (A)*, 702 (1971).

10. J. Chatt, J. R. Dilworth, and G. J. Leigh, *J. Chem. Soc., Dalton Trans.*, 612 (1973).
11. D. L. Hughes, A. J. L. Pombeiro, C. J. Pickett, and R. L. Richards, *J. Organomet. Chem.*, 248, C26 (1983).
12. G. J. Leigh, R. H. Morris, C. J. Pickett, and D. R. Stanley, *J. Chem. Soc., Dalton Trans.*, 800 (1981).
13. The authors wish to thank the National Science Foundation (EPSCoR) and the University of Wyoming (Basic Research Grant) for support of this work.

44. BIS(2,4-PENTANEDIONATO)IRON(II)
[IRON(II) BIS(ACETYLACETONATE)]

SUBMITTED BY J. M. MANRÍQUEZ,* E. E. BUNEL,† B. OELCKERS,*
and E. ROMÁN‡
CHECKED BY C. VÁSQUEZ§ and J. S. MILLER§

$$FeCl_2 + 2Hacac + 2C_5H_{10}N \xrightarrow{THF} Fe(acac)_2 + 2C_5H_{11}NCl$$

(Hacac = 2,4-pentanedione)

Iron(II) bis(acetylacetonate) has been prepared by reaction of iron(II) sulfate in aqueous solution with 2,4-pentanedione in the presence of bases such as pyridine and ammonia[1] and by reaction of iron(II) chloride tetrahydrate in aqueous solution with the stoichiometric amounts of 2,4-pentanedione and piperidine.[2] Neither publication reports yields and the lack of complete experimental details makes duplication difficult. The following procedure is a modification of that reported by Buckingham et al.[2] and employs either anhydrous iron(II) chloride or the corresponding adduct with THF. The reaction is carried out in diethyl ether as solvent instead of water.

Procedure

Conduct all operations under nitrogen or argon. The apparatus illustrated in Fig. 1 is used for the reaction sequence. A 1-L, two-necked, round-bottomed flask is fitted with a reflux condenser with a nitrogen inlet, a rubber septum, and an efficient stirring bar. The flask is evacuated and filled with nitrogen three times.

* Universidad Técnica Federico Santa María, Casilla 110-V, Valparaíso, Chile. Research supported by FONDECYT 799/90.
† Present address, E.I. du Pont de Nemours and Company, Wilmington, DE.
‡ Centro de Investigación Minera y Metalúrgica, Casilla 170, Santiago, Chile.
§ E.I. du Pont de Nemours and Company, Wilmington, DE.

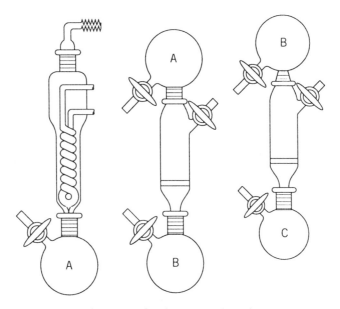

Figure 1. Apparatus for the preparation of Fe(acac)₂.

Anhydrous iron(II) chloride (9.23 g, 73 mmol; Strem Chemicals, Inc.), 450 mL of anhydrous diethyl ether,* and 12 mL (147 mmol) of anhydrous THF are introduced into the flask and the stirring of the light brown suspension is started. Alternatively, the FeCl₂·2THF adduct[3] (19.75 g, 73 mmol) can be used as starting material with similar results; in this case, no THF is added to the reaction flask.

2,4-Pentanedione* (15.40 g, 154 mmol) (Hacac) is added quickly to the flask via a syringe, and then piperidine* (13.10 g, 154 mmol) is added slowly to keep the temperature of the reaction below its boiling point. After completion of the addition, a fine precipitate appears and the mixture has a rusty color.

The mixture is stirred for 2 h at room temperature. Then the reflux condenser is removed and the flask is fitted with a frit (medium porosity, 5–6 cm diameter). The mixture is filtered to remove the piperidine hydrochloride, the solid is washed twice with 50-mL portions of diethyl ether and the filtrate, which is dark brownish red, is evaporated to dryness under vacuum to get a dark red solid.

---

* All solvents are distilled and dried under N₂. Diethyl ether, THF, and petroleum ether are distilled from Na/benzophenone. 2,4-Pentanedione and piperidine are distilled from P₂O₅ according to a literature procedure.[4]

The frit is removed, and the flask is fitted with another frit (medium porosity, 3 cm diameter). Then 60 mL of petroleum ether* (40–60°C) is added to the flask via a syringe. The solid is dislodged from the sides of the flask with the aid of an external magnet, and then it is filtered and washed twice with 15-mL portions of petroleum ether to remove traces of tris(2,4-pentanedionato)-iron(III).

The filtrate is discarded and the brownish-red solid is dried under vacuum to yield approximately 15–17 g of the crude product, which is transferred to a sublimator. The solid is covered with a glass wool disk to prevent the jumping of light particles from the residue. It is sublimed at 140°C (10^{-4} torr) to yield the product as a brownish-orange crystalline powder. Yield: 11.5–12.6 g (62–68%).[†]

Anal. Calcd. for $C_{10}H_{14}FeO_4$: C, 47.3; H, 5.55. Found: C, 47.1; H, 5.85.

Properties

Bis(2,4-pentanedionato)iron(II) is a polymeric crystalline material. It has a brownish-orange color when finely divided but darkens to almost black in larger clusters. The X-ray structure shows it to be a tetramer with six-coordinate iron(II) as a result of both oxygen bridges and weak Fe–C bonds.[5] It is paramagnetic and very air-sensitive, but there is a little visible change in color after oxidation. The compound is characterized by mass spectrometry ($M^+ = 254.028$) and IR spectroscopy [Nujol film, main bands at 1570(s), 1520(s), 1266(m), 1015(m), 926(m), 798(w), 770(w), 731(w), 660(w), 554(w), 433(m), and 407(m) cm^{-1}]. It melts at 170–171°C (sealed capillary). It sublimes without appreciable decomposition. It is slightly soluble in aromatic solvents and aliphatic hydrocarbons; it dissolves easily in coordinating solvents like THF with the formation of the monomeric adducts.

References

1. B. Emmert and R. Jarczynski, *Ber. Dtsch. Chem. Ges.*, **64**, 1072 (1931).
2. D. A. Buckingham, R. C. Gorges, and J. T. Henry, *Aust. J. Chem.*, **20**, 281 (1967).
3. G. Wilkinson, *Org. Synth.*, **4**, 473 (1963).
4. D. D. Perrin and W. L. F. Armarego, *Purification of Laboratory Chemicals*, Pergamon Press, Oxford, 1988, p. 261.
5. F. A. Cotton and G. W. Rice, *Nouv. J. Chim.*, **1**, 301 (1977).

[†] The checkers report a 90% yield when using the $FeCl_2 \cdot 2THF$ adduct as starting material.

45. SYNTHESIS OF
trans-TETRAAMMINEDICHLOROCOBALT(III)
CHLORIDE

SUBMITTED BY LONDA L. BORER* and HOWARD W. ERDMAN*
CHECKED BY CHARLES NORRIS,[†] JEROME WILLIAMS,[†]
and JAY WORRELL[†]

Coordination complexes of cobalt(III) are ideal for undergraduate work in inorganic chemistry because of the inert nature of the d^6 ion. The first synthesis of this *trans* complex was performed in 1896 by S. M. Jorgenson and took approximately 96 h.[1] The synthesis described herein takes approximately 15 min once the starting material, $[Co(NH_3)_4CO_3]NO_3$, has been prepared by known procedures.[2]

$$[Co(NH_3)_4CO_3]NO_3 + 3HCl \rightarrow$$

$$[Co(NH_3)_4Cl_2]Cl \cdot H_2O + CO_2 + H^+ + NO_3^-$$

Procedure

The compound, $[Co(NH_3)_4CO_3]NO_3$, is prepared according to the procedure described in *Inorganic Syntheses* using the hydrogen-peroxide method to save time.[2] A sand bath is heated such that a beaker of water could be heated to 80°C. Tetraammine(carbonato)cobalt(III) nitrate (1 g, 4 mmol) is dissolved in 5 mL of water in a 50-mL flask which contains a thermometer and magnetic stir bar. The solution is heated to 50–60°C in the preheated sand bath for 3 min with constant stirring. Concentrated hydrochloric acid (3.3 mL) is added as quickly as possible (over a period of 15–20 sec) with care being taken not to allow the solution to "boil" over. The solution is heated to 80°C with vigorous stirring for 5 min (*This temperature is critical!*). A dark green precipitate of *trans*-tetraamminedichlorocobalt(III) chloride is produced. The solution is cooled quickly to room temperature in an ice bath and the crude product is collected by vacuum filtration. The crude product is transferred to a 100-mL beaker, swirled in 30 mL of ice-cold H_2O to dissolve any *cis* cobalt complex crystals, and filtered immediately. The green *trans*-isomer crystals are washed with approximately 2 mL of ice-cold methanol and air dried to yield 0.45–0.55 g (45–55%) of product.

The purple *cis*-tetraammineaquachloro cobalt(III) chloride isomer can be obtained from the filtrate.

* Department of Chemistry, California State University, Sacramento, CA 95819.
† Department of Chemistry, University of South Florida, Tampa, FL 33620.

Anal. Calcd. for $[Co(NH_3)_4Cl_2]Cl \cdot H_2O$: H, 5.61; N, 22.28; Cl, 42.30; Co, 23.44. Found: H, 5.70; N, 22.16; Cl, 42.31; Co, 23.09.

Properties

trans-Tetraamminedichlorocobalt(III) chloride forms green crystals (dec. 240°C) which are soluble in aqueous solution. The IR spectrum of the complex (Nujol mull on polyethylene film) shows a Co–N stretch at 501 cm^{-1}, a Co–Cl stretch at 355.3 cm^{-1}, and bending vibrations at 169.7 and 283.7 cm^{-1}, which agree with those reported in literature.[3] The kinetics of aquation of the *trans* complex have been studied and have been reproduced by us.[4] An aquation of the *trans* complex at wavelength 253 nm, with HClO$_4$ at pH 2, provided a k value of 2.28×10^{-3} sec^{-1} and an extinction coefficient, ε, of 2.39×10^4 (lit. 1.90×10^{-3} sec^{-1} and 2.40×10^4 respectively).[5]

References

1. G. Brauer, ed. *Handbook of Preparative Inorganic Chemistry*, Vol. 2, Academic, New York, 1965, p. 1537.
2. E. G. Rochow, ed. *Inorganic Syntheses*, Vol. IV, McGraw-Hill, New York, 1960, p. 173.
3. I. Nakagawa and T. Simanouchi, *Spectrochim. Acta*, **22**, 759 (1956).
4. R. G. Linck, *Inorg. Chem.*, **8**, 1016 (1969).
5. R. G. Pearson, R. C. Boston, and F. J. Basolo, *J. Phys. Chem.*, **59**, 304 (1955).

46. [[3,3'-(1,3-PROPANEDIYLDIIMINO)BIS-[3-METHYL-2-BUTANONE]DIOXIMATE] (1−)-*N,N',N'',N'''*]NICKEL(II), NIOYL

$(CH_3)_2CH(CH_2)_2NO_2$ + 2HCl $\longrightarrow$ $(CH_3)_2CH(CH_2)_2Cl$ + NOCl + H_2O

$(CH_3)_2C{=}CHCH_3$ + NOCl $\longrightarrow$ $(CH_3)_2C(Cl)CH(NO)CH_3$

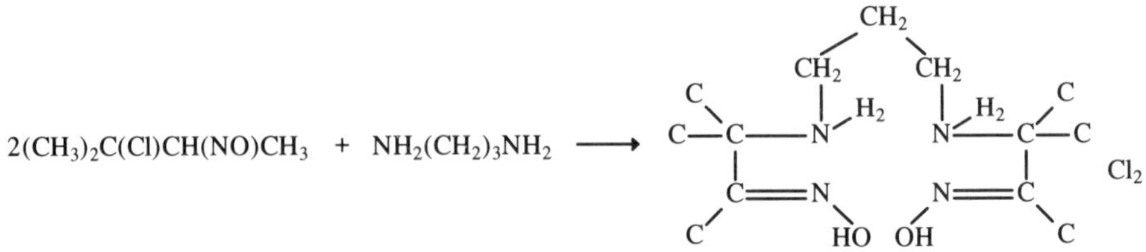

$2(CH_3)_2C(Cl)CH(NO)CH_3$ + $NH_2(CH_2)_3NH_2$ $\longrightarrow$

PnAO·2HCl

PnAO·2HCl + 2NaOH $\longrightarrow$ PnAO + 2NaCl + 2HOH

PnAO + $NiCl_2$ $\longrightarrow$ [Ni(PnAO)-H]Cl + HCl

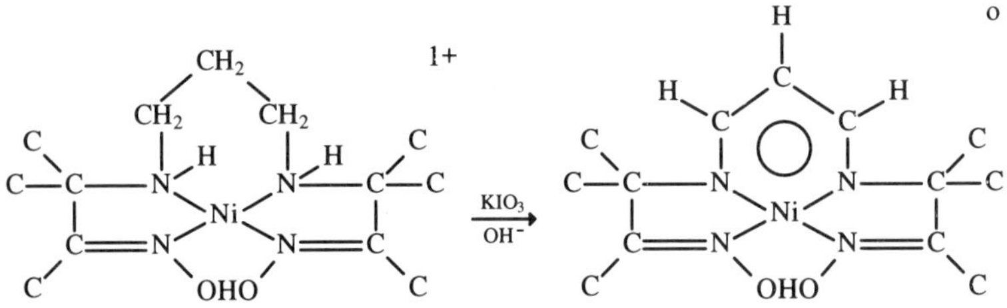

[Ni(PnAO)-H]⁺   NIOYL

SUBMITTED BY R. KENT MURMANN* and EDWIN G. VASSIAN[†]
CHECKED BY DARYLE H. BUSCH[‡] and ALEXANDER KOLCHINSKI[‡]

[[3,3'-(1,3-Propanediyldiimino)bis[3-methyl-2-butanone]dioximato] (1 −)-*N,N',N'',N'''*]nickel(II), Nioyl, is a neutral, planar, macrocyclic, quasi-aromatic heterocycle containing the heteroatom nickel, two nitrogens, and three carbons in the ring. It is prepared by treating nickel(II) ion with PnAO (3,3'-(1,3-pro-panediyldiimino)bis[3-methyl-2-butanone]dioxime) forming [Ni(PnAO)-H]Cl, [[3,3'-(1,3-propanediyldiimino)bis[3-methyl-2-butanone]dioximato] (1 −)-*N,N',N'',N'''*]nickel(II) which can then be oxidized by a variety of

* Department of Chemistry, University of Missouri, Columbia, MO 65211.
[†] Department of Chemistry, Western Carolina University, Cullowhee, NC 28723.
[‡] Department of Chemistry, University of Kansas, Lawrence, KS 66045-0046.

oxidizing agents (e.g., H_2O_2, PbO_2, O_2, or KIO_3) in basic aqueous media to Nioyl.[1] The unusual reactivity of Nioyl is shown by the facile manner in which electrophilic substitution of, for instance, halogen (Cl, Br, I), nitro, nitroso, acetyl, and benzoyl groups occurs on the central bridging carbon.[2] Although other quasi-aromatic macrocyclic complexes show reactivity toward electrophiles,[3] none have been shown to undergo as wide a variety of reactions as Nioyl. For example, the dimer Nioyl–Nioyl, linked through the central methine carbon, has been prepared as well as the methane-substituted derivatives $(Nioyl)_2CH_2$ and $(Nioyl)_3CH$. More recently, aromatic diazonium salts were found to react quantitatively to give, for example, Nioyl–$N=N$–$(C_6H_4SO_3H)$.[4] The scope of the area of interest has increased dramatically with the recent discovery that analogous aromatic-like macrocycles of Cu(II), Pd(II), and Pt(II) are stable and undergo the same reactions. Thus, these unique, well-behaved molecules can serve as models for the preparation of a wide variety of metal-containing aromatic ring complexes.

A. 3,3'-(1,3-PROPANEDIYLDIIMINE)BIS-[3-METHYL-2-BUTANONE]DIOXIME, PnAO

Procedure

A 1:1 mixture of cold 40% (v/v) H_2SO_4 and *t*-pentyl alcohol is distilled slowly and the volatile product consisting mainly of 2-methyl-2-butene is collected between 41 and 55°C. The alkene is cooled to -78°C in a dry ice–acetone bath to freeze out residual water and alcohol, which are removed on a filter. The filtrate is redistilled, collecting distillate that distills between 38 and 44°C. Yield: 90%. The rest of the procedure should be carried out in a fume hood. Into a 500-mL Erlenmyer flask equipped with a magnetic stirrer is added 100 mL (0.72 mol) of isopentylnitrite (3-methyl-1-butanol nitrite) which is cooled to -10°C using a dry ice–acetone bath. Seventy-five (75) mL (0.71 mol) of the alkene is added, followed by the dropwise addition of concentrated HCl. The reaction mixture is stirred rapidly and care should be taken to ensure that the temperature of the contents of the flask does not rise above 0°C. The temperature should be maintained between -5 and 0°C throughout the addition of the acid. The solution appears green after addition of approximately 20 mL of acid and white solid appears on the sides of the flask after 25–30 mL of acid has been added. After a total of 120 mL of acid is added, the mixture is allowed to remain overnight in the freezer at about -5°C. The bluish-white solid, 3-methyl-3-chloro-2-nitrosobutane, is collected in a 250-mL, sintered-glass funnel and then washed three times with 50-mL portions of cold (-5°C) anhydrous methanol. (■ **Caution.** *3-methyl-3-chloro-2-nitrosobutane is a lachrymator and it is essential that this* *operation be conducted in a fume hood. Also, improper or insufficient washing will result in subsequent decomposition resulting in a lower yield of the final product.*) The 3-methyl-3-chloro-2-nitrosobutane is placed under water-pump-vacuum for 30 min using a cold trap of dry ice–acetone and then held (covered) in a freezer at about − 20°C. It can be kept for about 1 wk under these conditions without significant decomposition. Yield: 76 g (79%).

The 3-methyl-3-chloro-2-nitrosobutane (76 g, 0.56 mol) is then placed in a 1-L round-bottomed flask equipped with a magnetic stirrer and immersed in a dry ice–acetone bath. Four hundred (400) mL of cold (− 5°C) anhydrous methanol is added and with continuous stirring, 1,3-diaminopropane (21 g, 0.28 mol) is added slowly over a 3- to 4-min period. The resulting light green solution is slowly allowed to come to room temperature, after which it is refluxed for 15 h, during which time the color changes to pale yellow.

In an alternate procedure, the last step is carried out in acetonitrile using the same quantities. During the addition of the diamine, the solution is maintained at 5°C, allowed to rise slowly to room temperature for about 15 min, and then warmed to about 75°C for one-half hour. (■ **Caution.** *The temperature must be monitored closely since the reaction is strongly exothermic.*) The relatively pure amine oxime hydrochloride (PnAO · 2HCl) is collected on a filter and washed with 50-mL portions of acetonitrile and dried in air.

When methanol is used, the solvent is removed under vacuum, leaving a creamy white solid or a yellow oil. The oil may be converted to the solid by dissolving it in methanol and then removing the methanol. The product obtained at this point is mainly (PnAO · 2HCl). This material is slurried with anhydrous benzene to assist removal of undesired by-products, collected on a filter, and dried in vacuo at 70°C.

The (PnAO · 2HCl) is added to 500 mL of water and a little dilute HCl is added until the mixture becomes slightly acidic (pH 3–4). After filtration, a saturated solution of K_2CO_3 is added slowly to a pH of 9, followed by several milliliters of 10 M NaOH to yield a pH of 10–11. The resulting white precipitate is collected on a filter. After a few hours, additional precipitate is recovered from the mother liquor. The combined precipitates are washed twice with small portions of cold water and dried in vacuo at 70°C. Yield: 34 g (45%). PnAO may be recrystallized from boiling methanol. Yield: 60%.

Anal. Calcd. for $C_{13}H_{28}O_2N_4$: C, 57.32; H, 10.36; N, 20.57. Found: C, 57.38; H, 10.28; N, 20.49.

Properties

PnAO is a white, air-stable, crystalline solid only slightly soluble in water, soluble in methanol, acetonitrile, and strong acids (with protonation).

Melting point range, 188–189°C. IR(KBr, cm^{-1}): 3400(s), 3280 and 3180 (doublet, sb), 1150(s), 1090(m), 990(m). ^{1}H-NMR (ppm in DMSO-d_6 relative to TMS): 1.097(4 CH$_3$) singlet; 1.685 (2 CH$_3$) singlet; 1.395 (CH$_2$) (average of five); 2.216(2 CH$_2$) (average triplet); 10.330 (2 oxime) singlet.

B. [[3,3'-(1,3-PROPANEDIYLDIIMINO)BIS[3-METHYL-2-BUTANONE]OXIMATO] (1 −)-N,N',N'',N''']NICKEL(II), NIOYL

Procedure

PnAO, 2,2'-(1,3-diaminopropane)bis(2-methyl-3-butanone)dioxime (1.59 g, 0.0058 mol) is weighed into a 150-mL beaker, followed by addition of 10 mL of water and 5.0 mL of 6.0 M HCl. The mixture is stirred until all of the ligand is dissolved, after which 1.30 g (0.0058 mol) of NiCl$_2$·6H$_2$O is added and dissolved. Then 6–7 mL of 5.0 M NaOH is added slowly until a pH of 9–10 is reached. The solution color changes from green to yellow to a final deep orange. The solution volume is adjusted to 30 mL using water. Twenty (20) mL of 5.0 M NaOH is added to obtain a solution that is 2.0 M in NaOH. Under magnetic stirring, 6.2 g KIO$_3$ (0.029 mol) is added to the beaker. Within 1 min the reaction mixture turns from yellow to beige and becomes very thick as the solid product forms. Sufficient water (10 mL) is added to allow stirring to proceed, which is continued for 5 h. The mixture is then centrifuged and the liquid portion is discarded. Methanol is used to extract the remaining solid, the mixture is thoroughly stirred to bring the Nioyl into solution, followed by centrifugation. This step is repeated until the liquid portion is almost colorless. A total of about 300 mL of reddish-orange solution is collected, which is filtered to remove KIO$_3$ particles. The volume is reduced to 80 mL by evaporation in air, followed by the addition of 80 mL of water, whereupon Nioyl comes out of solution. It is collected on a sintered-glass crucible and washed with water until the filtrate tests neutral. The solid is dried at 70°C under vacuum for 2 h. Yield: 1.2 g (63%).

Anal. Calcd. for C$_{13}$H$_{22}$O$_2$N$_4$Ni: C, 48.03; H, 6.82; N, 17.24; Ni, 18.06. Found: C, 48.00; H, 7.00; N, 17.05; Ni, 17.93.

Properties

Nioyl is an orange solid soluble in organic solvents but only slightly soluble in water. It is characterized by IR bands (CHCl$_3$, cm^{-1}): 2966(s), 1796(sb), 1598(s), and 1504(s) and by ^{1}H-NMR (ppm in DMSO-d_6 relative to TMS): 1.29 (4 CH$_3$), 1.89 (2 CH$_3$); 5.07 (average triplet), 7.21 (average doublet); 19.1 (OHO) singlet. UV-VIS (ethanol, ε_{max}): 486 nm, 145; 385 nm, 3.2 × 10^3; 363 nm, 5.2 × 10^3; 256 nm, 1.6 × 10^4.

References

1. E. G. Vassian and R. K. Murmann, *Inorg. Chem.*, **6**, 2043 (1967).
2. L. O. Urban and E. G. Vassian, *Inorg. Chem.*, **18**, 867 (1979).
3. J. P. Collman, in R. Gould (ed.), *Reactions of Coordinated Ligands and Homogeneous Catalysis*; Advanced Chemical Series, No. 37, American Chemical Society, Washington, DC, 1963, p. 78.
4. R. K. Murmann and E. G. Vassian, *Coord. Chem. Rev.*, **105**, 1 (1990).

47. PLATINUM COMPLEXES SUITABLE AS PRECURSORS FOR SYNTHESIS IN NONAQUEOUS SOLVENTS

SUBMITTED BY LARS I. ELDING*, ÅKE OSKARSSON*,
and VADIM YU. KUKUSHKIN†*
CHECKED BY GORDON K. ANDERSON‡

A prerequisite for preparative coordination chemistry in nonaqueous solvents is the availability of soluble complex compounds. The solubility in organic solvents of compounds of the type $M[X]$ (M^+ = the cation of an alkali metal) can be increased significantly if M^+ is substituted by a bulky organic cation, Q^+.[1]

The μ-nitrido-bis(triphenylphosphorus)(1+) ion,[2-4] $[\{(C_6H_5)_3P\}_2N]^+$, is one of the most convenient counterions for such purposes. Compounds of the $[\{(C_6H_5)_3P\}_2N][X]$ type are soluble in most commonly used organic solvents. They also often form good-quality single crystals, suitable for X-ray diffraction analysis.

Herein we describe simple procedures for the synthesis of the compounds $[\{(C_6H_5)_3P\}_2N]_2[PtCl_4]$ and $[\{(C_6H_5)_3P\}_2N]_2[Pt_2(\mu\text{-}Cl)_2Cl_4]$. Unlike $M_2[PtCl_4]$ and $M_2[Pt_2(\mu\text{-}Cl)_2Cl_4]$ (e.g., with $M^+ = K^+$), these compounds are soluble in different organic solvents and can be used as convenient starting materials for the nonaqueous synthesis of other platinum compounds. The $[Pt_2Cl_6]^{2-}$ complexes are suitable as starting material for syntheses involving bridge-splitting reactions (see ref. 5 and references therein).

* Inorganic Chemistry 1, Chemical Center, Lund University, P.O. Box 124, S-221 00 Lund, Sweden.
† On leave from Department of Chemistry, St. Petersburg State University, Universitetsky Pr. 2, 198904 Stary Petergof, Russian Federation. V. Yu. K. is grateful to the Royal Swedish Academy of Sciences and the Academy of Sciences of Russia for financial support for his stay at Lund University.
‡ Department of Chemistry, College of Art and Sciences, University of Missouri-St. Louis, 8001 Natural Bridge Road, St. Louis, MO 63121–4499.

The compound $[\{(C_6H_5)_3P\}_2N]_2[PtCl_4]$ is obtained by a metathesis reaction as described previously for $[(C_6H_5)_3PCH_2C_6H_5]_2[PtCl_4]$.[6] The $[\{(C_6H_5)_3P\}_2N]_2[Pt_2(\mu\text{-}Cl)_2Cl_4]$ compound is prepared by halide-abstraction from $[PtCl_4]^{2-}$ as in the synthesis of the $[(C_6H_5)_3PCH_2C_6H_5]_2$-$[Pt_2(\mu\text{-}Cl)_2Cl_4]$ [5]. In contrast to $[(C_6H_5)_3PCH_2C_6H_5]^+$, the counterion $[\{(C_6H_5)_3P\}_2N]^+$ is commercially available as the chloride salt.

■ **Caution.** *μ-nitrido-bis(triphenylphosphorus) (1 +) chloride is a harmful dust and irritant. Boron trifluoride diethyl etherate is a moisture-sensitive liquid. Small amounts of highly toxic HF and flammable diethyl ether can form in the synthesis of* $[\{(C_6H_5)_3P\}_2N]_2[Pt_2(\mu\text{-}Cl)_2Cl_4]$. *All organic solvents which are used here are toxic. Contact with the liquid or vapor should be avoided. Appropriate precautions must be taken, and an efficient hood must be used.*

A. μ-NITRIDO-BIS(TRIPHENYLPHOSPHORUS)(1 +) TETRACHLOROPLATINATE(II)

$$K_2[PtCl_4] + 2[\{(C_6H_5)_3P\}_2N]Cl \xrightarrow{(H_2O)}$$
$$[\{(C_6H_5)_3P\}_2N]_2[PtCl_4] + 2KCl$$

Procedure

A 150-mL beaker or Erlenmeyer flask is filled with a solution of 0.566 g (1.36 mmol) of potassium tetrachloroplatinate(II), $K_2[PtCl_4]$ (purchased from Aldrich), in 60 mL of water and placed on a magnetic stirrer. A solution of 1.878 g (3.27 mmol) of $[\{(C_6H_5)_3P\}_2N]Cl$ (purchased from Aldrich) dissolved in 10 mL of ethanol is added dropwise from a 25-mL dropping funnel over a period of 10 min. A pink precipitate appears immediately after addition of the first few drops. After complete mixing, the reaction mixture is stirred at 20–25°C for 15 min. The fine-grained pink precipitate is collected on a Millipore filter (type HV, 0.45 μm; other small-pore glass or paper filters can be used as well) or separated by centrifugation, washed three times with 3-mL portions of water, and dried in air at 80–90°C.

The crude precipitate (1.851 g, 96% yield) is dissolved in 10 mL of dichloromethane and filtered through a filter paper (Munktell Filter Paper V120H was used) into a 200-mL beaker and the filter is washed twice with 1-mL portions of CH_2Cl_2. Acetone (90 mL) is added and the beaker, covered with "Parafilm M", is left for 7 h at 20–25°C. The resulting brownish-red rhombic prisms are filtered off, washed on a filter three times with 0.5-mL portions of acetone and twice with 1-mL portions of diethyl ether and dried in air at 20–25°C. Yield of $[\{(C_6H_5)_3P\}_2N]_2[PtCl_4]$ is 1.427 g (74%) based on starting $K_2[PtCl_4]$.

Anal. Calcd. for $[\{(C_6H_5)_3P\}_2N]_2[PtCl_4]$: C, 61.2; H, 4.3; Cl, 10.0; N, 2.0. Found: C, 61.5; H, 4.3; Cl, 10.1; N, 1.9.

Properties

The complex is readily soluble in dichloromethane, chloroform, nitromethane, and N,N-dimethylformamide, but only slightly soluble in acetone, toluene, and diethyl ether. The melting point is 254–255°C (capillary). IR spectrum (KBr pellet, cm^{-1}): 310 m-s $\nu(PtCl)$. ^{195}Pt NMR spectrum in CH_2Cl_2, δ: $- 1679$ ppm relative to 0.7 M $\{(n\text{-}C_4H_9)_4N\}_2[PtCl_6]$ in CH_2Cl_2. Crystal data parameters obtained by X-ray single-crystal method: triclinic, space group $P\,1$ or $P\,\bar{1}$, $a = 10.894(10)$, $b = 12.262(69)$, $c = 13.614(23)$ Å, $\alpha = 105.5(3)$, $\beta = 96.1(2)$, $\gamma = 91.0(3)°$, $V = 1741(12)$ Å^3.

B. μ-NITRIDO-BIS(TRIPHENYLPHOSPHORUS)(1+) TETRACHLORO-DI-μ-CHLORODIPLATINATE(II)

$$2[\{(C_6H_5)_3P\}_2N]_2[PtCl_4] + 2BF_3 \cdot O(C_2H_5)_2 \xrightarrow{(CH_2Cl_2)}$$

$$[\{(C_6H_5)_3P\}_2N]_2[Pt_2(\mu\text{-}Cl)_2Cl_4] + 2[\{(C_6H_5)_3P\}_2N](BF_3Cl) + 2O(C_2H_5)_2$$

Procedure

Boron trifluoride diethyl etherate, $BF_3 \cdot O(C_2H_5)_2$ (purchased from Janssen Chimica), (0.162 mL, 1.32 mmol) is added to a solution of 0.686 g (0.05 mmol) $[\{(C_6H_5)_3P\}_2N]_2[PtCl_4]$ in 7 mL of dichloromethane in a 50-mL beaker. The beaker is covered with "Parafilm M" and left to stand for 24 h at 20–25°C. The resulting cherry-red rhombic prisms are filtered off, washed on a filter twice with 1-mL portions of dichloromethane, and dried in air at 20–25°C. Yield of $[\{(C_6H_5)_3P\}_2N]_2[Pt_2(\mu\text{-}Cl)_2Cl_4]$ is 0.339 g (83%) based on starting $[\{(C_6H_5)_3P\}_2N]_2[PtCl_4]$.

Anal. Calcd. for $[\{(C_6H_5)_3P\}_2N]_2$ $[Pt_2(\mu\text{-}Cl)_2Cl_4]$: C, 51.5; H, 3.6; Cl, 12.7; N, 1.7. Found: C, 51.1; H, 3.6; Cl, 12.9; N, 1.7.

Properties

The complex is soluble in N,N-dimethylformamide and propylene carbonate (1,2-propanediol cyclic carbonate), slightly soluble in acetone and dichloromethane, and insoluble in ethanol, diethyl ether, and toluene. The melting point is 274–276°C (capillary). IR spectrum (KBr pellet, cm^{-1}): 338 m-s $v(PtCl_t)$ and 315 m $v(PtCl_\mu)$. ^{195}Pt NMR spectrum in $(CH_3)_2NCHO$, δ: -1381 ppm relative to 0.7 M $\{(n\text{-}C_4H_9)_4N\}_2[PtCl_6]$ in CH_2Cl_2. Crystal data parameters obtained by X-ray single-crystal method: monoclinic, space group $P\ 2_1/c$, $a = 9.532(7)$, $b = 15.676(7)$, $c = 22.59(1)$ Å, $\beta = 101.16(5)°$, $V = 3311(3)$ Å^3.

References

1. V. Yu. Kukushkin and Yu. N. Kukushkin, *Sov. J. Coord. Chem. (Engl. Transl.)*, **14**, 893 (1988); *Koord. Khim.*, **14**, 1587 (1988).
2. J. K. Ruff and W. J. Schlientz, *Inorg. Synth.*, **15**, 84 (1974).
3. V. Yu. Kukushkin and A. I. Moiseev, *Inorg. Chim. Acta*, **176**, 79 (1990).
4. F. J. Lalor and S. Chaona, *J. Organomet. Chem.*, **344**, 163 (1988).
5. V. Yu. Kukushkin, V. M. Tkachuk, I. A. Krol, Z. A. Starikova, B. V. Zhadanov, and N. P. Kiseleva, *J. Gen. Chem. (Engl. Transl.)*, **61**, 42 (1991); *Zh. Obsch. Khim.*, **61**, 51 (1991).
6. V. Yu. Kukushkin, E. Yu. Pankova, T. N. Fomina, and N. P. Kiseleva, *Sov. J. Coord. Chem. (Engl. Transl.)*, **14**, 625 (1988); *Koord. Khim.*, **14**, 1110 (1988).

48. TETRAKIS(PROPANENITRILE)PLATINUM(II) TRIFLUOROMETHANESULFONATE AS A SUITABLE INTERMEDIATE IN SYNTHETIC Pt(II) CHEMISTRY

SUBMITTED BY VADIM YU. KUKUSHKIN,*† ÅKE OSKARSSON,†
and LARS I. ELDING†
CHECKED BY S. JONASDOTTIR‡

Solvated transition metal cations containing weakly bonded organic ligands are potentially useful as intermediates in preparative coordination chemistry. The coordinated solvent molecules provide solubility of the compounds in various organic solvents, and they can be replaced readily by other ligands with better donor properties.[1, 2] The preparation of mononuclear transition metal complexes with weakly bonded anions has been described previously[3] and preparative applications of such complexes have been reviewed.[4]

* On leave from Department of Chemistry, St. Petersburg State University, Universitetsky Pr. 2, 198904 Stary Petergof, Russian Federation. V. Yu. K. is grateful to the Royal Swedish Academy of Sciences and the Academy of Sciences of Russia for financial support of his stay at Lund University.
† Inorganic Chemistry 1, Chemical Center, Lund University, P.O. Box 124, S-221 00 Lund, Sweden.
‡ Department of Chemistry, The University of Michigan, Ann Arbor, MI 48109.

Herein we describe the preparation of $[Pt(C_2H_5CN)_4](SO_3CF_3)_2$ and propose its use as an intermediate for the synthesis of other platinum(II) complexes. The two-step synthesis includes the formation of $[PtCl_2(C_2H_5CN)_2]$[5] followed by abstraction of the two chloride ligands of this compound by the use of 2 equiv of $AgSO_3CF_3$, yielding the $[Pt(C_2H_5CN)_4]^{2+}$ solvated cation.[6] A similar procedure has been described earlier for preparation of the very moisture-sensitive $[Pt(CH_3CN)_4]^{2+}$ complex.[7, 8]

The useful properties of the $[Pt(C_2H_5CN)_4](SO_3CF_3)_2$ compound include excellent solubility in commonly used organic solvents, relative insensitivity to atmospheric moisture, a fairly high thermal stability, and good leaving properties of the coordinated propanenitrile. As an example of synthesis of other Pt(II) compounds via substitution of the weakly bound propanenitriles, we also describe the reaction of $[Pt(C_2H_5CN)_4](SO_3CF_3)_2$ with thioethers resulting in the formation of $[Pt(thioether)_4](SO_3CF_3)_2$ compounds. This method is more convenient than the previously used procedure starting from tetraaquaplatinum(II), $[Pt(H_2O)_4]^{2+}$.[9]

■ **Caution.** *Silver trifluoromethanesulfonate is an irritant. Propanenitrile, thioethers, and all organic solvents used are toxic. Contact with the liquids and vapors should be avoided. Appropriate precautions must be taken, and an efficient hood must be used.*

A. DICHLOROBIS(PROPANENITRILE)PLATINUM(II)

$$K_2[PtCl_4] + 2C_2H_5CN \rightarrow [PtCl_2(C_2H_5CN)_2] + 2KCl$$

Procedure

Propanenitrile, C_2H_5CN (10 mL, 140 mmol), is added to a solution of 10.00 g (24 mmol) $K_2[PtCl_4]$ in 80 mL of water in a 200-mL beaker. The beaker is covered with "Parafilm M" and left to stand for 3 days at 20–25°C. The resulting lemon-yellow rod-like crystals are filtered off, washed twice on a filter with 5-mL portions of water, and dried in air at 70°C. The yield of $[PtCl_2(C_2H_5CN)_2]$ is 7.36 g (81%, based on Pt).

Anal. Calcd. for $[PtCl_2(C_2H_5CN)_2]$: Cl, 18.9%; Pt, 51.9%. Found: Cl, 18.9%; Pt, 51.7%.

Properties

The compound obtained as described above is in fact a 6:1 mixture of *cis*- and *trans*-isomers as determined by TLC or by NMR spectroscopy. The TLC is carried out on SiO_2 Merck Kieselgel 60 F_{254} plates: $R_f(cis) = 0.55$ (larger spot), $R_f(trans) = 0.82$ [CHCl$_3$:(CH$_3$)$_2$CO = 1:1, in volume]. ^{1}H NMR spectrum of *cis*-[PtCl$_2$(C$_2$H$_5$CN)$_2$] in CD$_2$Cl$_2$ (99.6 + atom % D) δ, ppm: 1.41 [t, J$_{HH}$ = 7.6 Hz, CH$_3$], 2.86 [q, J$_{HH}$ = 7.6 Hz, CH$_2$]. ^{1}H NMR spectrum of *trans*-[PtCl$_2$(C$_2$H$_5$CN)$_2$] in CD$_2$Cl$_2$ (99.6 + atom % D) δ, ppm: 1.38 [t, J$_{HH}$ = 7.6 Hz, CH$_3$], 2.88 [q, J$_{HH}$ = 7.6 Hz, CH$_2$]. The compound [PtCl$_2$(C$_2$H$_5$CN)$_2$] is readily soluble in dichloromethane, acetone, nitromethane, and propanenitrile, sparingly soluble in chloroform, and insoluble in ethanol, diethyl ether, and toluene. The solubility in water at 20–25°C is $\sim 5 \cdot 10^{-3}$ M. The aqueous solution becomes blue-colored in a few days. Electronic absorption spectrum in chloroform: $\lambda_{max.}$ 319 nm (lg ε = 2.3).[10] Refractive indices: n$_p$ = 1.504, n$_g$ = 1.737.[11] IR spectrum in KBr pellet, cm^{-1}: 2314 m-w ν(CN) [lit. 2303 ν(CN)].[12, 13] The melting point in a capillary is 112–114°C (lit. 116°C[5]) with subsequent crystallization of the *trans*-[PtCl$_2$(C$_2$H$_5$CN)$_2$] from the melt. The newly formed solid melts at 179–182°C. DTA analysis: heating rate is 2.5 deg/min; 105°C, endopeak; 110°C, exopeak (lit.: 112°C, endopeak; 130°C, exopeak[11]).

B. TETRAKIS(PROPANENITRILE)PLATINUM(II) TRIFLUOROMETHANESULFONATE

[PtCl$_2$(C$_2$H$_5$CN)$_2$] + 2AgSO$_3$CF$_3$ + 2C$_2$H$_5$CN →

[Pt(C$_2$H$_5$CN)$_4$](SO$_3$CF$_3$)$_2$ + 2AgCl

Procedure

The reaction is carried out in a 100-mL Erlenmeyer flask. Silver trifluoromethanesulfonate (2.77 g, 10.8 mmol), AgSO$_3$CF$_3$, is added to a solution of 1.93 g (5.1 mmol) of [PtCl$_2$(C$_2$H$_5$CN)$_2$] in 35 mL of dichloromethane and 0.9 mL (12.8 mmol) of propanenitrile. The suspension is stirred vigorously on a magnetic stirrer for 5 h. The light yellow solution is filtered from the precipitated silver chloride on a Millipore filter (type HV, 0.45 μm) into a 100-mL, round-bottomed flask. The precipitate on the filter is washed twice with 2-mL portions of CH$_2$Cl$_2$. The solvent is removed under vacuum by use of a rotary evaporator at 20–25°C. The jelly-like residue is dissolved in 15 mL of propanenitrile and filtered through paper filter (Munktell Filter Paper V120H was used) into a 400-mL beaker. Diethyl ether (250 mL) is added and results in the immediate formation of a white precipitate. The precipitate is collected on a filter, washed three times with 5-mL portions of diethyl ether, and dried in a flow of dry argon at 20–25°C. Yield of [Pt(C$_2$H$_5$CN)$_4$]-(SO$_3$CF$_3$)$_2$ is 2.20 g (60%, based on Pt).

Anal. Calcd. for $[Pt(C_2H_5CN)_4](SO_3CF_3)_2$: C, 23.6%; H, 2.8%; N, 7.9%; Pt, 27.3%. Found: C, 23.3%; H, 2.8%; N, 7.6%; Pt, 27.2%.

Properties

The $[Pt(C_2H_5CN)_4](SO_3CF_3)_2$ is readily soluble in dichloromethane, chloroform, propanenitrile, nitromethane, ethanol, and water. In the latter case a light blue coloration of the solution is observed within 1 h after dissolution. The compound is sparingly soluble in diethyl ether and toluene and insoluble in hexane. The melting point in a capillary is 175°C. IR spectrum in Nujol, cm^{-1}: 2334 ν(CN). ^1H NMR spectrum in CD_2Cl_2 (99.6 + atom % D) δ, ppm: 1.41 [t, J_{HH} = 7.5 Hz, CH_3], 3.30 [q, J_{HH} = 7.5 Hz, CH_2].

C. TETRAKIS(DIMETHYLSULFIDE)PLATINUM(II) TRIFLUOROMETHANESULFONATE

$$[Pt(C_2H_5CN)_4](SO_3CF_3)_2 + 4(CH_3)_2S \rightarrow$$
$$[Pt\{(CH_3)_2S\}_4](SO_3CF_3)_2 + 4C_2H_5CN$$

Procedure

Dimethyl sulfide (0.32 mL, 4.40 mmol) is added to a 30-mL beaker containing a solution of 0.52 g (0.73 mmol) of $[Pt(C_2H_5CN)_4](SO_3CF_3)_2$ in 5 mL of dichloromethane. A colorless precipitate of the desired compound is formed immediately. Diethyl ether (5 mL) is added to the reaction mixture and the solid is filtered off, washed on a filter twice with 2 mL of $(C_2H_5)_2O$, and dried in air at 20–25°C. Yield of $[Pt\{(CH_3)_2S\}_4](SO_3CF_3)_2$ is 0.49 g (91%, based on Pt).

Anal. Calcd. for $[Pt\{(CH_3)_2S\}_4](SO_3CF_3)_2$: C, 16.2%; H, 3.3%; Pt, 26.3%. Found: C, 16.5%; H, 3.2%; Pt, 26.3%.

D. TETRAKIS(1,4-OXATHIANE)PLATINUM(II) TRIFLUOROMETHANESULFONATE AND BIS(1,4-DITHIANE)-PLATINUM(II) TRIFLUOROMETHANESULFONATE

$$[Pt(C_2H_5CN)_4](SO_3CF_3)_2 + 4S(C_2H_4)_2O \rightarrow$$
$$[Pt\{S(C_2H_4)_2O\}_4](SO_3CF_3)_2 + 4C_2H_5CN$$
$$[Pt(C_2H_5CN)_4](SO_3CF_3)_2 + 2S(C_2H_4)_2S \rightarrow$$
$$[Pt\{S(C_2H_4)_2S\}_2](SO_3CF_3)_2 + 4C_2H_5CN$$

Procedure

The same method as that described for the preparation of $[Pt\{(CH_3)_2S\}_4]$-$(SO_3CF_3)_2$ is used for the synthesis of tetrakis(1,4-oxathiane)platinum(II) trifluoromethanesulfonate, $[Pt\{S(C_2H_4)_2O\}_4](SO_3CF_3)_2$, and bis(1,4-dithiane)-platinum(II) trifluoromethanesulfonate, $[Pt\{S(C_2H_4)_2S\}_2](SO_3CF_3)_2$. The yields are 86%.

Anal. Calcd. for $[Pt\{S(C_2H_4)_2O\}_4](SO_3CF_3)_2$: C, 23.8%; H, 3.6%; Pt, 21.4%. Found: C, 23.7%; H, 3.6%; Pt, 21.4%.

Anal. Calcd. for $[Pt\{S(C_2H_4)_2S\}_2](SO_3CF_3)_2$: C, 16.4%; H, 2.2%; Pt, 26.6%. Found: C, 16.8%; H, 2.2%; Pt, 26.5%.

Properties

The $[Pt\{(CH_3)_2S\}_4](SO_3CF_3)_2$ compound is readily soluble in nitromethane and dimethylformamide, sparingly soluble in acetone, ethanol, and water, and insoluble in diethyl ether, toluene, and hexane. ^1H NMR spectrum in CD_3NO_2 (99% + atom % D) δ, ppm: 2.76 ($J_{PtH} = 45.3$ Hz).[14] The melting point in a capillary is 241–243°C (lit. 229–233°C[9]). The compound crystallized from water belongs to a space group $P\,2_1/c$ with $a = 8.5630(8)$, $b = 8.692(1)$, $c = 16.360(3)$ Å, $\beta = 96.064(9)°$, $V = 1208.4(2)$ Å^3, and $Z = 2$. The Pt–S bond lengths are: 2.317(3), 2.318(3), 2.319(4), and 2.321(4) Å.[9]

The $[Pt\{S(C_2H_4)_2O\}_4](SO_3CF_3)_2$ compound is readily soluble in nitromethane and dimethylformamide, slightly soluble in acetone, ethanol, and water and insoluble in diethyl ether, toluene, and hexane. ^1H NMR spectrum in CD_3NO_2 (99% + atom % D) δ, ppm: 3.04 (d), 4.05(t), $J_{HH} = 12.3$ Hz, 3.51(t), 4.30(d), $J_{HH} = 11$ Hz. The melting point in a capillary is 230–232°C (lit. 230–232°C[9]). The compound crystallizes from undried nitromethane as a monohydrate in space group $C\,2/c$ with $a = 14.438(2)$, $b = 13.016(2)$, $c = 21.696(3)$ Å, $\beta = 126.32(1)°$, $V = 3285.0(6)$ Å^3, and $Z = 4$. The Pt–S bond lengths are: 2.321(2) and 2.318(2) Å.[9]

The $[Pt\{S(C_2H_4)_2S\}_2](SO_3CF_3)_2$ compound is readily soluble in nitromethane and dimethylformamide, slightly soluble in acetone, ethanol, and water, and insoluble in diethyl ether, toluene, and hexane. ^1H NMR spectrum in CD_3NO_2 (99% + atom % D) δ, ppm: 3.19 ($J_{PtH} = 70.0$ Hz), 3.61 ($J_{HH} = 8.8$ Hz).[14] The compound has no characteristic melting point. On heating in a capillary it decomposes $> 250°C$. The $[Pt\{S(C_2H_4)_2S\}_2]$-$(SO_3CF_3)_2$ crystallizes from ethanol in space group *Pmnb* with $a = 11.300(2)$, $b = 12.759(1)$, $c = 14.712(2)$ Å, $V = 2122.6(3)$ Å^3, and $Z = 4$. The Pt–S bond lengths are: 2.299(2) and 2.302(2) Å.[15]

References

1. J. A. Davies and F. R. Hartley, *Chem. Rev.*, **81**, 79 (1981).
2. K. R. Dunbar, *Comments Inorg. Chem.*, **13**, 313 (1992).
3. W. L. Driessen and J. Reedijk, *Inorg. Synth.*, **29**, 111 (1992).
4. V. Yu. Kukushkin and Yu. N. Kukushkin, *Theory and Practice of Coordination Compounds Synthesis*, Nauka, Leningrad: 1990, p. 45.
5. V. V. Lebedinskii and V. A. Golovnya, *Ann. secteur platine, Inst. chim. gén.*, **18**, 38 (1945); *Chem. Abstr.*, **41**, 6187d.
6. V. Yu. Kukushkin, Å. Oskarsson, and L. I. Elding, *Zh. Obshch. Khim.*, **64**, 881 (1994).
7. A. de Renzi, A. Panunzi, A. Vitagliano, and G. Paiaro, *Chem. Comm.*, 47 (1976).
8. F. R. Hartley, S. G. Murray, and C. A. McAuliffe, *Inorg. Chem.*, **18**, 1394 (1979).
9. Z. Bugarcic, B. Norén, Å. Oskarsson, C. Stålhandske, and L. I. Elding, *Acta Chem. Scand.*, **45**, 361 (1991).
10. V. A. Golovnya and Ni Chia-Chiang, *Russ. J. Inorg. Chem. (Engl. Transl.)*, **6**, 61 (1961); *Zh. Neorg. Khim.*, **6**, 124 (1961).
11. V. A. Golovnya and Ni Tszya-Tszyan, *Russ. J. Inorg. Chem. (Engl. Transl.)*, **5**, 715 (1960); *Zh. Neorg. Khim.*, **5**, 1474 (1960).
12. Yu. A. Kharitonov, Ni Chia-Chiang, and A. V. Babaeva, *Russ. J. Inorg. Chem. (Engl. Transl.)*, **7**, 9 (1962); *Zh. Neorg. Khim.*, **7**, 21 (1962).
13. Yu. A. Kharitonov, Ni Tszya-Tzan', and A. V. Babaeva, *Proc. Acad. Sci. USSR (Engl. Transl.)*, **141**, 1185 (1961); *Dokl. Akad. Nauk. SSSR*, **141**, 645 (1961).
14. U. Frey, S. Elmroth, B. Moullet, L. I. Elding, and A. E. Merbach, *Inorg. Chem.*, **30**, 5033 (1991).
15. Z. Bugarcic, S. Elmroth, V. Yu. Kukushkin, B. Nore'n, Å. Oskarsson, and L. I. Elding, to be published.

49. [(1,2,5,6-η)-1,5-CYCLOOCTADIENE]DIMETHYL-PLATINUM(II)

SUBMITTED BY E. COSTA*, P. G. PRINGLE*, and M. RAVETZ*
CHECKED BY RICHARD J. PUDDEPHATT[†]

The complex $[Pt(CH_3)_2(1,5\text{-cod})]$ (1,5-cod = 1,5-cyclooctadiene) is widely used as a source of the $Pt(CH_3)_2$ moiety since the 1,5-cyclooctadiene is readily substituted by phosphines, isocyanides, pyridines, and other ligands.[1] Its volatility and clean thermal decomposition have led to its use in MOCVD experiments[2] and as a source of finely dispersed platinum for heterogeneous catalysis.[3] It can be made by treatment of suspensions of $[PtCl_2(1,5\text{-cod})]$ or $[PtI_2(1,5\text{-cod})]$ in diethyl ether with MeLi,[1] MeMgI,[4] or $LiCuMe_2$[5] in high

* School of Chemistry, University of Bristol, Bristol, UK.
[†] Department of Chemistry, University of Western Ontario, London, Ontario N6A 5B7, Canada.

reported yields. In the preparation described below, the reaction is carried out in dry dichloromethane in which the products and reactants are soluble so that a homogeneous reaction takes place.

$$[PtCl_2(1,5\text{-}cod)] + 2MeMgI \rightarrow [PtMe_2(1,5\text{-}cod)] + 2MgClI$$

The dichloromethane is freshly distilled from CaH_2 or P_2O_5. The reaction is carried out under nitrogen but the workup is in air. The $[PtCl_2(1,5\text{-}cod)]$ must be pure because poor yields and impure product were obtained if this starting material was slightly grey owing to the presence of platinum metal; it was made by treatment of $K_2[PtCl_4]$ with 1,5-cyclooctadiene in glacial acetic acid (Whitesides method[6]) and recrystallized from CH_2Cl_2/n-pentane. The diethyl ether solution of MeMgI (■ **Caution**: *extremely flammable*) was purchased from Aldrich or freshly prepared from MeI and Mg in diethyl ether.[7]

A two-necked, 250-mL, round-bottomed flask is fitted with a gas inlet adapter, a rubber septum, and a magnetic stirrer bar. The flask containing the $[PtCl_2(1,5\text{-}cod)]$ (1.0 g, 2.67 mmol) is evacuated and filled with nitrogen and then 50 mL of dry, nitrogen-saturated dichloromethane is introduced by syringe. The flask is then evacuated and refilled with nitrogen at least three times to ensure that the solvent is oxygen free. The resulting solution is stirred and cooled in an ice bath and then the solution of MeMgI (5.4 mL, 1 M in diethyl ether, 5.35 mmol) is added dropwise over a period of 10 min. As soon as the addition of the Grignard reagent begins, the solution becomes yellow and remains so until close to the end of the addition when it sharply returns to colorless. After the addition is complete, the mixture is stirred for a further 30 min. The mixture is then cooled to $-78°C$ in a dry-ice/acetone bath and 5 mL of nitrogen-saturated propan-2-ol are added. The bath is removed and the mixture is allowed to warm to 0°C before addition of 100 mL of nitrogen-saturated water. The mixture is stirred for a further 5 min and then transferred to a separating funnel (in air). The lower (dichloromethane) layer is separated. The aqueous layer is washed with 2×25 mL portions of dichloromethane and the combined organic extracts washed with 2×25 mL of distilled water and then dried over $MgSO_4$, filtered, treated with activated charcoal, and then filtered again to give a near-colorless solution. The solvent is removed on a rotary evaporator and the off-white solid product is collected in yields of 85–90%. The product thus obtained is analytically pure but, if required, it can be recrystallized readily by dissolving it in the minimum amount of boiling hexane (ca. 1 mL per 0.1 g) and then cooling the solution slowly to $-20°C$. The off-white needles are then filtered off and washed with a little ice-cold hexane; recovery is 80%. Similar yields are obtained when the reaction is scaled up using 2 g or scaled down using 0.2 g of $[PtCl_2(1,5\text{-}cod)]$.

Properties

Pure [PtMe$_2$(1,5-cod)] is an off-white, air-stable solid which is soluble in most common organic solvents (e.g., acetone, dichloromethane, toluene, diethyl ether, and pentane). Elemental analysis, before recrystallization: C, 35.65 (36.00); H, 5.75 (5.45). The ^{1}H NMR spectrum in CDCl$_3$ has signals at 4.82 (s, 4H, J_{PtH} 39 Hz), 2.34 (s, 8H), and 0.75 (s, 6H, J_{PtH} 83 Hz).

References

1. H. C. Clark and L. E. Manzer, *J. Organometal. Chem.*, **59**, 411 (1973).
2. R. Kumar, S. Roy, M. Rashidi, and R. J. Puddephatt, *Polyhedron*, **8**, 551 (1989).
3. T. R. Lee and G. M. Whitesides, *J. Am. Chem. Soc.*, **113**, 2576 (1991).
4. W. Gitzel, H. J. Keller, R. Lorentz, and H. H. Rupp, *Z. Naturforsch. B.*, **28**, 161 (1973).
5. R. Bassan, K. H. Bryars, L. Judd, A. W. G. Platt, and P. G. Pringle, *Inorg. Chim. Acta*, **121**, L41 (1986).
6. J. X. McDermott, J. F. White, and G. M. Whitesides, *J. Am. Chem. Soc.*, **98**, 6521 (1976).
7. L. F. Feiser and M. Feiser, *Reagents for Organic Synthesis*, Vol. 1, Wiley, New York, 1967, pp. 415–417.

50. BIS(2,2,6,6-TETRAMETHYL-3,5-HEPTANEDIONATO) COPPER

SUBMITTED BY WILLIAM S. REES, JR.* and MICHAEL W. CARRIS*
CHECKED BY GREGORY S. GIROLAMI† and YUJIAN YOU†

Introduction

Both elemental copper and copper oxide have been deposited by CVD from Cu(tmhd)$_2$, (tmhd = 2,2,6,6-tetramethyl-3,5-heptanedionato).^{1} The former is used widely as a conductor in electronic materials and the latter is the common component of the superconducting metal oxides YBa$_2$Cu$_3$O$_{7-\delta}$, BiSrCaCuO, and (TlO)BaCaCuO. In the CVD of complex compositions containing both copper and a group 2 element, fluorine-containing ligands such as tfac (CF$_3$COCHCOCH$_3$)$^{1-}$ and hfac (CF$_3$COCHCOCF$_3$)$^{1-}$ must be avoided because of the disadvantageous formation of EF$_2$ (E = Ca, Sr, Ba) by reaction of the group 2 element with the fluorine-containing ligand, rather than the desired formation of EO.^{2} Therefore, owing to the wide applicability

* Department of Chemistry and Biochemistry and Department of Materials Science and Engineering and Molecular Design Institute, Georgia Institute of Technology, Atlanta, Georgia 30332-0400.
† School of Chemical Sciences, University of Illinois at Urbana-Champaign, Urbana, IL 61801.

of Cu(tmhd)$_2$ to solving problems associated with alternate copper CVD source compounds, we now report its preparation, purification, and properties.

$$Cu(NO_3)_2 + 2Htmhd + 2KOH \xrightarrow[H_2O]{EtOH} 2KNO_3 + Cu(tmhd)_2$$

Procedure

A 1-L Erlenmeyer flask containing 98 g (0.53 mol) of Htmhd[3] and 300 mL of 95% ethanol is stirred rapidly with a magnetic stir bar. A solution of 30 g (0.53 mol) of KOH in 100 mL of distilled H$_2$O is added over a period of 15 min. Once the addition of the KOH solution is complete, a solution of 62 g (0.27 mol) of Cu(NO$_3$)$_2 \cdot$ 2.5 H$_2$O (Fisher) in 75 mL of distilled H$_2$O is added dropwise to the rapidly stirred mixture. A dark purple/blue precipitate forms immediately upon addition of the blue Cu^{2+} solution to the pale yellow alcohol solution.[4] After the addition is complete, the pH of the resulting solution is adjusted with 1.0 M KOH solution to 8. The solution is allowed to stir for 12 h.

The purple solid is collected by suction filtration and subsequently dissolved in diethyl ether (300 mL), and the resulting dark blue solution is washed with 5 × 200 mL portions of distilled water. The diethyl ether layer is dried over anhydrous MgSO$_4$, gravity filtered, and evaporated to dryness on a rotary evaporator. The resulting dark purple/blue crystals are allowed to air dry. Yield: 85 g (0.20 mol) of crude Cu(tmhd)$_2$, 75%.

Purification

A sublimator equipped with a dry ice cold finger is charged with 15 g (0.035 mol) of crude Cu(tmhd)$_2$. The system is evacuated to 5×10^{-3} mmHg. After 20 min at ambient temperature, the sublimator is heated to 110°C for 20 min without cooling the cold finger. The cold finger subsequently is cooled to -78°C and the Cu(tmhd)$_2$ is allowed to sublime for 1.5 h at 150°C. The sublimed product is removed from the cold finger in an inert-atmosphere glove box; yield, 14.1 g, (0.033 mol), 94%, of purified Cu(tmhd)$_2$.

Properties

A purple crystalline solid with a mp of 196.0–196.2°C. Solution IR: (cm^{-1}, in CCl$_4$ vs. CCl$_4$ reference) 2960(vs, sh) [CH], 1590(m, sh), 1560(s), 1550(s), 1530(s) [CO], 1495(vs, sh) [CuOC], 1450(m), 1410(vs), 1360(vs, sh) [CH], 1245(m, sh), 1220(m, sh), 1180(m, sh) [CCOC], 1145(s, sh) [CuOC], 960(w, sh), 870(s, sh) [OCC], 640(m, sh), 510(w, sh), 490(w, sh) [CuO]. Solid-state IR: (cm^{-1}, Nujol mull) 2960(vs, sh) [CH], 1590(m, sh), 1565(s), 1550(vs, sh) [CO], 1500(vs, sh) [CuOC], 1455(m), 1400(vs, sh), 1360(s, sh) [CH], 1250(m, sh), 1230(m, sh), 1180(m, sh) [CCOC], 1150(m, sh) [CuOC]. 880(m, sh) [OCC], 800(m, sh), 645(m, sh) [CuO]. ESR: the appearance of the X-band spectrum at room temperature in toluene is the same as that observed by Yokoi ($g = 2.04$, $a = 0.526$ cm^{-1}).[5] TGA (under a N_2 atmosphere)–onset of weight loss: 200°C; 2 wt% residue to 267°C. Magnetic susceptibility: ground to a powder, 2.57×10^{-6} cgs at 23°C. The ^{1}H and ^{13}C spectra were difficult to interpret because of the paramagnetic nature of the compound. ^{1}H NMR [300 MHz, positive δ downfield referenced to Si(CH$_3$)$_4$ = 0 ppm utilizing residual CHCl$_3$ = 7.24 ppm in solvent CDCl$_3$]: 4.72 [s, fwhm = 230 Hz]. UV/VIS (nm, hexane vs. hexane reference): $\lambda_{max} = 297$, $c = 3.57 \times 10^{-5}$ M, $\varepsilon = 1.58 \times 10^4$ M^{-1} cm^{-1}; $\lambda_2 = 251$. MS: [EI, 70 eV, m/e^+ (fragment); L = tmhd, M = parent ion]: 372 (M-tBu), 315 (M-2tBu), 189 (M-L-tBu), 127 (L-tBu), 57 (tBu).

Anal. Calcd. for C$_{22}$H$_{38}$O$_4$Cu: C, 61.43%; H, 8.92%. Found (average of four analyses): C, 61.16%; H, 8.80%.

References

1. (a) F. A. Houle, C. R. Jones, R. Wilson, and T. H. Baum, in *Laser Chemical Processing of Semiconductor Devices*, F. A. Houle, T. F. Deutsch, and R. H. Osgood, Jr., (eds.), Extended Abstracts of Proceedings of Symposium B, 1984 Fall Materials Research Society Meeting, Materials Research Society, Pittsburgh, 1984, pp. 64–66. (b) B. Raveau and C. Michel, *Annu. Rev. Mater. Sci.*, **19**, 319 (1989). (c) W. S. Rees, Jr. and C. R. Caballero, *Mater. Res. Soc. Symp. Proc.*, **250**, 297 (1992). (d) W. S. Rees, Jr. and C. R. Caballero, "Metal—Organic Chemical Vapour Deposition (MOCVD) Growth Utilising Cu(acac)$_2$ (acac = pentane-3,5-dionato) as a Source for Copper-containing Materials: Influence of Carrier Gas on Surface Morphology," *Advanced Materials for Optics and Electronics*, **1**, 59 (1992).
2. L. M. Tonge, D. S. Richeson, T. J. Marks, J. Zhao, J. Zhang, B. W. Wessels, H. O. Marcy, and C. R. Kannewurf, in *Electron Transfer in Biology and the Solid State: Inorganic Compounds with Unusual Properties*, M. K. Johnson, R. B. King, D. M. Kurtz, Jr., C. Kutal, M. L. Norton, and R. A. Scott, (eds.), Advances in Chemistry Series 226, American Chemical Society, Washington, DC 1990, pp. 351–368.
3. (a) G. S. Hammond, D. C. Nonhebel, and C. S. Wu, *Inorg. Chem.*, **2**, 73 (1963). (b) Commercially available from Aldrich Chemical Company, Milwaukee, WI.
4. (a) For the preparation of tmhd complexes of the rare-earth elements see K. J. Eisentraut and R. E. Sievers, *Inorganic Syntheses*, Vol. XI, W. L. Jolly, (ed.), McGraw-Hill, New York, 1968, pp. 94–98. (b) For the preparation of another β-diketonato complex of copper see T. J. Wenzel, E. J. Williams, and R. E. Sievers, *Inorganic Syntheses*, Vol. 23, S. Kirschner, (ed.), Wiley-Interscience, New York, 1985, pp. 144–149.
5. H. Yokoi, *Inorg. Chem.*, **17**, 538 (1978).

51. LEWIS BASE ADDUCTS OF 1,1,1,5,5,5-HEXAFLUORO-2,4-PENTANDIONATO-COPPER(I) COMPOUNDS

SUBMITTED BY K.-M. CHI,* H.-K. SHIN,* M. J. HAMPDEN-SMITH,*
and T. T. KODAS[†]
CHECKED BY MALCOLM A. DE LEO,[‡] SUBHASH C. GOEL,[‡] and
WILLIAM E. BUHRO[‡]

Metal-organic compounds that are to be used as precursors for the formation of metal films via CVD must fulfill a number of molecular design criteria. These criteria include high vapor pressure, thermal stability at room temperature, low thermal decomposition temperature (ideally $< 400°C$) and the formation of high-purity films. It has been demonstrated recently that in a series of copper(I) compounds of general formula (hfac)CuL, where hfac = 1,1,1,5,5,5-hexafluoro-2,4-pentandionate, and L = triorganophosphine, olefin or alkyne fulfill these criteria for the CVD of copper.[1-12] These species exhibit reasonable vapor pressures [e.g., 100 mtorr for Cu(hfac)-(PMe₃) at 60°C[13]], are thermally stable at room temperature, and deposit high-purity copper films by virtue of a thermally induced disproportionation reaction according to the stoichiometry of eq. (1), where the products Cu(hfac)₂[14] and L are thermally inert at the disproportionation temperature.

$$2Cu(hfac)^{(I)}L \rightarrow Cu^{(0)} + Cu^{(II)}(hfac)_2 + 2L \qquad (1)$$

In particular, CVD of the derivatives Cu(hfac)(PMe₃),[1,2] Cu(hfac)(1,5-cod),[3-6] Cu(hfac)(2-butyne),[7,8] and Cu(hfac)(vtms),[9-12] where 1,5-cod = 1,5-cyclooctadiene and vtms = vinyltrimethylsilane, has been studied in detail. These species can be used to deposit copper films either selectively or nonselectively on various surfaces depending on the nature of the precursor, the deposition conditions, and the substrate surface pretreatment. The syntheses of these species from a general salt elimination reaction according to eq. (2) is described here in detail.[10,13,15-17] It should be noted that other general methods of preparation of this class of compounds have been reported elsewhere.[18]

$$CuCl + L + Na(hfac) \rightarrow Cu(hfac)L + NaCl \qquad (2)$$

* Department of Chemistry and Center for Micro-Engineered Ceramics, University of New Mexico, Albuquerque, NM 87131.
†Department of Chemical Engineering and Center for Micro-Engineered Ceramics, University of New Mexico, Albuquerque, NM 87131.
‡Department of Chemistry, Washington University, St. Louis, MO 63130.

In general, nitrogen-atmosphere Schlenk techniques are used throughout these syntheses.[19] All hydrocarbon and ethereal solvents are refluxed over sodium benzophenone and stored over 4-Å molecular sieves prior to use. The NMR data presented here are obtained in benzene-d_6 at 23°C at resonance frequencies of 250.13 MHz (^{1}H), 62.5 MHz (^{13}C), or 101.3 MHz (^{31}P). Infrared spectra are recorded in the solid state in KBr pellets and in the gas phase in a 10-cm cell with KBr windows. The reagent Na(hfac) is prepared as described in the literature[13] and [CuCl(PMe$_3$)] is prepared by addition of PMe$_3$ to CuCl by a method analogous to that reported for other triorgano-phosphine adducts of CuCl.[20]

A. 1,1,1,5,5,5-HEXAFLUORO-2,4-PENTANDIONATO (TRIMETHYL-PHOSPHINE) COPPER(I), [Cu(hfac)(PMe$_3$)]

$$CuCl + PMe_3 + Na(hfac) \rightarrow Cu(hfac)PMe_3 + NaCl$$

Procedure

■ **Caution.** *Trimethylphosphine is a toxic, pyrophoric substance and should be used in an efficient fume hood. It should be destroyed by reaction with bleach.*

The solids [CuClPMe$_3$] (6.00 g, 34.3 mmol) and Na(hfac) (10.30 g, 44.8 mmol) are placed in a 250-ml Schlenk flask and 200 mL of dry n-pentane are added at room temperature. This mixture is stirred for 3 h at room temperature, during which time the solution turns orange. The stirrer is stopped, and the white solid is allowed to settle. The solution is then filtered and the solid is washed with three 50-ml portions of n-pentane. The orange filtrates are combined and the volatile components are removed in vacuo to give 10.60 g of an orange powder, which sublimes at 60°C and 10^{-2} torr, in high yield, to give a crystalline orange solid that analyzes correctly as Cu(hfac)(PMe$_3$)—a yield of 89.2%.

Characterization Data. ^{1}H NMR data: 6.33 ppm, s, 1H, C$\underline{H}$; 0.38 ppm, d, 8.0 Hz, 9H, P$\underline{Me}_3$. ^{13}C NMR data: 177.2 ppm, q, 33 Hz, C$\underline{O}$; 118.5 ppm, q, 289 Hz, C$\underline{F}_3$; 89.2, s, C$\underline{H}$; 14.0 ppm, d, 28 Hz, P$\underline{Me}_3$. ^{31}P NMR: -42.7 ppm, s, br. IR data (KBr disc, cm^{-1}): 2983(m), 2915(m), 1672(s), 1646(s), 1548(s), 1525(s), 1490(s), 1327(m), 1255(s), 1200(s), 1138(s), 965(s), 946(s), 848(w), 785(s), 761(m), 737(m), 669(s), 662(s), 583(m), 526(w). Gas phase: 2978(w), 2906(w), 1672(s), 1642(m), 1524(s), 1485(m), 1310(w), 1252(s), 1202(s), 1141(s), 960(m), 794(m), 737(m), 661(s), 576(w). Mass spectral data (70 eV, m/e): 487, 19%, [Cu$_2$(hfac)(PMe$_3$)$_2$]$^+$·; 346, 38%, [Cu(hfac)PMe$_3$]$^+$·; 277, 12%, [Cu$_2$(PMe$_3$)$_2$]$^+$·; 215, 95%, [Cu(PMe$_3$)$_2$]$^+$·; 201, 4%, [Cu(O$_2$C$_2$

CHCF$_3$)]$^+$; 139, 100%, [CuPMe$_3$]$^+$; 124, 4%, [CuPMe$_2$]$^+$; 109, 2%, [CuPMe]$^+$; 76, 30%, [PMe$_3$]$^+$; 69, 6%, [CF$_3$]$^+$; 61, 14%, [PMe$_2$]$^+$; 45, 7%, [PMe]$^+$.

Elemental Analysis Data. Calcd. for C$_8$H$_{10}$O$_2$F$_6$PCu: C, 27.72%; H, 2.91%. Found: C, 28.28%; H, 2.80%. Cryoscopic molecular weight (C$_6$H$_6$)—measured: 340 ± 14 Daltons (0.040 M); 338 ± 7 Daltons (0.078 M); calculated for (hfac)Cu(PMe$_3$), 347 Daltons. Mp = 67°C.

B. 1,1,1,5,5,5-HEXAFLUORO-2,4-PENTANDIONATO-(1,5-CYCLO-OCTADIENE) COPPER(I), [(hfac)Cu(1,5-cod)]

$$\text{CuCl} + 1,5\text{-cod} + \text{Na(hfac)} \rightarrow \text{Cu(hfac)}1,5\text{-cod} + \text{NaCl}$$

Procedure

[CuCl(1,5-cod)]$_2$ (5.20 g, 12.55 mmol) and Na(hfac) (5.90 g, 25.64 mmol) are placed in a 250-mL Schlenk flask. The addition of 150 mL of diethyl ether immediately gives a yellow solution. After 2 h stirring and removal of solvent in vacuo, the greenish-yellow solid is obtained. Extraction of the product with pentane (150 mL), followed by filtration and removal of the volatile components in vacuo, provided bright yellow crystalline Cu(hfac)(1,5-cod) in 70% yield (6.67 g). The compound can be purified further by sublimation at 65°C under reduced pressure (100 mtorr).

Characterization Data. NMR data (C$_6$D$_6$, 23°C): ^{1}H δ 1.80 (s, br, 8H, C$\underline{\text{H}}_2$ in COD), 5.07 (s, br, 4H, C$\underline{\text{H}}$ in COD), 6.17 (s, 1H, C$\underline{\text{H}}$ in hfac) ppm. ^{13}C{^{1}H}: δ 28.0 (s, $\underline{\text{C}}$H$_2$ in COD), 89.0 (s, $\underline{\text{C}}$H in hfac), 115.0 (s, $\underline{\text{C}}$H in COD), 118.5 (q, J$_{\text{C–F}}$ = 286.2 Hz, $\underline{\text{C}}$F$_3$ in hfac), 177.8 (q, J$_{\text{C–F}}$ = 33.6 Hz, CF$_3$$\underline{\text{C}}$O in hfac) ppm. IR data (KBr disc, cm^{-1}): 2953(w), 2892(w), 2842(w), 1637(s), 1593(m), 1553(m), 1526(m), 1483(s), 1432(m), 1345(w), 1259(s), 1210(s), 1142(s), 1097(m), 997(w), 810(w), 796(m), 747(m), 725(w), 671(m), 585(m), 528(w). Mass spectral data: 477, 7%, [Cu(hfac)$_2$]$^+$; 441, 2%, [Cu(hfac) (cod)]$^+$; 408, 23%, [Cu(hfac) (CF$_3$COCHCO)]$^+$; 378, 46%, [(hfac)Cu(COD)]$^+$; 339, 23%, [Cu(CF$_3$COCHCO)$_2$]$^+$; 279, 5%, [Cu(cod)$_2$]$^+$; 201, 51%, [Cu(CF$_3$CO-CHCO)]$^+$; 171, 98%, [Cu(cod)]$^+$; 117, 26%, [Cu(C$_4$H$_6$)]$^+$; 108, 8%, [COD]$^+$; 93, 10%, [C$_7$H$_9$]$^+$; 80, 35%, [C$_6$H$_8$]$^+$; 79, 37%, [C$_6$H$_7$]$^+$; 69, 8%, [CF$_3$]$^+$; 67, 100%, [C$_5$H$_7$]$^+$; 63, 35%, [Cu]$^+$; 54, 95%, [C$_4$H$_6$]$^+$.

Elemental Analysis Data. Calcd. for C$_{13}$H$_{13}$O$_2$CuF$_6$: C, 41.22%, H, 3.46%. Found: C, 41.10%, H, 3.48%, Mp = 109°C.

C. 1,1,1,5,5,5-HEXAFLUORO-2,4-PENTANDIONATO(2-BUTYNE) COPPER(I), [Cu(hfac)(2-butyne)]

$$CuCl + 2\text{-butyne} + Na(hfac) \rightarrow Cu(hfac)2\text{-butyne} + NaCl$$

Procedure

A diethyl-ether (50 mL) solution of 2-butyne (2.85 g, 52.78 mmol) is added to a slurry of CuCl (3.34 g, 33.74 mmol) in diethyl ether (50 mL) at 0°C. After the mixture is stirred for 30 min, a solution of Na(hfac) (8.01 g, 34.83 mmol) in 150 mL of diethyl ether is added. The mixture turns yellow immediately and is stirred continuously at 0°C for 10 h. The volatile species are removed from the reaction mixture in vacuo to give a crude pale yellow solid which is extracted into *n*-pentane (150 mL). After filtration and removal of *n*-pentane in vacuo, 4.83 g of pale yellow Cu(hfac)(2-butyne) is obtained (44% yield). The compound can be purified further by sublimation at 35°C under reduced pressure (100 mtorr).

Characterization Data. NMR data (C_6D_6, 23°C): ^1H δ 1.55 (s, 6H, C$\underline{H}_3$ in butyne), 6.14 (s, 1H, C$\underline{H}$ in hfac) ppm. $^{13}C\{^1H\}$: δ 6.0 (s, $\underline{C}H_3$ in butyne), 81.1 (s, $CH_3\underline{C}$ in butyne), 90.0 (s, $\underline{C}H$ in hfac), 118.4 (q, $J_{C-F} = 286$ Hz, $\underline{C}F_3$ in hfac), 178.1 (q, $J_{C-F} = 34$ Hz, $CF_3\underline{C}O$ in hfac) ppm. IR data (KBr disc, cm^{-1}): 2055(w), 1638(s), 1555(m), 1530(s), 1486(s), 1374(w), 1258(s), 1210(s), 1146(s), 1029(w), 801(m), 798(m), 746(w), 674(m), 589(m), 527(w), 468(w). Mass spectral data: 477, 3%, [Cu(hfac)$_2$]$^+$; 408, 10%, [Cu(hfac)(CF$_3$COCHCO)]$^+$; 339, 8%, [Cu(CF$_3$COCHCO)$_2$]$^+$; 324, 5%, [Cu(hfac)(2-butyne)]$^+$; 270, 1%, [Cu(hfac)]$^+$; 255, 6%, [Cu(CF$_3$COCHCO)(2-butyne)]$^+$; 201, 37%, [Cu(CF$_3$COCHCO)]$^+$; 117, 22%, [Cu(2-butyne)]$^+$; 104, 2%, [Cu(C$_3$H$_5$)]$^+$; 91, 2%, [Cu(C$_2$H$_4$)]$^+$; 69, 42%, [CF$_3$]$^+$; 63, 55%, [Cu]$^+$; 54, 100%, [2-butyne]$^+$; 53, 63%, [C$_4$H$_5$]$^+$; 39, 46%, [C$_3$H$_3$]$^+$.

Elemental Analysis Data. Calcd. for $C_9H_7O_2CuF_6$: C, 33.29%; H, 2.17%. Found: C, 33.04%; H, 1.92%. Mp = 81°C (dec).

D. 1,1,1,5,5,5-HEXAFLUORO-2,4-PENTANDIONATO (VINYLTRIMETHYLSILANE) COPPER(I), [Cu(hfac)(vtms)]

$$CuCl + vtms + Na(hfac) \rightarrow Cu(hfac)vtms + NaCl$$

Procedure

The addition of a solution of vinyltrimethylsilane (3.03 g, 30.3 mmol) in 30 mL of diethyl ether to a cold (0°C) slurry of CuCl (2.38 g, 24.0 mmol) in diethyl ether (50 mL) in a 250-mL Schlenk flask gives a pale pink solution. After the mixture is stirred for 20 min, a solution of Na(hfac) (5.75 g, 25.0 mmol) in 100 mL of diethyl ether is added and the slurry immediately turns yellow. After stirring for 3 h, removal of solvent in vacuo gives a pale yellow solid. Extraction of the product with 50 mL of pentane, followed by filtration and removal of volatile materials in vacuo at 0°C, gives 6.06 g (68% yield) of yellow liquid Cu(hfac)(vtms).

Characterization Data. NMR data (C_6D_6, 23°C): ^{1}H: δ −0.01 (s, 9H, C$\underline{H}_3$ in vtms), 4.00–4.35 (m, 3H, vinyl protons in vtms), 6.13 (s, 1H, C$\underline{H}$ in hfac) ppm. ^{13}C{^{1}H}: δ − 1.6 (s, $\underline{C}H_3$ in VTMS), 88.8 (s, H$_2\underline{C}$ in vtms), 90.0 (s, $\underline{C}$H in hfac), 99.3 (s, $(CH_3)_3Si\underline{C}$ in vtms), 118.3 (q, $J_{C-F} = 285$ Hz, $\underline{C}F_3$ in hfac), 178.3 (q, $J_{C-F} = 34.5$ Hz, $CF_3\underline{C}O$ in hfac) ppm. IR data (KBr disc, cm^{-1}): 2962(w), 1668(s), 1642(s), 1606(m), 1529(s), 1514(m), 1485(s), 1257(s), 1213(s), 1194(s), 1149(s), 1118(s), 943(w), 843(m), 803(m), 790(m), 765(m), 737(w), 662(s), 579(m), 524(w).

Elemental Analysis Data. Calcd. for $C_{10}H_{13}O_2F_6CuSi$: C, 32.39%; H, 3.53%. Found: C, 32.37%; H, 3.33%. Mp: $\sim$ −30°C

Properties

The species described above are volatile, pale yellow, or orange solids, except Cu(hfac)(vtms), which is a pale yellow liquid at room temperature. The vapor pressures of Cu(hfac)(PMe$_3$) and Cu(hfac)(1,5-cod) have been measured as a function of temperature. At 60°C, they exhibit vapor pressures of ~ 100 mtorr[13] and ~ 62 mtorr[4], respectively. All the compounds are quite soluble in hydrocarbon solvents such as *n*-pentane, benzene, toluene, and ethers. The compound Cu(hfac)PMe$_3$ is monomeric in benzene solution[13] and all the compounds are monomeric in the solid state by single-crystal X-ray diffraction.[8, 13, 15, 21] These species are sensitive to atmospheric oxygen, which leads to slow oxidation (green discoloration), except Cu(hfac)(1,5-cod), which may be exposed to air for long periods without appreciable detectable decomposition. The compound Cu(hfac)(2-butyne) is thermally unstable and is best stored in a refrigerator. Some thermal decomposition is observed even during sublimation at 35°C.

References

1. H. K. Shin, K. M. Chi, M. J. Hampden-Smith, T. T. Kodas, M. F. Paffett, and J. D. Farr, *Angew. Chem. Adv. Mater.*, **3**, 246 (1991).

2. H. K. Shin, M. J. Hampden-Smith, T. T. Kodas, M. F. Paffett, and J. D. Farr, *Chem. Mater.*, **4**, 788 (1991).
3. A. Jain, K. M. Chi, M. J. Hampden-Smith, T. T. Kodas, M. F. Paffett, and J. D. Farr, *J. Mater. Res.*, **7**, 261 (1992).
4. S. K. Reynolds, C. J. Smart, E. F. Baran, T. H. Baum, C. E. Larson, and P. J. Brock, *Appl. Phys. Lett.*, **59**, 2332 (1991).
5. S. L. Cohen, M. Liehr, and S. Kasi, *Appl. Phys. Lett.*, **60**, 1585 (1992).
6. R. Kumar, A. W. Maverick, F. R. Fronczek, G. Lai, and G. L. Griffin, 200th American Chemical Society Meeting, Atlanta, April 1991, Abstract INOR 256.
7. A. Jain, K. M. Chi, M. J. Hampden-Smith, T. T. Kodas, M. F. Paffett, and J. D. Farr, *Chem. Mater.*, **3**, 995 (1991).
8. T. H. Baum and C. E. Larson, *Chem. Mater.*, **4**, 365 (1992).
9. J. A. T. Norman, B. A. Muratore, P. N. Dyer, D. A. Roberts, and A. K. Hochberg, *J. Phys.* **IV** (1), C2-271 (1992).
10. J. A. T. Norman and B. A. Muratore, US patent, 5085731, Feb. 4, 1992.
11. A. Jain, K. M. Chi, M. J. Hampden-Smith, T. T. Kodas, M. F. Paffett, and J. D. Farr, *J. Electrochem. Soc.*, in press, 1992.
12. J. Farkas, K. M. Chi, J. Farkas, M. J. Hampden-Smith, and T. T. Kodas, *J. Appl. Phys.*, in press, 1992.
13. H. K. Shin, K. M. Chi, J. Farkas, M. J. Hampden-Smith, T. T. Kodas, and E. N. Duesler, *Inorg. Chem.*, **31**, 424 (1992).
14. W. G. Lai, Y. Xie, and G. L. Griffin, *J. Electrochem. Soc.*, **138**, 3499 (1991).
15. H. K. Shin, M. J. Hampden-Smith, T. T. Kodas, and E. N. Duesler, *Polyhedron*, **6**, 645 (1991).
16. K. M. Chi, H. K. Shin, M. J. Hampden-Smith, T. T. Kodas, and E. N. Duesler, *Polyhedron*, **10**, 2293 (1991).
17. K. M. Chi, M. J. Hampden-Smith, T. T. Kodas, and E. N. Duesler, *Inorg. Chem.*, **30**, 4293 (1991).
18. G. Doyle, K. A. Eriksen, and D. Van Engen, *Organometallics*, **4**, 830 (1985).
19. D. F. Shriver and M. A. Drezden, *The Manipulation of Air-Sensitive Compounds*, 2nd ed., Wiley-Interscience, New York, 1986, p. 78.
20. K. M. Chi, M. J. Hampden-Smith, T. T. Kodas, and E. N. Duesler, *J. Chem. Soc., Dalton Trans.*, 3111 (1992).
21. J. A. T. Norman, B. A. Muratore, P. N. Dyer, D. A. Roberts, and A. K. Hochberg, European Materials Research Society Meeting, Strasburg, Spring, 1992.

52. COPPER(II) ALKOXIDES

SUBMITTED BY SUBHASH C. GOEL* and WILLIAM E. BUHRO*
CHECKED BY KAI-MING CHI[†] and MARK J. HAMPDEN-SMITH[†]

Soluble and volatile alkoxides of copper are precursors in sol-gel-like routes to cuprate-oxide superconductors[1] and in the CVD of copper and copper oxides,[2] respectively. Copper(II) alkoxides are typically prepared by alcohol-interchange reactions with dimethoxycopper(II); however, most are insoluble

* Department of Chemistry, Washington University, St. Louis, MO 63130-4899.
[†] Department of Chemistry, The University of New Mexico, Albuquerque, NM 87131.

and nonvolatile.[3] Dimethoxycopper(II) has been prepared by the decomposition of methylcopper(I) in MeOH in the presence of air,[4] and by reaction of CuX_2 (X = Cl, Br) and MOMe (M = Li, Na).[4,5] The following procedures describe dimethoxycopper(II) and soluble and volatile copper(II) alkoxides derived from it by alcohol interchange.

A. DIMETHOXYCOPPER(II)

$$CuCl_2 + 2LiOMe \rightarrow Cu(OMe)_2 + 2LiCl$$

Procedure[5]

All ambient-pressure operations are carried out under purified nitrogen using oxygen- and moisture-free solvents. Anhydrous copper(II) chloride[6] (10.00 g, 74.4 mmol) and MeOH (70 mL) are combined in an oven-dried, 250-mL Schlenk flask equipped with a magnetic stirring bar. A LiOMe solution is prepared separately as follows. The impurity coating is removed from lithium wire by scraping the surface of the wire in a glove box. A 200-mL Schlenk flask is charged with cleaned lithium wire (1.04 g, 150 mmol) and a magnetic stirring bar, removed from the glove box, attached to a Schlenk line, and cooled in a wet-ice/acetone bath. Methanol (125 mL) is added to the cooled flask in approximately 2 min, resulting in a controlled exothermic reaction. The stirred mixture is allowed to warm to room temperature over a period of 1 h to complete the disappearance of the lithium and the generation of LiOMe. The LiOMe solution is transferred to the stirred copper(II) chloride solution via cannula. An exothermic reaction ensues and a blue precipitate forms (ca. 5 min).

The reaction mixture is stirred for 6 h at room temperature, and the blue solid is then collected in a 200-mL Schlenk filter funnel. The LiCl by-product and other side products are extracted from the blue dimethoxycopper(II) by vigorously shaking the collected precipitate with 100 mL of MeOH in the filter funnel, followed by removal of the MeOH extract. This process is repeated until the silver–nitrate test for chloride ion[7] in the extract is negative. During each extraction, the mixture must be shaken vigorously enough to break up the solid cake. Six such washings are typically enough to remove the LiCl and side products; however, when the synthesis is performed on larger scales, more and larger-volume MeOH washings are required. The dimethoxycopper(II) product is finally dried under reduced pressure. Yield: 8.80 g (94%).

Anal. Calcd. for $C_2H_6O_2Cu$: C, 19.12; H, 4.81; Cl, 0; Cu, 50.57; Li, 0. Found: C, 18.87; H, 4.65; Cl, 0.38; Cu, 49.97; Li, ≪0.01 (undetectable).

Properties[4, 5]

Dimethoxycopper(II) is a moisture-sensitive blue compound that is insoluble in common organic solvents. It can be recrystallized from $MeOH/NH_3$ to give a microcrystalline solid. Analysis of copper by iodometric titration provides a quick routine purity determination for dimethoxycopper(II).[8] The complete removal of residual chloride from dimethoxycopper(II) is not easily achieved; the most likely impurity is CuCl(OMe), which is a green compound.[5] Dimethoxycopper(II) must be washed thoroughly as described above to minimize contamination by CuCl(OMe), and is obtained in >98% purity by this procedure (on the basis of the C, H, Cl, Cu, and Li analyses). IR (KBr, cm^{-1}): 2917(vs), 2885(vs), 2806(vs), 1436(w), 1150(w), 1052(vs), 528(vs), 438(s).

B. BIS[1-(DIMETHYLAMINO)-2-PROPANOLATO]COPPER(II)

$$Cu(OMe)_2 + 2HOCH(Me)CH_2NMe_2 \rightarrow$$

$$Cu(OCH(Me)CH_2NMe_2)_2 + 2MeOH$$

■ **Caution.** *Benzene is a toxic substance. This procedure should be conducted in an efficient fume hood taking care to avoid inhalation of benzene vapors.*

Procedure[9]

All ambient-pressure operations are carried out under purified nitrogen using oxygen- and moisture-free solvents. A suspension of $Cu(OMe)_2$ (2.16 g, 17.2 mmol) is prepared in 50 mL of benzene in an oven-dried, 100-mL Schlenk flask equipped with a magnetic stirring bar. 1-(Dimethylamino)-2-propanol (4.25 mL, 3.55 g, 34.5 mmol, Aldrich) is added with stirring and the flask is fitted with a 6-in. fractionating column and a variable-return distillation head (see Fig. 1). A deep-purple homogeneous solution is formed within 30 min at room temperature; however, the subsequent procedure need not be delayed for this event to occur. The solution is heated gently (50–60°C) for 1 h, and then is heated to reflux. The MeOH liberated during the reaction is removed as a benzene/MeOH azeotrope (bp 58.3–80.2°C) and is collected at a rate of 1–2 drops per minute over 4–6 h. The cessation of MeOH liberation is determined by the following test. An aliquot of distillate (ca. 1 mL) is combined (in air) with a solution of 1 N potassium dichromate in 12.5% (by volume) H_2SO_4 (ca. 1 mL), which turns brown to green when MeOH is present in the distillate.[10] The liberation of MeOH has ceased when the color of the test mixture remains unchanged (orange-red) for 10 min. The solution is then cooled and the solvent is removed under reduced pressure. The

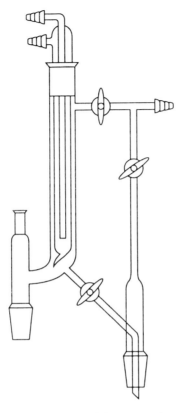

Figure 1. Variable-return distillation head for separation of the benzene–MeOH azeotrope.

residual dark purple solid is purified by sublimation (60°C bath/10^{-4} torr) to give bis[1-(dimethylamino)-2-propanolato]copper(II) as dark-purple crystals. Yield: 3.45 g (75%).

Anal. Calcd. for $C_{10}H_{24}N_2O_2Cu$: C, 44.84; H, 9.03; N, 10.45; Cu, 23.72. Found: C, 44.69; H, 9.04; N, 10.41; Cu, 23.54.

Properties[9]

Bis[1-(dimethylamino)-2-propanolato]copper(II) is a moisture-sensitive purple compound that is soluble in common organic solvents such as benzene, toluene, and hexane. It has a *trans*, square-planar molecular structure in the solid state and is mononuclear in solution. Bis[1-(dimethylamino)-2-propanolato]copper(II) is among the most volatile copper(II) complexes known. It undergoes thermal decomposition ($<300°C$) under N_2 to give copper metal (as the only copper-containing phase) and organic by-products. Mp, 134–135°C dec. IR (KBr, cm^{-1}): 3002(m), 2969(vs), 2947(vs), 2911(vs), 2865(s), 2835(s), 2727(s), 2765(s), 2719(m), 2581(m), 1458(s), 1354(m), 1332(s), 1135(vs), 1107(m), 1089(vs), 1031(s), 1011(s), 944(vs), 855(m), 836(s), 622(vs), 541(s), 497(w), 481(w).

C. BIS[2-(2-METHOXYETHOXY)ETHANOLATO]COPPER(II)

$$Cu(OMe)_2 + 2HOCH_2CH_2OCH_2CH_2OMe \rightarrow$$

$$Cu(OCH_2CH_2OCH_2CH_2OMe)_2 + 2MeOH$$

■ **Caution.** *Benzene is a toxic substance. This procedure should be conducted in an efficient fume hood taking care to avoid inhalation of benzene vapors.*

Procedure[1c]

All ambient-pressure operations are carried out under purified nitrogen using oxygen- and moisture-free solvents. A suspension of Cu(OMe)$_2$ (4.00 g, 31.8 mmol) is prepared in 70 mL of benzene in an oven-dried, 250-mL Schlenk flask equipped with a magnetic stirring bar. 2-(2-Methoxyethoxy)-ethanol) (7.6 mL, 7.67 g, 63.9 mmol, Aldrich) is added with stirring and the flask is fitted with a 6-in. fractionating column and a variable-return distillation head (see Fig. 1). The reaction mixture is heated gently (50–60°C) for 2 h and then under reflux. The reaction temperature must be increased with caution during this procedure because the mixture has a tendency to foam and to carry unreacted Cu(OMe)$_2$ up into the condenser. [If excessive foaming occurs, Cu(OMe)$_2$ may be returned to the flask by carefully shaking the apparatus or by washing its walls with fresh benzene.] Liberated MeOH is collected over 6–8 h as described above in Section 52.B. When the MeOH formation is complete (see Section 52.B), the mixture is cooled and filtered. The volume of the blue-green filtrate is reduced to ca. 30 mL in vacuo. Hexane (50–60 mL) is then added and bis[2-(2-methoxyethoxy)ethanolato]-copper(II) precipitates as a blue solid, which is collected by filtration, washed with 2×20 mL portions of hexane, and finally dried in vacuo. Yield 8.50 g (89%).

Anal. Calcd. for $C_{10}H_{22}O_6Cu$: C, 39.80; H, 7.35; Cu, 21.05. Found: C, 39.41; H, 7.41; Cu, 20.95.

Properties[1c]

Bis[2-(2-methoxyethoxy)ethanolato]copper(II) is a moisture-sensitive, non-crystalline, nonvolatile solid that is soluble in common organic solvents such as benzene, toluene, and THF. It is an oligomer in solution; in benzene it exists as $[Cu(OCH_2CH_2OCH_2CH_2OMe)_2]_n$, where n is ≥ 5. Unlike most copper(II) complexes, bis[2-(2-methoxyethoxy)ethanolato]copper(II) exhibits a well-resolved (paramagnetically shifted) ^1H NMR spectrum (δ, C_6D_6): $+110$ (v br s, 2H), $+6.8$ (br s, 2H), $+6.3$ (br s, 2H), $+5.1$ (br s, 2H), and $+4.00$ (s, 3H) ppm. Mp, 183–184°C dec. IR (KBr, cm^{-1}): 2982(m), 2884(vs, br), 1456(m), 1351(w), 1252(m), 1201(m), 1123(vs, br), 1073(vs, br), 1027(m), 923(m), 875(w), 846(w), 537(w sh), 499(m).

References

1. (a) S. Hirano, T. Hayashi, M. Miura, and H. Tomonaga, *Bull. Chem. Soc. Jpn.*, **62**, 888 (1989).
   (b) H. S. Horowitz, S. J. McLain, A. W. Sleight, J. D. Druliner, P. L. Gai, M. J. VanKavelaar, J. L, Wagner, B. D. Biggs, and S. J. Poon, *Science*, **243**, 66 (1989). (c) S. C. Goel, K. S. Kramer, P. C. Gibbons, and W. E. Buhro, *Inorg. Chem.*, **28**, 3619 (1989).
2. P. M. Jeffries and G. S. Girolami, *Chem. Mater.*, **1**, 8 (1989).
3. J. V. Singh, B. P. Baranwal, and R. C. Mehrotra, *Z. Anorg. Allg. Chem.*, **477**, 235 (1981).
4. G. Costa, A. Camus, and N. Marsich, *J. Inorg. Nucl. Chem.*, **27**, 281 (1965).
5. C. H. Brubaker, Jr. and M. Wicholas, *J. Inorg. Nucl. Chem.*, **27**, 59 (1965).
6. A. R. Pray, *Inorg. Synth.*, **5**, 153 (1957).
7. A. I. Vogel, *Vogel's Qualitative Inorganic Analysis*, 6th ed., revised by G. Svehla, Wiley, New York, 1987, p. 174.
8. A. I. Vogel, *Vogel's Textbook of Quantitative Inorganic Analysis*, 4th ed., revised by J. Bassett, R. C. Denney, G. H. Jeffery, and J. Mendham, Longman, Essex, 1978, p. 379.
9. S. C. Goel, K. S. Kramer, M. Y. Chiang, and W. E. Buhro, *Polyhedron*, **9**, 611 (1990).
10. D. C. Bradley, F. M. A. Halim, and W. Wardlaw, *J. Chem. Soc.*, 3450 (1950).

53. PYRAZOLATO COPPER(I) COMPLEXES

G. A. ARDIZZOIA* and G. LA MONICA*
CHECKED BY C.-W. LIU[†] and JOHN P. FACKLER, JR.[†]

The pyrazolato copper(I) complex, $[Cu(pz)]_n$ (Hpz = $1H$-pyrazole), and related copper(I) complexes having different substituents on the pyrazole ring are well known.[1,2] For these pyrazolato derivatives, various synthetic routes have been reported. However, the methods are, in general, complex and do

* Dipartimento di Chimica Inorganica, Metallorganica, e Analitica e Centro CNR, Via G. Venezian, 21, I-20133 Milano (Italy).
† Department of Chemistry, Texas A and M University, College Station, TX 77843.

not allow pure products to be obtained in large amounts.[2-4] Moreover, a general synthetic method for pyrazolato copper(I) complexes has not been reported to date. The pyrazolato ligand, owing to its capability to coordinate to metal centers in an *exo*-bidentate fashion, appears to be highly suitable for stabilizing di- and polynuclear systems. Owing to their potential role in multimetal-centered catalysis, the synthesis of such systems, in which metal centers are maintained in close proximity, is an important objective in transition metal chemistry.

The following procedure represents a convenient preparative method leading to the synthesis of copper(I) pyrazolates, namely, $[Cu(pz)]_n$ (Hpz = 1*H*-pyrazole), $[Cu(dmpz)]_3$ (Hdmpz = 3,5-dimethyl-1*H*-pyrazole), $[Cu(dcmpz)]_n$ (Hdcmpz = dimethyl-1*H*-pyrazole-3,5-dicarboxylate), $[Cu(dctpz)]_n$ (Hdctpz = di-*tert*-butyl-1*H*-pyrazole-3,5-dicarboxylate), and $[Cu(dppz)]_4$ (Hdppz = 3,5-diphenyl-1*H*-pyrazole). In all cases, high yields (>90%) of the pure products are obtained.

Procedure

$$[Cu(CH_3CN)_4]BF_4 + Hdmpz + N(C_2H_5)_3 \rightarrow$$

$$1/3[Cu(dmpz)]_3 + [(C_2H_5)_3NH]BF_4 + 4CH_3CN$$

A 250-mL, two necked, round-bottomed flask equipped with a magnetic stirrer is connected to a nitrogen line. The flask is purged thoroughly with nitrogen and charged with 3.04 g (31.6 mmol) of Hdmpz* and 100 mL of acetone. Under nitrogen, $[Cu(CH_3CN)_4]BF_4$ (prepared as described in ref. 5, but with aqueous HBF_4 used instead of HPF_6), 5.00 g (15.9 mmol) is added under stirring. When a clean solution is obtained (a few minutes), degassed triethylamine (10 mL) is added over a period of 1 min. The product precipitates as a colorless solid. The suspension is stirred for 30 min and then filtered; the product is washed with acetone (50 mL) and hexane (20 mL) and dried in vacuo. Yield 2.42 g (96%).

Anal. Calcd. for $C_5H_7CuN_2$: C, 37.8; H, 4.4; N, 17.7. Found: C, 37.8; H, 4.4; N, 17.7

---

* 1*H*-pyrazole, 3,5-dimethyl-1*H*-pyrazole and 1*H*-pyrazole-3,5-dicarboxylic acid were purchased from Aldrich Chemical Co. 3,5-Diphenyl-1*H*-pyrazole is also a commercially available product (Lancaster Synthesis). Dimethyl-1*H*-pyrazole-3,5-dicarboxylate has been obtained as described in ref. 6, di-*tert*-butyl-1*H*-pyrazole-3,5-dicarboxylate has been prepared from the corresponding acyl chloride[7] by esterification with *tert*-butyl alcohol in 1,2-dichloroethane in the presence of triethyl amine.

Analogous Complexes

The aforementioned pyrazolato copper(I) complexes, [Cu(pz)]$_n$, [Cu(dcmpz)]$_n$, and [Cu(dppz)]$_4$ are prepared analogously. [Cu(dctpz)]$_n$ is prepared by using ethanol instead of acetone, because of its high solubility in the latter solvent.[#]

Properties

[Cu(pz)]$_n$, [Cu(dmpz)]$_3$, and [Cu(dcmpz)]$_n$ are insoluble in the common organic solvents. [Cu(dctpz)]$_n$ and [Cu(dppz)]$_4$ are soluble in dichloromethane and toluene. Although the complexes may be handled in air (in the solid state), they are best stored under nitrogen to ensure their purity.

The nuclearity of the reported copper(I) complexes is well documented only for [Cu(dmpz)]$_3$[4] and [Cu(dppz)]$_4$.[8] Employing a different synthetic route,[3] Fackler and Raptis obtained a trimeric copper(I) complex, [Cu(dppz)]$_3$. The significant IR absorptions of these copper(I) complexes are reported in Table I.

Copper(I) pyrazolates react with neutral ligands (L), such as 1,10-phenanthroline and cyclohexyl isocyanide, giving binuclear complexes of general formula [Cu(pz)(L)]$_2$.[9,10] Solutions of [Cu(dmpz)]$_3$ in pyridine rapidly absorb oxygen in the presence of water, giving the blue octanuclear hydroxo-

TABLE I. Significant IR Absorptions of Pyrazolato Copper(I) Complexes

Compound	Absorptions[a]
[Cu(pz)]$_n$	3114(w), 1490(w), 1410(m), 1181(m), 1059(m), 764(s), 740(s)
[Cu(dmpz)]$_3$	1524(s), 1346(m), 1148(m), 1058(m), 767(s)
[Cu(dcmpz)]$_n$	3154(m), 1736(sh), 1732(s), 1718(sh), 1713(s), 1522(m), 1348(m), 1258(s,br), 1081(s), 820(s), 759(s)
[Cu(dctpz)]$_n$	1723(sh), 1715(s), 1708(sh), 1699(s), 1509(w), 1416(w), 1263(s), 1161(s), 1080(m), 1064(m), 845(m), 767(m)
[Cu(dppz)]$_4$	3064(w), 3017(w), 1605(w), 1538(m), 1121(m), 1008(m), 908(w), 800(w), 754(s), 695(s), 689(s), 686(s)

[a] Nujol mull, cm^{-1}, w: weak, m: medium, s: strong, sh: shoulder.

[#] Crystal data for [Cu(dppz)]$_4$: triclinic, space group $P\bar{1}$ (no. 2), $Z = 2$, $a = 19.423(9)$ Å, $b = 11.627(1)$ Å, $c = 11.905(2)$ Å, $\alpha = 85.75(1)°$, $\beta = 72.92(2)°$, $\gamma = 88.92(2)°$. X-Ray powder-diffraction data confirmed the single-crystal results given above.

copper(II) complex $[Cu_8(dmpz)_8(OH)_8]$,[11] which catalyzes the oxidation of triphenylphosphine, carbon monoxide, isocyanides, and primary or secondary amines.[11]

References

1. S. Trofimenko, *Chem. Rev.*, **72**, 497 (1972).
2. H. Okkersen, W. L. Groeneveld, and J. Reedijk, *Recl. Trav. Chim. Pays-Bas*, **92**, 945 (1973).
3. R. G. Raptis and J. P. Fackler, *Inorg. Chem.*, **27**, 4149 (1988).
4. M. K. Ehlert, S. J. Rettig, A. Storr, R. C. Thompson, and J. Trotter, *Can. J. Chem.*, **68**, 1444 (1990).
5. G. J. Kubas, *Inorg. Synth.*, **19**, 90 (1979).
6. R. V. Rothenburg, *Chem. Ber.*, **27**, 1098 (1894).
7. C.V. Greco and F. Pellegrini, *J. Chem. Soc. Perkin I*, 720 (1972).
8. G. A. Ardizzoia, S. Cenini, G. La Monica, N. Masciocchi, and M. Moret, *Inorg. Chem.*, **33**, 1458 (1994).
9. G. A. Ardizzoia, G. La Monica, M. Angaroni, and F. Cariati, *Inorg. Chim. Acta*, **158**, 159 (1989).
10. G. A. Ardizzoia, M. Angaroni, G. La Monica, N. Masciocchi, and M. Moret, *J. Chem. Soc. Dalton Trans.*, 2277 (1990).
11. G. A. Ardizzoia, M. Angaroni, G. La Monica, F. Cariati, S. Cenini, M. Moret, and N. Masciocchi, *J. Chem. Soc. Chem. Commun.*, 1021 (1990); G. A. Ardizzoia, M. Angaroni, G. La Monica, F. Cariati, S. Cenini, M. Moret, and N. Masciocchi, *Inorg. Chem.*, **30**, 4347 (1991).

54. *TRIS*(2,2,6,6-TETRAMETHYL-3,5-HEPTANEDIONATO)-YTTRIUM

SUBMITTED BY WILLIAM S. REES, JR.* and MICHAEL W. CARRIS*
CHECKED BY YUJIAN YOU† and GREGORY S. GIROLAMI†

Introduction

There is much current interest in volatile compounds of some of the metallic elements which may be useful for CVD of thin films of electronic or structural materials.[1] Among the presently utilized precursors for metal oxide-containing films, the β-diketonate family has been employed by numerous researchers.[2] A previous volume of *Inorganic Synthesis* described the general

* Department of Chemistry and Biochemistry and Department of Materials Science and Engineering and Molecular Design Institute, Georgia Institute of Technology, Atlanta, Georgia 30332-0400.
† School of Chemical Sciences, University of Illinois at Urbana-Champaign, Urbana, IL 61801.

preparation of Ln(tmhd)$_3$.[3] It may not be necessary to employ anhydrous precursors for CVD.[1,2] Because of the current interest in Y$_2$O$_3$,[4] yttria-stabilized zirconia (YSZ),[5] and YBa$_2$Cu$_3$O$_{7-\delta}$,[6] we now detail the specific preparations and purifications of Htmhd, Y(tmhd)$_3 \cdot$ H$_2$O, and Y(tmhd)$_3$, the latter two compounds having proved successful for the CVD of yttrium-containing thin films. Compilations of their properties also are presented.[3,7]

$$Y_2O_3 + 6HNO_3 \rightarrow 2Y(NO_3)_3 \cdot 6H_2O$$

Procedure

■ **Caution.** *The reaction shown above is rather exothermic. The yttrium oxide must be added slowly to maintain control over the reaction rate.*

To a 400-mL beaker containing 188 mL of 6 M nitric acid and a magnetic stirring bar, 50 g (0.22 mol) of 99.9% Y$_2$O$_3$ (Strem Chemicals) is added over a period of 30 min. Once the addition is complete, the beaker is covered with a watch glass and heated on a steam bath until all of the solid is dissolved. At this point, the watch glass is removed and the solution is concentrated by evaporation on a steam bath at ambient pressure to 100 mL. Upon cooling, a mass of white solid precipitates.

The solid is collected by suction filtration on a coarse-porosity, fritted-glass disk and transferred to a clean 400-mL beaker and dissolved in a minimum amount of 1 M HNO$_3$. This solution is concentrated as described above to 100 mL. The beaker is placed in an ice-water bath and, with constant manual stirring, a white solid begins to precipitate. Before the solid hardens, it is collected by vacuum filtration on a coarse-porosity frit. The highly hydrated solid is dried for 2 days in a desiccator over concentrated H$_2$SO$_4$. The reaction affords 80 g (0.24 mol) of Y(NO$_3$)$_3 \cdot$ 3.35H$_2$O, 55% yield based on Y$_2$O$_3$. The product is identified by its TGA.[8]

$$2.2 (CH_3)_3CCOOCH_2CH_3 + (CH_3)_3CCOCH_3 + 2.9NaH \xrightarrow[(2) \ HCl_{(aq)}]{(1) \ DME}$$

$$(CH_3)_3CCOCH_2COC(CH_3)_3 + NaCl + CH_3CH_2OH + H_2$$

A. 2,2,6,6-TETRAMETHYL-3,5-HEPTANEDIONE (Htmhd)[9]

Under N$_{2(g)}$, 11.0 g (0.46 mol) of NaH and 400 mL of DME are added to a three-necked, 1-L, round-bottomed flask. The flask is equipped with a reflux condenser, an overhead stirrer, and an addition funnel charged with 15.7 g (0.157 mol) of pinacolone and 44.6 g (0.343 mol) of ethyl pivalate diluted with 30 mL of DME. The flask is heated at reflux and the contents of the addition funnel are added dropwise (≈ 3 drop/min). Upon completion of the addition, the solution is allowed to reflux for an additional 7 h. The reaction mixture is cooled to 0°C and 50 mL of 12 M HCl are added slowly with rapid stirring, followed by the addition of 100 ml of H_2O and 100 mL of pentane. The resulting mixture is warmed to ambient temperature and allowed to stir for 0.5 h. The solution is transferred to a 1-L separatory funnel, the organic layer is washed with 3×75 mL aliquots of H_2O, dried over $MgSO_4$ for 12 h, gravity filtered, and distilled at 6 torr. The fraction collected at 72–73°C is identifiable as pure (GC), authentic Htmhd (lit. bp[10a] 100–102°C, 36 torr, 63% yield; lit. bp[10c] 96–97°C, 20 torr, 28% yield). Yield: 22.8 g (0.124 mol), 79% based on pinacolone.

Characterization Data. ^1H NMR: [300 MHz, positive δ downfield referenced to $Si(CH_3)_4 = 0$ ppm utilizing residual $CHCl_3 = 7.24$ ppm in solvent $CDCl_3$] 5.70 [s, 2H, CH_2], 1.24 [s, 18H, CH_3]. $^{13}C\{^1H\}$: [75 MHz, positive δ downfield referenced to $Si(CH_3)_4 = 0$ utilizing solvent $CDCl_3 = 77.0$ ppm] 201.9 [CO], 90.7 [CH_2], 39.3 [ipso C], 27.2 [CH_3]. MS: {EI, 70 eV, m/e^+, (fragment); LH = Htmhd} 184(LH), 169(LH–CH_3), 127(LH–tBu), 111(LH-tBu–O), 97(tBu–CO–C), 85(tBuCO), 69(tBuC–O), 57(tBu). Solution IR: (4000–400 cm^{-1}, in CCl_4 vs. CCl_4 reference, NaCl plates) 2960(vs) [CH], 1685(vs, sh), 1600(vs) [CO], 1460(s) [CH], 1440(s), 1350(s, sh) [CH_3], 1275(m) [CCH_3], 1210(m, sh), 1190(m, sh), 1120(m) [$CHCCH_3$], 940(w) [CH_3], 860(m, sh) [$CCCO$]. UV/VIS: (750–200 nm, hexane vs. hexane reference) $\lambda_{max} = 275.4$, $c = 2.22 \times 10^{-5}$ M, $\varepsilon = 1.01 \times 10^4$ M^{-1} cm^{-1}.

$$Y(NO_3)_3 \cdot 3.35H_2O + 3Htmhd + 3NaOH \xrightarrow[H_2O]{EtOH} Y(tmhd)_3 \cdot H_2O + 3NaNO_3$$

B. Y(tmhd)$_3$

To a 1-L flask containing a magnetic stir bar, 115 g (0.62 mol) of Htmhd and 400 mL of 95% ethanol, 25 g (0.62 mol) of NaOH in 100 mL of distilled H_2O is added over a period of 30 min with rapid stirring. A solution of 80 g (0.24 mol) of $Y(NO_3)_3 \cdot 3.35H_2O$ dissolved in 125 mL of distilled H_2O then is added dropwise. Upon this addition a white precipitate forms immediately. After stirring for 12 h, the solution is transferred to a 2-L, round-bottomed, flask and the majority of the ethanol is removed by rotary evaporation. Then 500 mL of distilled H_2O is added to the remaining solution and it is again allowed to stir for 12 h.

The white/yellow solid that precipitates is collected on a medium-porosity, glass-suction frit and air dried for 3 h. The solid is ground to a powder with a mortar and pestle and placed in a desiccator over Drierite® at ambient temperature and 10^{-3} torr for 24 h. The semidry solid is dissolved in 600 mL of hexane (freshly distilled under $N_{2(g)}$ from $Li[AlH_4]$) and dried further over anhydrous Na_2SO_4 for 24 h. The pale-yellow solution is gravity filtered and the filtrate is evaporated to dryness on a rotary evaporator. The resulting white powder is redissolved in 450 mL of hexane at room temperature and allowed to stand for 24 h.

The white needle-like crystals that precipitate are collected on a suction frit and allowed to air dry. A second crop of crystals is harvested from the mother liquor by further concentration on a rotary evaporator to one-fourth of the initial volume at ambient temperature. The combined solids again are ground to a powder and placed in a round-bottomed flask. The flask is evacuated to 0.01 torr and heated at 111°C for 7 h to remove all of the hexane and any remaining Htmhd. Yield: 105 g (0.16 mol) of $Y(tmhd)_3 \cdot H_2O$, 77%.

Dehydration

A sublimator equipped with a dry-ice cold finger is charged with 3.0 g (0.0046 mol) of $Y(tmhd)_3 \cdot H_2O$. After evacuation to 5×10^{-5} torr, the sublimator is heated to 187°C, the cold finger is cooled to $-78°C$ and the $Y(tmhd)_3$ is allowed to sublime for 70 min. The white solid is scraped off the cold finger in an inert-atmosphere glove box. Yield: 2.75 g (0.0043 mol), 93%, of $Y(tmhd)_3$, based on the initial charge of $Y(tmhd)_3 \cdot H_2O$.

Properties

$Y(tmhd)_3 \cdot H_2O$. The solid collected is a white powder which melts at 170.9°C. ¹H NMR: [300 MHz, positive δ downfield referenced to $Si(CH_3)_4 = 0$ ppm utilizing residual $CHCl_3 = 7.24$ ppm in solvent $CDCl_3$] 5.71 [s, 1H, CH], 1.16 [s, 18H, CH_3]. Solution IR: (4000–400 cm⁻¹, in CCl_4 vs. CCl_4 reference, NaCl plates) 2960(vs) [CH], 1590(s), 1570(vs), 1550(s) [CO], 1500(vs, sh) [YOC], 1445(s), 1400(vs), 1355(vs, sh) [CH], 1285(w, sh), 1245(m,sh), 1220(s, sh), 1175(m, sh) [CCOC], 610(m, sh), 490(m, sh), 475(m, sh) [YO]. TGA: (under a $N_{2(g)}$ atmosphere, 10°C/min) onset of weight loss: 204°C, 4 wt% residue to 500°C. ¹³C{¹H} NMR: [75 MHz, positive δ downfield referenced to $Si(CH_3)_4 = 0$ utilizing solvent $CDCl_3 = 77.0$ ppm] 201.7 [CO], 91.7 [CH], 40.5 [ipso C], 27.9 [CH_3]. UV/VIS: (750–200 nm, hexane vs. hexane reference) $\lambda_{max} = 275$, $c = 2 \times 10^{-5}$ M, $\varepsilon = 4.02 \times 10^4$ M⁻¹cm⁻¹. MS: [EI, 70 eV, m/e^+, (fragment); L = tmhd, M = parent ion-H_2O] 638(M), 623(M–Me), 581(M–*t*Bu), 455(M–L), 313(YLC₂HO), 184(LH), 146(Y*t*Bu), 127(LH–*t*Bu), 57(*t*Bu), 42(*t*Bu–Me). Analysis: Calcd. C, 60.3; H, 9.1. Found: C, 60.6; H, 8.9.

*Y(tmhd)*₃. The solid collected is a white powder which melts at 168.4°C. Analysis: Calcd. C, 62.1; H, 9.0. Found: C, 61.9; H, 9.2. MW: (benzene cryoscopy) Calcd. (monomer) 639 g/mol. Found 661 g/mol.

References

1. (a) G. B. Stringfellow, *Organometallic Vapor-Phase Epitaxy: Theory and Practice*, Academic, New York, 1989. (b) W. S. Rees, Jr., A. R. Barron, *Adv. Mat. Opt. Electron.*, **2**, 271 (1993). (c) W. S. Rees, Jr., "Superconductors: An Overview of the Present and Potential Materials and Markets," *Ceramic Industries International*, **1993**, 22–26.
2. (a) R. C. Mehrotra, R. Bohra, and D. Gaur, *Metal β-Diketonates and Allied Derivatives*, Academic, New York, 1978. (b) W. S. Rees, Jr. and A. R. Barron, "Group IIA β-diketonate Compounds as CVD Precursors for High T_c Superconductors," *Materials Science Forum*, Volume 137–139, A. Hepp, S. A. Alterovitz, J. J. Pouch, and R. R. Romanofsky, Eds., **1993**, 473–494.
3. Ln = Sc, Y, La, Pr, Nd, Sm, Eu, or Gd. K. J. Eisentraut and R. E. Sievers, *Inorganic Syntheses*, Vol. 11, W. L. Jolly, (ed.), McGraw-Hill, New York, 1968, pp. 94–98.
4. C. Brecher, G. C. Wei, and W. H. Rhodes, *J. Am. Ceram. Soc.*, **73**, 1473 (1990).
5. E. Lilley, *Ceram. Trans.*, **10**, 387 (1990).
6. (a) L. M. Tonge, D. S. Richeson, T. J. Marks, J. Zhao, J. Zhang, B. W. Wessels, H. O. Marcy, and C. R. Kannewurf, in *Electron Transfer in Biology and the Solid State: Inorganic Compounds With Unusual Properties*, M. K. Johnson, R. B. King, D. M. Kurtz, Jr., C. Kutal, M. L. Norton, and R. A. Scott, (eds.), Advances in Chemistry Series 226, American Chemical Society, Washington, DC, 1990, pp. 351–368. (b) W. S. Rees, Jr., L. R. Testardi and Y. S. Hascicek, in *Better Ceramics Through Chemistry V*, M. J. Hampden-Smith, W. G. Klemperer, and C. J. Brinker, (eds.), Materials Research Society, Symposium Proceedings, Vol. 271, 1992, pp. 925–931. (c) W. S. Rees, Jr., "Growth of YBa₂Cu₃O₇₋δ Thin Films on LaAlO₃, in "Proceedings of the Fourth Florida Microelectronics Conference," E. J. Claire, Ed.; University of South Florida, **1992**, 83–89.
7. W. S. Rees, Jr., H. A. Luten, M. W. Carris, E. J. Doskocil, and V. L. Goedken, in *Better Ceramics Through Chemistry V*, M. J. Hampden-Smith, W. G. Klemperer, and C. J. Brinker, (eds.), Materials Research Society, Symposium Proceedings, Vol. 271, 1992, pp. 141–147.
8. (a) TGA: (10°C/min, N₂₍g₎ atm) onset 123°C, 34 wt% residue to 600°C. (b) H. Bergmann, in *Gmelin Handbuch der Anorganischen Chemie: Sc, Y, La, und Lanthanide, Teil C 2*, Springer-Verlag, New York, 1974, 233 pp.
9. (a) Modified procedure from G. S. Hammond, D. C. Nonhebel, and C. S. Wu, *Inorg. Chem.*, **2**, 73 (1963). (b) Also available commercially from Aldrich Chemical Company (Milwaukee, WI); 1993 cost: $ 18.90/5 g. (c) J. T. Adams and C. R. Hauser, *J. Am. Chem. Soc.*, **66**, 1220 (1944).

55. LEWIS BASE ADDUCTS OF URANIUM TRIIODIDE AND TRIS[BIS(TRIMETHYLSILYL)AMIDO]URANIUM*

SUBMITTED BY DAVID L. CLARK[†] and ALFRED P. SATTELBERGER[†]

CHECKED BY RICHARD A. ANDERSEN[‡]

The nonaqueous chemistry of trivalent uranium is relatively unexplored compared to the nonaqueous chemistry of tetravalent uranium.[1] We[2] and others[3] have argued that the paucity of molecular uranium(III) compounds is due to the virtual lack of suitable starting materials. During the course of our studies of trivalent uranium we have developed a simple procedure for the preparation of organic-solvent-soluble Lewis-base adducts of uranium triiodide, UI_3L_4, on a synthetically useful scale (~ 50 g).[2] A key to these preparations has been the development of a technique for generating amalgamated, essentially "oxide-free" uranium metal. The UI_3L_4 compounds are exceptionally good starting materials for the high-yield preparation of trivalent uranium complexes via metathetical reactions employing sodium or potassium salts.[2] We describe here the details of our procedures for the preparation of "oxide-free" metal and the adducts $UI_3(THF)_4$ and $UI_3(py)_4$, and then illustrate the utility of $UI_3(THF)_4$ as a reagent in the synthesis of the trivalent uranium amido complex $U[N(SiMe_3)_2]_3$.[4]

General Procedures and Techniques

Because of the oxygen and moisture sensitivity of the starting materials and products, all operations are carried out under rigorously dry and oxygen-free atmospheres using Schlenk or inert-atmosphere glove-box techniques. We prefer glove-box techniques[5,6] for the transfer of air-sensitive solids and the workup of air-sensitive reaction mixtures, and employ a Vacuum Atmospheres Company glove box equipped with a high-capacity purification system (MO-40-2H) and a Dry-Cold freezer maintained at $-40°C$. The helium atmosphere[7] of our glove box system is maintained at or below 1.0 ppm O_2 and is monitored continuously with a Teledyne oxygen analyzer (Model 316). The checkers, working on smaller scales with appropriately adjusted quantities of reagents, used Schlenk and cannula techniques[5,6] for all manipulations except the transfer and storage of dry solids (glove box) and

* Research performed under the auspices of the Office of Energy Research, Division of Chemical Sciences, US Department of Energy, Washington, DC.
† Isotope and Nuclear Chemistry Division, Los Alamos National Laboratory, Los Alamos, NM 87545.

‡ Department of Chemistry, University of California, Berkeley, Berkeley, CA 94720.

obtained comparable yields. Toluene, diethyl ether, THF and hexane are degassed and distilled from sodium–potassium alloy or sodium benzophenone ketyl under argon. Pyridine (py) is predried by stirring for 48 h over KOH pellets. After decanting from the KOH and degassing with argon, the pyridine is refluxed over and subsequently distilled from CaH_2 or sodium under argon. Solvents are stored in the glove box in 500-mL bottles. Benchtop manipulations are performed using standard Schlenk techniques[5,6] in glassware equipped with 15- or 25-mm Solv-Seal joints (Fisher-Porter). After preparation, the trivalent uranium compounds are handled in the glove box and stored in glass scintillation vials at $-40°C$, or stored on the benchtop in sealed glass ampoules under argon at room temperature. Uranium residues remaining after the preparation of $UI_3(THF)_4$ or $UI_3(py)_4$ are covered with water, and then destroyed by cautious addition of 6 M nitric acid. These nitrate solutions are discarded as low-level radioactive waste.

Precautions for Handling Radioactive Materials

Uranium-238 is a weak α-emitter (4.195 MeV) with a half-life of 4.51×10^9 years. All of the manipulations described below are carried out in monitored fume hoods or in an inert-atmosphere glove box in a radiation laboratory equipped with α- and β-counting equipment.[8] (■ **Caution.** *Lab coats, eye protection, and surgical gloves should be worn at all times when handling uranium turnings and compounds.[9] Every precaution should be taken to avoid inhalation or ingestion of uranium-containing dust. Aqueous, nonaqueous, and solid radioactive waste should be disposed of in locally approved containers.*)

Starting Materials

Sublimed elemental iodine and reagent-grade mercury(II) iodide are purchased from Aldrich Chemical Company, stored under ultrahigh-purity (UHP) grade argon, and used without further purification. Concentrated nitric acid and reagent grade acetone are purchased from J. T. Baker Chemical Company and used without purification. 1,1,1,3,3,3-Hexamethyldisilazane is purchased from Aldrich Chemical Company, distilled, and stored under argon prior to use. Sodium hydride was purchased from Aldrich as a 60% dispersion in mineral oil. The mineral oil dispersion is taken into the glove box, washed with hexane, and vacuum filtered on a coarse-porosity, sintered-glass frit. The resulting solid is washed with hexane until the filtrate is colorless and then dried in vacuo to yield an off-white pyrophoric powder. The hexane filtrate contains some reactive NaH which is destroyed outside the glove box by careful addition of isopropyl alcohol with stirring. Depleted uranium metal turnings (99.9%) are obtained from Los Alamos National

Laboratory stock or purchased from Cerac Incorporated (99.7%).[10] No difference in yield is found between reactions employing uranium turnings obtained from laboratory stock and those performed using turnings purchased from Cerac. Although the commercial turnings are sold packaged in airtight glass containers under argon, removal of the oxide coating from the uranium metal remains the key step in the preparation of UI_3L_4 compounds in high yield and purity.

A. AMALGAMATED, "OXIDE-FREE" URANIUM METAL TURNINGS

Procedures

In a typical procedure, approximately 30 g of uranium metal turnings are cut into small strips of approximately 0.5–1.5 in. in length using steel scissors and transferred to a 250-mL Pyrex beaker inside a well-ventilated fume hood. *Concentrated* nitric acid (16 M) is added to just cover the uranium turnings. The turnings are swirled with the acid approximately 1 min.

■ **Caution.** *The reaction of uranium metal with concentrated HNO_3 is exothermic, is accompanied by the liberation of toxic NO_2 gas, and should only be carried out in a well-ventilated hood. The nitric acid for this step must be concentrated. Acid-resistant gloves and a face shield are highly recommended safety additions.*

The reaction of nitric acid with uranium metal is accompanied by the evolution of heat and brown NO_2 gas as the metal turnings lose the black oxide coating to expose a shiny, metallic surface. The acid solution is decanted carefully from the turnings into an acid waste container, and the procedure is repeated. The resulting shiny uranium turnings are washed three times with copious amounts of distilled water to remove all traces of acid. After the third water wash, the turnings are rinsed three times with reagent-grade acetone to remove water.

■ **Caution.** *Never discard the acetone washes in the same container as the concentrated nitric acid waste. This will result in a violent reaction.*

The shiny turnings are dumped from the beaker onto a paper towel, patted dry, and transferred, using tweezers, to a 250-mL, one-neck, Schlenk reaction vessel. The vessel is evacuated to 10^{-3} torr, and then back-filled with UHP argon. The evacuate/refill cycle is repeated twice. Under an argon purge, approximately 2 g of mercury(II) iodide are added, and the evacuate/refill cycle is repeated three more times. The Schlenk reaction vessel is then taken into the inert atmosphere glove box and 100 mL of distilled THF are added. The solution is shaken vigorously by hand for 10 min to produce a dark blue solution of $UI_3(THF)_4$ and uranium metal turnings coated with a thin film of uranium-mercury amalgam. The amalgamated turnings are vacuum filtered on a coarse-porosity, sintered-glass frit, washed with THF, and dried in vacuo. This procedure yields very shiny, essentially oxide-free, amalgamated uranium turnings which should be used immediately. Attempts to store the turnings in the glove box result in the formation of a brown-black oxide coating on the metal surface.

B. TRIIODOTETRAKIS(TETRAHYDROFURAN)- URANIUM, UI₃(THF)₄

$$U_{\text{"oxide-free"}} + 1.5I_2 \xrightarrow[-10°C]{\text{THF}} UI_3(THF)_4$$

Procedure

Inside the glove box, an oven-dried 500-mL, one-neck Schlenk reaction vessel is charged with 23.93 g (100.5 mmol) of freshly amalgamated, "oxide-free" uranium turnings (see Section 55.A), a 1-in., Teflon-coated magnetic stirring bar, and approximately 200 mL of THF. The vessel is removed from the glove box, connected to the argon manifold of a conventional Schlenk line,[5, 6] and cooled to approximately $-10°C$ in an ice/salt bath. Under an argon purge, 17.5 g (68.9 mmol, 45% of theoretical) of sublimed I_2 is added, in one portion, to the cooled THF solution. The Schlenk reaction vessel is removed from the ice bath and shaken vigorously by hand until the vessel is warm to the touch, and then immersed back into the ice bath to cool without shaking. The suspension should not be allowed to reflux because this will decrease the yield of product through ring opening of THF by uranium(III). This shaking/cooling procedure is repeated 4 or 5 times, until no further heat evolution is detected. A second 17.5-g portion of iodine is added under an argon purge to the cooled suspension and the entire procedure is repeated until enough uranium metal has dissolved to allow for magnetic stirring of the dark-red suspension. The latter is stirred for 3 h at $-10°C$, after which time the ice bath is removed, and the solution is allowed to warm to room temperature and stir for an additional 10–12 h. At this time a dark-blue solution with much dark-blue microcrystalline precipitate is observed together with some unreacted metal turnings. A 10% excess of uranium is used in the reaction to insure oxidation state integrity. The Schlenk vessel is taken into the glove box and stirred vigorously to suspend the $UI_3(THF)_4$ precipitate in the THF. The stirring is stopped and the uranium turnings are allowed to settle quickly to the bottom of the flask before the suspension is decanted and vacuum filtered on a 150-mL, medium-porosity, sintered-glass frit. This procedure of suspending the product in THF, allowing the metal to settle, decanting the product onto a frit, and vacuum filtration is repeated several times. Finally, the remaining metal turnings are washed with THF and the washings decanted from the metal and onto the frit. The dark blue solid is dried in vacuo to yield 42.86 g of microcrystalline $UI_3(THF)_4$. The blue-green filtrate is reduced in vacuo to approximately 30 mL, layered with 60 mL of diethyl ether, and cooled to $-40°C$. After 12 h, $UI_3(THF)_4$ is collected by filtration, washed with 25 mL of cold ($-40°C$) ether, and dried in vacuo. Yield: 10.40 g; combined yield: 53.26 g (63.9% based on I_2). The checkers started with 2.0 g of uranium turnings and report an average yield of 68% for three syntheses.

Anal. Calcd. for $UI_3O_4C_{16}H_{32}$: C, 21.18; H, 3.56; I, 41.97. Found: C, 20.68; H, 3.38; I, 41.20.

Properties

Triiodotetrakis(tetrahydrofuran)uranium is an extremely air- and moisture-sensitive dark-blue solid. We have stored this compound in glass scintillation vials in the glove-box freezer, or in sealed-glass ampoules at room temperature for 2 years without noticeable signs of decomposition. The compound is soluble in THF to approximately 3 g per 100 mL THF at 20°C. $UI_3(THF)_4$ is initially soluble in toluene, but over a period of several hours a dark-green solid, presumably an oligomer, precipitates from solution. Toluene solutions spiked with approximately 5% THF (by volume) are stable for long periods of time. Freshly prepared toluene-d_8 NMR samples show two broad ^{1}H NMR (250 MHz, 20°C) resonances at δ 9.5 and 5.4 corresponding to the α- and β-protons of coordinated THF.[11] The IR spectrum, recorded as a Nujol mull between KBr plates, has absorptions at 1342(w), 1312(w), 1292(w), 1246(w), 1172(w), 1034(2), 1011(s), 922(m), 853(s), 833(s), 721(m), 667(m), and 574(w) cm^{-1}.

C. TRIIODOTETRAKIS(PYRIDINE)URANIUM, $UI_3(py)_4$

$$U_{\text{"oxide-free"}} + 1.5I_2 \xrightarrow[-10°C]{py} UI_3(py)_4$$

Procedure

In the glove box, 8.1 g (34.0 mmol) of freshly amalgamated uranium turnings (see Section 55.A) are placed in an oven-dried, 250-mL, one-neck Schlenk reaction vessel, along with a 1-in. Teflon-coated magnetic stirring bar, and ~ 200 mL of distilled, degassed pyridine. The reaction vessel is removed from the dry box, connected to the argon manifold of a conventional Schlenk line,[5, 6] and cooled to approximately $-10°C$ in an ice/salt bath. Under an argon purge, 8.2 g (32.3 mmol, 90% of theoretical) of sublimed I_2 is added, in one portion, to the cooled pyridine solution.

■ **Caution.** *If a larger scale preparation is attempted, the iodine should be added portionwise to prevent a violent reaction.*

It is essential that a slight excess (approximately 10%) of metal be used to ensure that all the iodine is consumed and that the oxidation state of uranium is maintained at $+3$. The Schlenk reaction vessel is removed from the ice bath and shaken vigorously by hand until the vessel is warm to the touch, and then immersed back into the ice bath to cool without shaking. This shaking/cooling procedure is repeated 4 or 5 times until enough uranium metal has dissolved to allow magnetic stirring. The reaction mixture is stirred for 2 h at $-10°C$, and then allowed to warm to room temperature, where it is stirred for an additional 24 h to produce a deep-purple, almost black, solution. The reaction vessel is taken into the glove box, where unreacted uranium metal is removed by vacuum filtration through a coarse-porosity frit. The glassware and turnings are washed with pyridine until the filtered washings are colorless. The filtrate volume is reduced in vacuo to approximately 30 mL, a 1-in., Teflon-coated magnetic stirring bar and 30 mL of toluene are introduced, and the solution is stirred vigorously for 1 h to produce a black microcrystalline precipitate. The product is collected by vacuum filtration on a 60-mL, medium-porosity, sintered-glass frit and dried in vacuo to yield 17.3 g (80% based on I_2) of analytically pure $UI_3(py)_4$. The checkers started with 2.0 g of uranium turnings and report a yield of 93%.

Anal. Calcd. for $UI_3N_4C_{20}H_{20}$: C, 25.69; H, 2.16; N, 5.99; I, 40.70. Found: C, 25.81; H, 2.25; N, 5.89; I, 40.68.

Properties

Triiodotetrakis(pyridine)uranium is an extremely air- and moisture-sensitive, purple-black solid, which may be stored for months in the absence of air and moisture. It is appreciably soluble in pyridine: approximately 17 g in 30 mL pyridine at 20°C. The compound is sparingly soluble in benzene and toluene, but readily dissolves in THF with ligand substitution. Freshly prepared samples of $UI_3(py)_4$ in benzene-d_6 show three broad ^1H NMR resonances (250 MHz, 20°C) at δ 17.8, 15.4, and 12.0 in a 2:1:2 ratio corresponding to the protons of coordinated pyridine.[11] The IR spectrum, recorded as a Nujol mull between KBr plates, has absorptions at 1597(m), 1304(w), 1215(m), 1169(w), 1150(m), 1061(m), 1034(m), 999(m), 976(w), 756(s), 725(w), 698(s), 621(s), and 425(w) cm^{-1}.

D. SODIUM BIS(TRIMETHYLSILYL)AMIDE, Na[N(SiMe₃)₂]

$$NaH + HN(SiMe_3)_2 \xrightarrow[\Delta]{\text{toluene}} Na[N(SiMe_3)_2] + H_2(g)$$

Procedure

This preparation is a modification of the procedure of Krüger and Nieder-prüm.[12] In the glove box, an oven-dried, 250-mL, one-neck Schlenk reaction vessel is charged with 4.0 g (166.7 mmol) of sodium hydride powder, and a 1-in., Teflon-coated magnetic stirring bar. After 120 mL of toluene are introduced, the vessel is removed from the glove box and connected to the argon manifold of a conventional Schlenk line. Under an argon purge, a slight excess (38 mL, 29.1 g, 180.1 mmol) of hexamethyldisilazane is added to the NaH suspension using a predried glass syringe. The Schlenk reaction vessel is then fitted with a reflux condenser and heated to gentle reflux under argon. After 48 h the heating source is removed, and the vessel is allowed to cool to room temperature, resulting in the formation of some gelatinous brown precipitate and a yellow solution. The vessel is taken into the glove box, and the solution is filtered through a 60-mL medium-porosity, sintered glass frit to give a clear yellow filtrate. The solvent is removed from the filtrate to yield off-white Na[N(SiMe₃)₂]. The solid is loaded into a one-neck Schlenk vessel and pumped to dryness on a high-vacuum (10^{-6} torr) manifold for 8 h. This procedure ensures the removal of any residual hexamethyldisilazane. Yield: 25.1 g (82% based on NaH).

Properties

Sodium bis(trimethylsilyl)amide is an air- and moisture-sensitive white solid which is soluble in toluene, diethyl ether, THF, and pyridine. An alternative procedure for its synthesis may be found in an earlier volume of this series.[12]

E. TRIS[BIS(TRIMETHYLSILYL)AMIDO]URANIUM, U[N(SiMe₃)₂]₃

$$UI_3(THF)_4 + 3NaN(SiMe_3)_2 \xrightarrow[20°C]{\text{THF}} U[N(SiMe_3)_2]_3 + 3\,NaI$$

Procedure

In the glove box, 18.2 g (99.3 mmol) of NaN(SiMe₃)₂ are placed in an oven-dried, 250-mL Erlenmeyer flask, 100 mL of THF are added, and the solution is gently swirled until all of the sodium bis(trimethylsilyl)amide is dissolved. Also in the glove box, a 1-L, oven-dried Erlenmeyer flask, topped with a 24/40 standard-taper joint, is charged with 30.0 g (33.1 mmol) of purple $UI_3(THF)_4$, a 1-in., Teflon-coated magnetic stirring bar, and 300 mL of THF. The uranium triiodide solution is stirred magnetically. The sodium bis(trimethylsilyl)amide solution is added slowly dropwise to the uranium triiodide solution using a Pasteur pipet. The total addition requires just over 1 h. The reaction proceeds smoothly with the formation of a deep wine-red solution and the precipitation of some white sodium iodide. The flask is stoppered and, after stirring for 12 h in the glove box, the solution is vacuum filtered through Celite on a 150-mL, coarse-sintered glass frit. The Celite pad and glassware are washed with THF until the filtrate is colorless. The combined filtrate is then removed in vacuo to provide a purple-red micro-crystalline product. A significant amount of sodium iodide remains in this crude product. The desired uranium product is extracted into 100 mL of hexane, filtered through a 60-mL, fine-porosity, sintered-glass frit and cooled to $-40°C$ for 12 h to produce large purple needles. The crystalline product is isolated by vacuum filtration and dried in vacuo. Reduction of the filtrate volume to ~ 20 mL and cooling to $-40°C$ produces a second crop of purple needles which are isolated by vacuum filtration. The combined yield is 19.5 g [82% based on $UI_3(THF)_4$]. The checkers started with 2.0 g of $UI_3(THF)_4$ and report a yield of 79%.

Properties

Tris[bis(trimethylsilyl)amido]uranium is an extremely air- and moisture-sensitive, red-purple solid, which can be stored for months in the absence of air and moisture without noticeable signs of decomposition. It is soluble in both aliphatic and aromatic hydrocarbons. The ^1H NMR spectrum in benzene-d_6 (250 MHz, 20°C) shows a single broad resonance at δ -11.5.[11] The compound sublimes readily at 80°C under good vacuum (10^{-6} torr). The checkers report a melting point of 137–140°C. The IR spectrum, recorded as a Nujol mull between KBr plates, has absorptions at 1248(s), 1170(w), 990(s), 860(s), 828(s), 764(m), 676(m), 654(m), and 598(m) cm^{-1}. Other physicochemical properties are described in the literature.[4]

References

1. J. J. Katz, L. R. Morss, and G. T. Seaborg, *The Chemistry of the Actinide Elements*, Vols. 1 and 2, Chapman and Hall, New York, 1986, and references therein.
2. D. L. Clark, A. P. Sattelberger, S. G. Bott, and R. N. Vrtis, *Inorg. Chem.*, **28**, 1771 (1989); L. R. Avens, S. G. Bott, D. L. Clark, A. P. Sattelberger, J. G. Watkin, B. D. Zwick, *Inorg. Chem.*, **33**, 2248 (1994).

3. P. J. Fagan, J. M. Manriquez, T. J. Marks, C. S. Day, S. H. Vollmer, and V. W. Day, *Organometallics*, **1**, 170 (1982).

4. R. A. Andersen, *Inorg. Chem.*, **18**, 1507 (1979).

5. D. F. Shriver and M. A. Drezdzon, *The Manipulation of Air-Sensitive Compounds*, 2nd ed., Wiley Interscience, New York, 1986.

6. A. L. Wayda and M. Y. Darensbourg, *Experimental Organometallic Chemistry*, ACS Symposium Series **357**, American Chemical Society, Washington, DC, 1987.

7. The use of a helium atmosphere eliminates the static electricity problems that are often encountered in glove boxes operating under nitrogen or argon atmospheres.

8. It is much easier to detect uranium using a β-counter because of the β-activity associated with the short-lived daughter products $_{90}Th^{234}$ (24.1 days), $_{91}Pa^{234m}$ (1.14 min), and $_{91}Pa^{234}$ (6.75 h).

9. At Los Alamos National Laboratory, we operate our radiological areas under the guidelines and specifications of DOE Order 5480.11.

10. Natural abundance uranium contains $\sim$99.3% $_{92}U^{238}$ and $\sim$0.7% $_{92}U^{235}$. "Depleted uranium" is a term used to describe uranium metal that contains $\sim$0.2% of the 235 isotope. It is also referred to as D-38.

11. The checkers noted some slight differences (< 1 ppm) in the chemical shifts they obtained versus those reported here. The chemical shifts of paramagnetic uranium(III) complexes are temperature and concentration dependent. The data reported here were obtained on dilute solutions ($\sim$0.05 M) at 20 $\pm$ 1°C at 250 MHz.

12. C. R. Krüger and H. Niederprüm, *Inorg. Synth.*, **8**, 15 (1966).

CONTRIBUTOR INDEX
Volume 31

SUBJECT INDEX

Prepared by DAVID BOWER
and JASON WEISGERBER of
Chemical Abstracts Service

The Subject Index for *Inorganic Syntheses, Vol. 31* is based on the Chemical Abstracts Service (CAS) Registry nomenclature. Each entry consists of the CAS Registry Name, the CAS Registry Number and the page reference. The inverted form of the CAS Registry Name (parent index heading) is used in the alphabetically ordered index. Generally one index entry is given for each CAS Registry Number. Some less common ligands and organic rings may have a separate alphabetical listing with the same CAS Registry Number as given for the index compound, e.g. *3,10,14,18,21,25-Hexaazabicyclo[10.7.7] hexacosane*, cobalt (2+) deriv., [73914-18-8]. Simple salts, binary compounds and ionic lattice compounds, including nonstoichiometric compounds, are entered in the usual uninverted way, e.g. *Chromium chloride*, (*CrCl₂*) [10049-05-5]. Salts of oxo acids are entered at the acid name, e.g. *Sulfuric acid*, disodium salt [7757-82-6].

FORMULA INDEX
Prepared by DAVID BOWER
and JASON WEISGERBER of
Chemical Abstracts Service

The formulas in the *Inorganic Syntheses, Vol. 31* Formula Index are for the total composition of the entered compound. In many cases, especially ionic complexes, there are significant differences between the Inorganic Syntheses Formula Index entry and the CAS Registry formula, e.g. Sodium tetrahydroborate(1-) [16940-66-2], the I.S. Formula Index entry is BH_4Na while the CAS Registry formula is $BH_4 \cdot Na$. The formulas consist solely of the atomic symbols (abbreviations for atomic groupings or ligands are not used) and are arranged in alphabetical order with carbon and hydrogen always given last, e.g. $Br_3CoN_4C_4H_{16}$. To enhance the utility of the I.S. Formula Index, all formulas are permuted on the atomic symbols except for carbon. $FeO_{13}Ru_3C_{13}H_3$ is also listed at $O_{13}FeRu_3C_{13}H_3$, and $Ru_3FeO_{13}C_{13}H_3$. Ligands are not given separate formula entries in this I.S. Formula Index.

Water of hydration, when so identified, and other components of clathrates and addition compounds are not added into the formulas of the constituent compound. Components of addition compounds (other than water of hydration) are entered at the formulas of both components.

AlC_3H_9, Aluminum, trimethyl-, [75-24-1], 31:48

$AlNC_4H_{14}$, Aluminum, (*N,N*-dimethylethanamine)trihydro-, (*T*-4)-, [124330-23-0], 31:74

$Al_2N_2C_{23}H_{40}$, Aluminum, hexamethyl[μ-[4,4'-methylenebis[*N,N*-dimethylbenzenamine]-*N*:*N*']]di-, [118518-30-2], 31:49

$AsCl_3$, Arsenous trichloride, [7784-34-1], 31:148

$AsCl_3OPtC_{25}H_{20}$, Tetraphenylarsonium trichlorocarbonylplatinate(II), [108756-21-4], 31:244

$AsLiSi_2C_6H_{19}$, Arsine, bis(trimethylsilyl)-, lithium salt, [76938-15-3], 31:157

$AsSi_3C_9H_{27}$, Arsine, tris(trimethylsilyl)-, [17729-30-5], 31:151

$As_2F_{12}S_4$, Arsenate(1-), hexafluoro-, (tetra-sulfur)(2+) (2:1), [74775-27-2], 31:108

$As_2F_{12}Se_4$, Arsenate(1-), hexafluoro-, selenium ion (Se_4^{2+}) (2:1), [53513-64-7], 31:110

$BaF_{12}O_9C_{20}H_{24}$, Barium, bis(1,1,1,5,5,5-hexafluoro-2,4-pentanedionato-*O,O'*)(2,5,8,11,14-pentaoxapentadecane-*O,O',O'',O''',O''''*)-, [130691-53-1], 31:5

$BaN_2O_2Si_4C_{20}H_{52}$, Barium, bis(tetrahydrofuran) bis[1,1,1-trimethyl-*N*-(trimethylsilyl) silanaminato]-, (*T*-4)-, [131380-78-4], 31:8

$BaO_6TiC_{18}H_{12}$, Titanate(2-), tris[1,2-benzene-diolato(2-)-*O,O'*]-, barium (1:1), (*OC*-6-11)-, [121241-16-5], 31:13

$BiCo_3O_9C_9$, Cobalt, μ$_3$-bismuthylidynetri-μ-carbonylhexacarbonyltri-, *triangulo*, [110899-93-9], 31:224

$BiCo_3O_{12}C_{12}$, Cobalt, μ$_3$-bismuthylidynedo-decacarbonyltri-, [43164-49-4], 31:223

325

CHEMICAL ABSTRACTS SERVICE
REGISTRY NUMBER INDEX